Endemism in Vascular Plants

PLANT AND VEGETATION

Volume 9

Series Editor: M.J.A. Werger

For further volumes:
http://www.springer.com/series/7549

Carsten Hobohm
Editor

Endemism in Vascular Plants

Editor
Carsten Hobohm
Ecology and Environmental Education
Working Group
Interdisciplinary Institute of Environmental,
Social and Human Studies
University of Flensburg
Flensburg
Germany

ISSN 1875-1318 ISSN 1875-1326 (electronic)
ISBN 978-94-007-6912-0 ISBN 978-94-007-6913-7 (eBook)
DOI 10.1007/978-94-007-6913-7
Springer Dordrecht Heidelberg New York London

Library of Congress Control Number: 2013942502

Printed on acid-free paper

Springer is part of Springer Science+Business Media (www.springer.com)

Preface

What is the aim of a book about endemism in vascular plants – biogeography and evolution of vascular plants that have small ranges, vegetation ecology, a little bit social sciences, politics, and nature conservation? Does it mainly review and comment on some new scientific results including spotlights of the revolution in genetics, systematics and phylogeography?

It will be impossible not to review important publications, and scientific experts may excuse the review character of many parts in this book. And, yes, the relevance of endemism also concerns nature conservation. Endemism is not only a scientific game of characters and numbers. Endemism matters. On the other hand we present many scientific results which have not been published before.

What is the relationship between endemism and global warming? We want to show that the amount of recent climate change means very little in terms of extinction risks for vascular plants. If a single rock in Central Europe represents a climate of northern Italy on its southern face and of southern Norway on the opposite side, if the variability in a single forest represents many different micro- and mesoclimates, then a changing temperature of 1 °C plus or minus perhaps means almost nothing for a population of plants living close to the rock or in the forest. If both, the rock and the forest, are becoming destroyed by man, then the populations of the rocky habitat or forest will probably not survive. In this book, we estimate the impact of climate and climate change on the composition of endemic taxa in different regions. However, it is clear that we are talking about other dimensions than a single degree centigrade in average.

What is the impact of invading species? Are they competing with the endemics? Competition or introgression are keywords in this context. Both processes can result in extinction of genes and taxa and we don't want to play down the extinction risk for endemic plant taxa. However, we also want to show that the extinction risk for endemic vascular plant taxa is relatively low in most mainland areas because competition and introgression between invading and endemic plants do not play an important role compared to the effects of habitat disturbance or resource use. We are aware that the situation becomes more complicated on islands or habitat isolates particularly due to invading animals which are able to destroy habitat structures and change environmental conditions or regional food web structures. However, neither competition nor introgression seem to be the main problems in this case.

What is new? We want to focus on the relationships between endemism in vascular plants and habitat. This topic is relatively new to science. And it is relevant because one of the most important factors for the loss of biodiversity in the past was habitat destruction by man including domestic livestock. We assume that this factor in the future will persist to be important.

To secure ecosystem services and goods is a modern feature of recent strategies and payments in biodiversity policies (CBD). Therefore, the question arises what the relationship between the occurrence of endemic taxa and a modern strategy securing ecosystem goods and services could be. An example: *Cephalaria radiata* is a beautiful flowering plant scattered in semi-natural grasslands and endemic to Transylvania in Romania, Europe. What are the goods and services of this plant? And if there are no such, why should we lower the extinction risk of this species? In this moment grasslands and semi-natural pastures and meadows all over the world are declining in quantity and quality.

Last but not least we will point out gaps in our knowledge. For many regions in the world, especially mainland regions, we don't have adequate information about the amount and ecology of endemism. For example, the number of publications focussing on range sizes of plant species in tropical forests is still very limited. Modern molecular analyses on systematics will change our knowledge and earlier ideas in systematics. This process has just begun. We still do not know if most vascular plant species or most endemics on earth live in forests or in open landscapes. We do not know all the threats in every part of the world. Even the very important Red List database of the IUCN covers only a small percentage of the vascular plants living on earth. Many ecological conditions and events in the past which have been influencing the species composition and endemism in a region presently cannot be reconstructed. Thus, we know that we are far away from understanding the interferences of all the basic processes. We will ask these questions in our book. Nevertheless, many of them cannot be answered at the moment.

The structure of the book is artificial because it is impossible to describe the time axis independent of space or to understand spatial patterns without discussing underlying processes. Thus, we did not try to absolutely avoid overlaps but to pronounce patterns on the one hand and processes on the other. Both levels are important for an understanding of the relating complex phenomena and for the understanding of nature conservation purposes.

As a result of the process in which we organized the development of this book the collection of regions is a little bit patchwork-like. We tried to describe and analyse both patterns and processes in endemic-rich regions and also in regions with fewer endemic taxa. We wanted to represent mainland regions, continental and oceanic islands in different parts of the world, because of the different evolutionary and climatic histories. And we tried to present new results. However, in a scientific world which is evaluated and financed on the basis of scientific indicators created by a private company, it becomes more and more difficult to find experts writing a chapter in a book. Hopefully this trend is not the end of the road.

Acknowledgements

I am thankful to the editor of the series *Plant and Vegetation* Marinus Werger, Utrecht, who initiated the book-project in the year 2009 while we discussed the landscape-structures, plant compositions including endemism, cultural influences and other geobotanical features of different regions in Southwest China,

to our publishing editor Valeria Rinaudo and her assistant Elisabete Machado of Springer, Dordrecht, for all the help and kind communication,

to Richard Pott, Hannover, for the organization of many excursions and congresses, for information, questions and discussions related to ecology, geobotany, systematics and biogeography,

to my family, especially Uta, Merel and Till Herdeg, Lüneburg, for securing best conditions including transition zones between private and scientific environments,

to my University in Flensburg for local warming, financial support, for the opportunity to visit Madagascar twice in 2011 and 2012 and for allowing me to have a sabbatical semester in 2012/2013.

We invited many experts in the field of biogeography and endemism to write or to correct a chapter of this book. It is wonderful that many of them consented to take part. Therefore, I thank the coauthors Ines Bruchmann, Monika Janišová, Cindy Q. Tang, Caroline M. Tucker, Sula E. Vanderplank, Nigel P. Barker, V. Ralph Clark, Uwe Deppe, Sergio Elórtegui Francioli, Jihong Huang, Jan Jansen, Keping Ma, Andres Moreira-Muñoz, Masahiko Ohsawa, Jalil Noroozi, Gerhard Pils, Miguel de Sequeira, Marinus J.A. Werger, Wenjing Yang and Yongchuan Yang.

Flensburg, Germany Carsten Hobohm

Contents

Contributors

Nigel P. Barker Department of Botany, Rhodes University, Grahamstown, South Africa

Ines Bruchmann Ecology and Environmental Education Working Group, Interdisciplinary Institute of Environmental, Social and Human Studies, University of Flensburg, Flensburg, Germany

V. Ralph Clark Department of Botany, Rhodes University, Grahamstown, South Africa

Uwe Deppe Ecology and Environmental Education Working Group, Interdisciplinary Institute of Environmental, Social and Human Studies, University of Flensburg, Flensburg, Germany

Sergio Elórtegui Francioli Facultad de Ciencias de la Educación, Pontificia Universidad Católica de Chile, Santiago, Chile

Carsten Hobohm Ecology and Environmental Education Working Group, Interdisciplinary Institute of Environmental, Social and Human Studies, University of Flensburg, Flensburg, Germany

Jihong Huang Institute of Botany, Chinese Academy of Sciences, Beijing, China

Monika Janišová Institute of Botany, Slovak Academy of Sciences, Banská Bystrica, Slovakia

Jan Jansen Institute for Water and Wetland Research, Radboud University, Nijmegen, The Netherlands

Keping Ma Institute of Botany, Chinese Academy of Sciences, Beijing, China

Andrés Moreira-Muñoz Instituto de Geografía, Pontificia Universidad Católica de Chile, Santiago, Chile

Jalil Noroozi Department of Conservation Biology, Vegetation and Landscape Ecology, Faculty Centre of Biodiversity, University of Vienna, Vienna, Austria

Plant Science Department, University of Tabriz, Tabriz, Iran

Masahiko Ohsawa Institute of Ecology and Geobotany, Kunming University in China, Kunming, China

Gerhard Pils HAK Spittal/Drau, Kärnten, Austria

Miguel Pinto da Silva Menezes de Sequeira Centro de Ciências da Vida, Universidade da Madeira, Funchal, Portugal

Cindy Q. Tang Institute of Ecology and Geobotany, Yunnan University, Kunming, China

Caroline M. Tucker Department of Ecology and Evolutionary Biology, University of Toronto, Toronto, ON, Canada

Sula E. Vanderplank Department of Botany & Plant Sciences, University of California, Riverside, CA, USA

Marinus J.A. Werger Department of Plant Ecology, University of Utrecht, Utrecht, The Netherlands

Wenjing Yang Institute of Botany, Chinese Academy of Sciences, Beijing, China

Yongchuan Yang Faculty of Urban Construction and Environmental Engineering, Chongqing University, Chongqing, China

Introduction

If you want to cross a river without a boat or without having to swim then you have to look for a ford. Thousand years ago Hamburg developed as a crossroad-settlement near a ford across the River Elbe. The river is naturally shallow in this area because of the decreasing influence of the North Sea tides.

This section of the Elbe is subject to a combination of two major influences: tides from the west, on the one hand and freshwater from the east, on the other, making it a freshwater tidal area. And only here is the vascular plant species *Oenanthe conioides* to be found. This plant occurs in the Elbe area as a pioneer on the muddy soil of tidal flats with no salt water influence.

A shallow river with freshwater influenced by tides were optimal conditions for *Oenanthe conioides* to evolve and grow and for the first people to settle in this area. *Oenanthe conioides* is an endemic plant almost entirely restricted to the area of Hamburg with only few individuals occurring outside the city boundaries.

When we started our study of this plant very little was known about it. Is this taxon really a genetically isolated species or are there also hybrids with other species of the genus? Where exactly does the plant grow? What are its ecological and community characteristics? What do we know about its biological traits such as dispersal type or the possibility of building soil seed banks? What human influences pose a risk of extinction?

The combination of the very limited distribution of this species and a very limited knowledge of its biology including environmental conditions was surprising. Even in rich European countries, we realised, there is no guarantee that ignorance and extinction can be ruled out.

Today, in good years a few hundred individuals of *Oenanthe conioides* flower worldwide – that is, in Hamburg.

After analysing the ecology and biology of *Oenanthe conioides* we began to investigate other vascular plants with limited distributions. We established that plants which occur only in a specific nation are often relatively well-known because countries are normally proud to present such endemics. For many countries, such checklists can be found on the internet.

However, if the distribution area of plant species with a small range of occurrence covers two or more countries, people less often are aware of their special status. If, for example, half the populations of a particular plant species occur in Austria and the other half in Switzerland, the species is neither endemic nor subendemic to one of these countries; nevertheless, it is endemic to the Alps. For this reason we decided to begin compiling a checklist (EvaplantE) of all vascular plants which are restricted to Europe. Our aim was to pay attention to rare and endemic species, independent of whether these are country-endemics or not.

Such a list will never be complete. It is like a phone book. A great deal of effort is required to update the information or, in our case, to rework the taxonomy, as well as ecological and distribution data.

Our knowledge is growing rapidly, not only our knowledge of the ecology and genetics of vascular plant taxa, but also of methodologies of analyses in the field, from satellites, or statistics, respectively.

Endemism is a phenomenon with so many features that a single book could never give answers to all the questions concerning the biology of endemic plants including genetics and population dynamics, biogeography and geobotany, ecology, etc.

This book is an attempt to integrate information which helps to explain the occurrence of plants that are restricted to a single country, mountain range, mountain top, island or estuary, for example.

We describe and analyse important patterns and processes. We focus on the biogeography of endemic, rare and threatened vascular plants. We discuss where and why endemics are concentrated or missing in a region. We focus on the relationship between endemism and habitat, which we consider to be important for species conservation policies.

Almost all countries in the world signed the Convention on Biological Diversity and agreed to its main goal, which was to lower the extinction risk for species. We assume that endemism as an obligatory stage before extinction is one of the best indicators and predictors world-wide of the necessity for conservation activities.

> One important criterion for determining the optimal design for conservation units is the degree of biotic endemism.
> Gentry (1986: 153)

We hope to support the attempt to protect and conserve rare, threatened and endemic species and their habitats by publishing this book.

Flensburg, Germany Carsten Hobohm

Part I
The Meaning of Endemism

Chapter 1
The Increasing Importance of *Endemism*: Responsibility, the Media and Education

Carsten Hobohm and Caroline M. Tucker

1.1 What Is Endemism?

While the term *endemic* (from Greek *en demos* = in people) has connotations today for both medicine (disease) and biogeography, the term *endemism* is used in biogeography to refer to taxa that have small ranges. De Candolle (1820: 54) first defined the term *genres endémiques* (endemic taxa) in a biogeographical context (Fig. 1.1). He adapted the term from the medical meaning, where *endemic* described diseases that are present continuously in a certain area, to refer to the analogous concept of taxon restricted to a particular geographic region. If a disease is spreading through human populations across large regions then the disease becomes a *pandemic* disease (from Greek *pan demos* = across peoples) (Photo 1.1).

As the antonym to endemic taxa in biogeography, De Candolle used the term *genres sporadiques* for the more wide-spread taxa. He felt that locally or regionally restricted taxa normally occur in higher densities than wide-spread taxa with a more scattered distribution – which occasionally is observed (e.g. Rabinowitz et al. 1986).

Endemism is a function of spatial scale (cf. Laffan and Crisp 2003, and many others). Many small regions do not harbour any endemic taxon but solely non-endemics. At the scale of the earth, all taxa are of course endemic (Lu et al. 2007; Hobohm 2003). The necessary spatial context means that for every taxon, one can find a geographical unit where it exists as non-endemic (except when the taxon has become reduced to a last individual) and another larger area where it is endemic. Today, there are two general groups of biogeographic definitions of *endemism* or *endemic taxon*, and within each group the differences in practice are

C. Hobohm (✉)
Ecology and Environmental Education Working Group, Interdisciplinary Institute of Environmental, Social and Human Studies, University of Flensburg, Flensburg, Germany
e-mail: hobohm@uni-flensburg.de

C.M. Tucker
Department of Ecology and Evolutionary Biology, University of Toronto, Toronto, ON, Canada

C. Hobohm (ed.), *Endemism in Vascular Plants*, Plant and Vegetation 9,
DOI 10.1007/978-94-007-6913-7_1,

Parmi les phénomènes généraux que présente l'habitation des plantes, il en est un qui me paroît plus inexplicable encore que tous les autres : c'est qu'il est certains genres, certaines familles, dont toutes les espèces croissent dans un seul pays (je les appellerai, par analogie avec le langage médical, *genres endémiques*), et d'autres dont les espèces sont réparties sur le monde entier (je les appellerai, par un motif analogue, *genres sporadiques*).

Fig. 1.1 First book section with a definition of the term *genres endémiques* (endemic taxa) in a biogeographical context (De Candolle 1820: 54)

small. In the first group, endemism refers to taxa restricted to a certain sized area (e.g. 10,000 km^2) or number of cells within a geographic grid. The second group of definitions, which are independent of any artificial maximum size of a region, refer to taxa restricted to a defined geographic area or habitat type. Henceforth we will use the term endemic to correspond to the second group of definitions. This makes analyses such as Endemics-Area-Relationships (EARs), which include taxa with a variety of range sizes, possible.

What is the difference between endemism and rarity? Rabinowitz et al. (1986) noted that plants may be rare in several ways. They distinguished three traits that all species possess – a geographic range (broad area vs. endemic to a particular small area), habitat specificity (occurs in a variety of habitats vs. restricted to one or a few sites with special environmental characteristics) and local population size (large populations somewhere within its range vs. only small populations). According to Rabinowitz et al. (1986) a taxon can be rare in a region without being endemic (it has populations outside the region). It could also be endemic to the region but not rare, having large population sizes and/or broad habitat tolerance within its restricted range.

Cowling and Lombard (2002) stated that rarity or endemism *"is associated with the early (post-speciation) and late (pre-extinction) phases of the taxon cycle"*, which means that an obligatory stage in the distribution patterns of every taxon must be a phase of local endemic existence both at the beginning and the end of the taxon's existence. This would mean that every taxon originates and ends as endemic. Is this theoretically necessary? One could imagine special cases in which a wide-spread species is dividing into two species by long-term diverging genomes which ultimately are genetically isolated; the new taxa could each occupy half of the former range. In this case both "new" species do not start the taxon cycle in a small region or with a few individuals only. However, it is perhaps more likely that speciation events initially involve new taxa that are limited to only small portions of the original range, or to small areas of new habitat, and thus are endemic.

Regions that are rich in endemic taxa are referred to as having *high endemism.* The global distribution of regions with high and low endemism is incompletely

Photo 1.1 *Alstroemeria magnifica* near La Serena, Chile (photographed by Marinus Werger). Almost all c. 120 *Alstroemeria* spp. are restricted to one of two centres, one in central Chile, the other in E Brazil

recorded and the causes not fully understood. Yet, it is recognized that these designations are important to conservation planning and activities. Though we have not fully explained how and why regions of high endemicity arise, and how they are maintained, we know many processes and conditions in the past and present that are likely to be relevant. Current hypotheses and models on endemism are useful even in the face of this uncertainty: as Gentry (1986) showed, different explanations for diversity patterns may have very different implications for nature conservation strategies and practices. For example: the *habitat specialisation model* explains distribution patterns of local endemics by differences in soil, water or microclimate parameters (soft boundaries), whereas the *refugia model* relates endemism to a strong relationship between the location of (Pleistocene) refugia. The implication of the habitat specialisation model would be to protect landscapes, including all different habitat types, whereas the refugia model would primarily mean that as many of the refugial centers as possible should be protected by establishing reserves. We do not know much about distribution patterns of local endemics (e.g. in diverse tropical forest zones), and thus we must rely on the theories available. As a consequence, as many representative regions as possible of both habitat types and refugia should be protected.

The species concept that is mainly based on morphological, phenotypic similarities and on the idea of a putative reproduction barrier that disrupts gene-flow between distinct taxa was largely contradicted by data obtained from molecular analyses (genome analyses, allozyme analyses). Several phylogenetic studies have shown that speciation is to be understood as a more continuous process rather than a process which occurs at a discrete specific moment in time.

A comprehensive assessment which applied methods of morphological and genetical analyses (DNA barcoding) to species numbers of Madagascar's amphibian inventory, for example, resulted in an almost two-fold increase in species numbers (Vieitesa et al. 2009). Authors found large genetic variations that could no longer be interpreted as simple intraspecific genetic variations. In the light of modern genetic analyses it becomes clear that evaluating endemism is also a matter of the underlying species concept (see e.g. Hawksworth and Kalin-Arroyo 1995; Bruchmann 2011).

1.2 The Increasing Importance of Endemism in the Media

The growing attention on *endemism* and *endemic species* can be visualized based on the rising number of scientific publications including these terms. Though in 1820 De Candolle defined endemism in a biogeographical context, the term was not regularly used before the beginning of the twentieth century. Since that time the words *endemism* and *endemic* taxa began to occur more or less regularly in scientific publications, particularly in relation to the scientific description of new species with a restricted distribution or to the recognition of threats. In the last quarter of the twentieth century, especially in the late 1980s and 1990s, the term endemism began exploding. Many taxa in checklists and Red Lists became labeled as endemics. For scientists and students in the field of ecology and biogeography, endemism became a central issue.

The concept of *Biodiversity Hotspots* was coined by Norman Myers in 1988 and 1990. Biodiversity hotspots are regions of the highest conservation priority, based on a number of characteristics determining their contribution to global diversity and the level of threat they face: one requirement is that these regions have high endemism. Today these biodiversity hotspots play a leading role in international and national nature conservation strategies. In 1988, the United Nations Environmental Programme (UNEP) convened an Ad Hoc Working Group of Experts on Biological Diversity to explore the need for an international Convention on Biological Diversity (CBD). As a result this Convention was opened for signing in 1992 at the United Nations Conference on Environment and Development in Rio de Janeiro. The concepts of endemism and endemic species were included within this declaration (see e.g. Annex I of the CBD). Today endemism and endemic species are part of an uncountable number of scientific publications, national and international laws and conventions, internet sites (e.g. www.iucn.org), and many documentary films on TV, that use these terms for characterizing globally rare and threatened animals and plants. The terms endemism and endemic, in former times only used in science, are now becoming more and more common to non-scientists in many countries, many languages, environmental protection laws and nature conservation measures all over the world.

Furthermore, endemics are not only entities in nature conservation, scientific or political efforts. Endemism in general or particular endemic plants and animals are topics included for advertisement in tourism industry, especially in ecotourism.

Endemic plants may be used regionally as ornamental and medicinal plants (cf. El-Darier and El-Mogaspi 2009; Latheef et al. 2008). We saw different endemics used as ornamental plants in gardens, e.g. *Adansonia* and *Aloe* spp. on Madagascar, orchids in Colombia, *Proteas* in South Africa, and palms on Mauritius. Many plants widely if not globally used in horticulture, such as *Rhododendron* spp., *Metasequoia glyptostroboides* or *Ginkgo biloba*, were once restricted to small regions (cf. Zhao 2003). In a few cases, some plant families such as Cactaceae are almost completely represented in botanic gardens, many others, e.g. Orchidaceae, Asteraceae or Melastomataceae, are only represented by a small proportion of their total number of species (cf. Rauer et al. 2000).

Where endemic species have economic value, there can be positive or negative consequences. Endemic plants, e.g. cultivars of *Argyranthemum* spp. which are native on Canary Islands, are legally traded by florists on markets in Europe. Additionally, illegal trade in plants and animals is a serious and growing crime, and affects endemic as well as non-endemic species. The revenues generated by trafficking in endangered species (including, but not limited to endemic species) are estimated at 18–26 billion dollars per year (Europol 2011; Alacs and Georges 2008; Flores-Palacios and Valencia-Diaz 2007, see also Maggs et al. 1998).

1.3 Endemism, Responsibility and Education

Though the meaning of endemic species for nature conservation management is well recognized, it is also recognized that there is a paucity of information about diversity and endemism: "*educational value of biological diversity*" and the "*general lack of information and knowledge regarding biological diversity*" is pronounced in different chapters of the Convention on Biological Diversity (see e.g. Preamble). According to Article 13 – *Public Education and Awareness* – of the CBD

> The Contracting Parties shall:
>
> (a) Promote and encourage understanding of the importance of, and the measures required for, the conservation of biological diversity, as well as its propagation through media, and the inclusion of these topics in educational programmes; and
> (b) Cooperate, as appropriate, with other States and international organizations in developing educational and public awareness programmes, with respect to conservation and sustainable use of biological diversity.

Public preferences play an important role in prioritizing species for nature conservation activities. Stating that species with small ranges are important indicators and subject to the risk of extinction should be picked out as central theme in the education of children at schools and curricula. Meuser et al. (2009) found a preference of the public for endemism over other conservation-related species-attributes such as economic importance, regionally at risk but common elsewhere, cultural and traditional importance.

Endemism, the risk of extinction, and our responsibility in halting the loss of biodiversity might be topics in school subjects such as biology, philosophy, geography, economy, social studies. However, the reality seems to be far away from the education-related goals of the CBD. The term *biodiversity* has entered the schoolbooks in biology in different languages and countries (e.g. in Europe and North America). Furthermore, it is very easy to find teaching concepts on biodiversity on the internet in different languages. However, the term "biodiversity" does not automatically imply the meaning *halting the loss of biodiversity, nature conservation* or *sustainability* as is stated by the CBD. One can talk and teach about the diversity of life without any connection to questions of nature conservation.

Since children are responsible for tomorrow, it is clear that the discussion about how endemism is linked to aspects of species' survival and nature conservation policies and that this discussion should be intensified in schools and other programs (cf. Meuser et al. 2009; Ugulu et al. 2008). Teaching material about endemism, endemic species, the risk of extinction and/or histories of species which have been rescued, might easily be prepared for all types of school. For example, stories of species which were rescued from extinction due to activities in zoos or botanical gardens can result in positive outlooks on conservation by children at primary schools. More complex situations, including economic, social and ecological constraints, might be discussed in secondary schools or universities.

And in any case, there should be a focus at all levels on the target of halting the loss of biodiversity by protecting landscapes and habitats that harbour endemics.

References

Alacs E, Georges A (2008) Wildlife across our borders: a review of the illegal trade in Australia. Aust J Forensic Sci 40(2):147–160

Bruchmann I (2011) Plant endemism in Europe: spatial distribution and habitat affinities of endemic veascular plants. Dissertation, University of Flensburg, Flensburg. URL: www.zhb-flensburg.de/dissert/bruchmann

Cowling RM, Lombard AT (2002) Heterogeneity, speciation/extinction history and climate: explaining regional plant diversity patterns in the Cape Floristic Region. Divers Distrib 8: 163–179

De Candolle AB (1820) Essai elementaire de geographie botanique. In: Dictionnaire des sciences naturelles, vol 18. Flevrault, Strasbourg, pp 1–64

El-Darier SM, El-Mogaspi FM (2009) Ethnobotany and relative importance of some endemmic plant species at El-Jabal El-Akhdar Region (Libya). World J Agric Sci 5(3):353–360

Europol (ed) (2011) EU organised crime threat assessment. Europol, The Hague, File No. 2530-274

Flores-Palacios A, Valencia-Diaz S (2007) Local illegal trade reveals unknown diversity and involves a high species richness of wild vascular epiphytes. Biol Conserv 136:372–387

Gentry AH (1986) Endemism in tropical versus temperate plant communities. In: Soulé ME (ed) Conservation biology: the science of scarcity and diversity. Sinauer Associates, Inc.-Publisher, Sunderland, pp 153–182
Hawksworth DL, Kalin-Arroyo MT (1995) Magnitude and distribution of biodiversity. In: Heywood VH (ed) Global biodiversity assessment. UNEP – United Nations Environment Programm, Cambridge, pp 107–191
Hobohm C (2003) Characterization and ranking of biodiversity hotspots: centres of species richness and endemism. Biodivers Conserv 12:279–287
Laffan SW, Crisp M (2003) Assessing endemism at multiple spatial scales, with an example from the Australian vascular flora. J Biogeogr 30:511–520
Latheef SA, Prasad B, Bavaji M, Subramanyam G (2008) A database on endemic plants at Tirumala hills in India. Bioinformation 2(6):260–262
Lu H-P, Wagner HH, Chen X-Y (2007) A contribution diversity approach to evaluate species diversity. Basic Appl Ecol 8(1):1–12
Maggs GL, Craven P, Kolberg HH (1998) Plant species richness, endemism, and genetic resources in Namibia. Biodivers Conserv 7:435–446
Meuser E, Harshaw HW, Mooers AÖ (2009) Public preference for endemism over other conservation-related species attributes. Conserv Biol 23(4):1041–1046
Myers N (1988) Threatened biotas: hotspots in tropical forests. Environmentalist 8:1–20
Myers N (1990) The biodiversity challenge: expended hotspots analysis. Environmentalist 10: 243–256
Rabinowitz D, Cairns S, Dillon T (1986) Seven forms of rarity and their frequency in the flora of the British Isles. In: Soulé ME (ed) Conservation biology: the science of scarcity and diversity. Sinauer Associates, Inc.-Publisher, Sunderland, pp 182–204
Rauer G, Ibisch PL, von den Driesch M, Lobin W, Barthlott W (2000) The convention on biodiversity and botanic gardens. In: Bundesamt für Naturschutz (ed) Botanic gardens and biodiversity. Landwirtschaftsverlag, Münster, pp 25–64
Ugulu I, Aydin H, Yorek N, Dogan Y (2008) The impact of endemism concept on environmental attitudes of secondary school students. Natura Montenegrina Podgorica 7(3):165–173
Vieitesa DR, Wollenberg KC, Andreone F, Köhlerd J, Glawe F, Vencesb M (2009) Vast underestimation of Madagascar's biodiversity evidenced by an integrative amphibian inventory. PNAS 106:8267–8272
Zhao L (2003) Ornamental plant resources from China. In: Lee JM, Zhang D (eds) Acta Horticulture 620: Asian plants with unique horticultural potential – genetic resources, cultural practices, and utilization. International Society for Horticultural Science, Leuven

Chapter 2
How to Quantify Endemism

Carsten Hobohm and Caroline M. Tucker

2.1 Number of Individuals (N)

Counting or estimating the number of individuals in a population is only possible if the whole population is rather small and well documented. Because of the high extinction risk faced by small populations, the number of individuals should be continuously monitored. The IUCN Red List of threatened plants (Baillie et al. 2004; Walter and Gillett 1998, see also IUCN on the internet) uses information about the number of individuals to categorize level of threat experienced. In practice, as the examples that follow will show, almost all taxa with low numbers of individuals are categorized as critically endangered. Some spectacular examples of plant species which at the moment are represented by only a few individuals have been published, and we detail some of these below. Many but not all examples of plant species represented only by few individuals are restricted to marine islands. Unfortunately, some other plant taxa recently disappeared altogether from the globe – no longer having a single living individual. Avoiding extinction cannot be guaranteed, regardless how intensive the efforts to rejuvenate the species are.

The following examples are primarily in 2011 and 2012 adopted from the very important IUCN Red List (www.iucnredlist.org). However, some of the assessments are 10 or 20 years old and, therefore, cannot be assumed to represent recent conditions and should be updated. These examples are ordered in a geographical manner: from west to east and/or north to south, beginning with North America.

C. Hobohm (✉)
Ecology and Environmental Education Working Group, Interdisciplinary Institute of Environmental, Social and Human Studies, University of Flensburg, Flensburg, Germany
e-mail: hobohm@uni-flensburg.de

C.M. Tucker
Department of Ecology and Evolutionary Biology, University of Toronto, Toronto, ON, Canada

C. Hobohm (ed.), *Endemism in Vascular Plants*, Plant and Vegetation 9,
DOI 10.1007/978-94-007-6913-7_2, © Springer Science+Business Media Dordrecht 2014

2.1.1 California

Cercocarpus traskiae is a shrub or small tree, living on Catalina Island. Only a single wild population still exists, consisting of seven individuals in a canyon covering an area of approximately 250 km^2 (World Conservation Monitoring Centre 1998; Oldfield et al. 1998).

2.1.2 Mexico

Diospyros johnstoniana (*syn. D. xolocotzii*) is endemic to Michoacan de Ocampo, Mexico (Madrigal Sánchez and Guridi Gómez 2002; Madrigal Sánchez and Rzedowski 1988). It is known only from an area of 25 ha. Despite many surveys since 1998, no other locations have been found. In 2005, a census for this species found 36 trees, but in 2006 only 34 individuals remained. One was felled for agricultural expansion, and the other was severely damaged by human-caused fire in 2006 and died in the same year. The remaining population is fragmented.

D. johnstoniana grows in subtropical dry forest and woodland. It is a plant with a low number of fruits observed and a low rate of pollination success. It reaches sexual maturity at the age of about 25 years old. There is no trade in this species. But, the fruit is commonly eaten, and the species can be used as an ornamental plant. The plant can also be used for the genetic improvement of other species in the same genus. The species is severely impacted by habitat loss through agriculture, which directly threatens the remaining trees. Clear-cutting in the area is common practice to open new areas for agriculture. Furthermore, the presence of cattle and goats in the area threatens the remaining trees: livestock eat young plants and near-ground level foliage, severely affecting the surrounding habitat and compacting the ground. The human population is increasing in the area, and urban expansion is ongoing. This has several impacts on *D. johnstoniana*, as there is an increase in solid waste, human-caused fire, wood collecting, and the introduction of alien plant species, e.g. *Eucalyptus* trees. The species is included in Mexico's official list of species at risk, in the category of Special Protection. However, this is not enforced for this species and there is still no specific programme to protect this species (Villaseñor Gómez 2005; www.iucnredlist.org).

Mammillaria sanchez-mejoradae is a critically endangered cactus endemic to a single area in Mexico. Since the discovery of the species more than 20 years ago, the population has diminished by an estimated 75 %. The wild population currently is estimated at less than 500 plants and, despite legal protection of the species, the population continues to be highly threatened by illegal collecting (www.iucnredlist.org).

Acharagma aguirreanum is a cactus which occurs in semi-desert on calcareous rocks at an altitude of about 1,500 m in Sierra de la Paila over a range of 1 km^2, in Coahuila, Mexico. The total population numbers less than 100 individuals. Illegal collecting is a major threat (Anderson et al. 2002; Anderson 2001; Hunt 1999; Glass 1998).

2.1.3 *Costa Rica*

The present adult population of the palm *Cryosophila cookii* is estimated to number less than 100 individuals. The species is living in atlantic lowland rainforest near Limon, Costa Rica. Habitat conversion to arable land has caused a major population decline. Logging, increasing settlements and decline in the populations of dispersal/pollination agents have also contributed to losses. The palm heart is eaten locally for medicinal purposes (Evans 1998; Oldfield et al. 1998).

2.1.4 *Ecuador*

Centropogon cazaletii is an endemic herb or subshrub in Ecuador where it occurs in high Andean forest (3,500–4,000 m). It is known from two collections in Napo Province, both made inside the Reserva Ecológica Cayambe-Coca. One was in the surroundings of the Laguna de San Marcos; another from Oyacachi. After two field trips of experts to search for the species in 1998, no individual was recorded around the Laguna and only four individuals were found in Oyacachi.

Although the taxon occurs inside a protected area, it might be threatened by fire set by humans around the Laguna de San Marcos and deforestation around Oyacachi.

Another species of the same genus, *Centropogon pilalensis* is also endemic to Ecuador. It is known only from one population of less than 50 individuals in Cotopaxi province where it is living in high Andean forest and dry paramos (Moreno and Pitman 2003; Valencia et al. 2000).

In the Red List of the IUCN (www.iucnredlist.org) many critically endangered species of Ecuador are only known from the type collection and nobody knows how many of them have meanwhile become extinct (cf. Valencia et al. 2000).

2.1.5 *Juan Fernandez Islands, Chile*

In the following (Fig. 2.1), Skottsberg tells the story about a journey to the last living individual of *Santalum fernandezianum* which he saw and photographed on Robinson Crusoe Island (also known as Masatierra) in the year 1908.

2.1.6 *Cerrado, Brazil*

Dimorphandra wilsonii is a critically endangered tree, found only in Minas Gerais State in southeast Brazil. Before 2010 (last assessment 2006), there were only some ten mature trees and six juveniles growing in the wild, in the middle of pastures of *Brachiaria*, an alien invasive grass species, in a strongly deforested and fragmented

140 THE WILDS OF PATAGONIA

IN MEMORY OF
ALEXANDER SELKIRK,
Mariner,
A native of Largo in the county of
Fife, Scotland.
Who lived on this island in complete solitude, for four years
and four months.
He was landed from the Cinque
Ports galley, 96 tons, 16 guns, a.d.
1704, and was taken off in the
Duke, privateer, 12th Feb. 1709.
He died lieutenant of H.M.S. Weymouth, a.d. 1723, aged 47 years.
This tablet is erected
near Selkirk's lookout by
Commodore Powell and the
officers of H.M.S. Topaze, a.d. 1868.

This is the historical basis of Defoe's work. It may look somewhat meagre, but one can understand that poor Selkirk had to work to preserve his life. What a mental trial, not to hear a word spoken by another, not to see a human soul for four years and four months! Thus his fate was pretty adventurous even if told without embellishment. On the other hand, he left his ship at his own request, discontented with the life on board. Besides, he might have chosen a worse place. The climate is very mild, it rains just enough, snow or frost is unknown. A few plants are edible, and the goats, which were much more numerous in Selkirk's time than they are now, provided him with fresh meat.

Through a walk lined with marvellous trees and precious ferns we pass the natural gate and are on the south side of the island. Down it goes, almost as precipitous as on the other side. We have a magnificent view of the coast and Santa Clara, where a tremendous surf roars. Soon we came out of the forest,

ROBINSON CRUSOE'S ISLAND 141

and continued on to the barren slopes near the sea. The vegetation here is more like that of a steppe, with short grass and some heath-plants; only along the streams is there a bright green strip, a mosaic of gigantic pangue-leaves. And we bent the thick stalks at the side and drank to the health of Masatierra and Robinson and the whole world. There is only one way back, the way we had come; it was getting dark and we hurried on through showers of rain; large drops splashed on the heads of the rosette-trees, the soil emitted strong, peculiar scents. The last part of the way we slid down in the slippery clay.

Above I happened to mention the sandal-wood. The discovery of this kind of wood, famous since the days of Solomon, on Juan Fernandez most surely attracted notice. We have no reports of it previous to 1624, when, according to Burney, L'Heremite reported sandal-trees in great number. According to another authority ships used to visit the place as early as 1664 to bring the valuable wood to the coast, where it was highly appreciated. One did not think of preserving anything; a hundred years later it was hardly possible to find a living tree, and in the beginning of last century it was regarded as extinct. No botanist had ever seen the leaves or flowers. Suddenly F. Philippi in Santiago got some fresh twigs brought to him in 1888; he found them to belong to the genus *Santalum;* the species being new, it received the name of *S. fernandezianum.* The general interest in the tree was increased, but nobody told where the branches came from; a living tree was still unknown. Only in 1892 did Johow

142 THE WILDS OF PATAGONIA

get news of one; a colonist had found it in Puerto Ingles, high up in the valley. He was the first botanist who saw this plant. It is easily understood that I was anxious to become the second. How many people had looked for other specimens! All their efforts were fruitless; as far as we knew Johow's tree was the very last. If it were still there!

The man who brought Johow to the spot still lived, and after we had explained our purely scientific interest he promised to send his son with us. It would have been more than uncertain for us alone to look for a single tree in a valley clad with virgin forest.

It is possible to climb across the ridge that separates Cumberland Bay from the English Harbour, but we preferred to go there with a well-manned boat. The landing is, as in most places on the islands, performed with some risk; one must jump just at the right moment, and there has to be a good crew in the yawl, or the boat would be thrown on the rocks and capsized. Perhaps I ought to mention that the place in question only has the *name* of a harbour. We walked up the valley and made an ascent of the western side; the place is so steep that one is forced to grasp the trees and shrubs to get a foothold. Our guide stopped, looked round for a minute, down a few hundred yards, and we had reached our destination. The last sandal-tree. Absolutely the last descendant of *Santalum fernandezianum.* It is so queer to stand at the death-bed of a species; probably we were the last scientists who saw it living. We look at the old tree with a religious respect, touch the stem and the firm, dark green leaves—it is not only

ROBINSON CRUSOE'S ISLAND 143

an individual, it is a species that is dying. It cannot last very long. There is only one little branch left fresh and green; the others are dead. We cut a piece to get specimens of the peculiar red, strongly scented wood. A photo was taken, I made some observations on the place, and we said good-bye. Should I happen to go there once more I shall not see the sandal-tree; it will be dead and its body cut up into precious pieces—curiosities taken away by every stranger.

Fig. 2.1 Travelogue of Skottsberg about a journey to Robinson Crusoe Island where he visited the last individual of *Santalum fernandezianum* (Skottsberg 1911: 140–143)

region. The species is threatened by deforestation for charcoal production, which is the most important threat to the Cerrado Biome. There is also deforestation for pasture establishment and any seedlings face competition from *Brachiaria*. People also deliberately eradicate this species because its seeds can be harmful to pregnant cattle (cf. Moreira Fernandez 2006; Alves 2004, Mendonça).

2.1.7 Puerto Rico

Some ten trees of *Auerodendron pauciflorum* are recorded on Puerto Rico where a single small population is found in woodland on a limestone cliff. Most individuals occur on land owned by a development company. The trees have not been seen to seed but efforts are being made to cultivate the plant from tissue culture. Information needs to be updated because the last year assessed was 1998 (cf. World Conservation Monitoring Centre 1998; Oldfield et al. 1998).

2.1.8 Anegada, Virgin Islands

Anegada belongs to the British Virgin Islands and is located northwest of Puerto Rico. The population of *Acacia anegadensis*, endemic to Anegada, has been reduced by past exploitation for resin. The potential extent for *A. anegadensis* is approximately 25 km^2 because fieldwork has determined that the species is found primarily on limestone and scattered in sand dune habitats. Different numbers of individuals have been recorded, from a few mature trees to locally common (Baillie et al. 2004; Clubbe et al. 2003). However, many human and natural effects such as habitat destruction, livestock, invasive species, or hurricanes cause a risk of extinction. Anegada is under severe development pressure resulting in both loss of habitat to residential and tourism infrastructure, and further fragmentation due to upgrading and construction of new roads. Loose cattle, goats and donkeys roam the island and damage the habitat through trampling and grazing. Natural disasters are a current and on-going threat e.g., hurricanes and coastal inundation (Clubbe et al. 2003; Smith-Abbott et al. 2002; Proctor and Fleming 1999; D'Arcy 1971).

2.1.9 St. Helena

Trochetiopsis erythroxylon (Redwood) is a tree endemic to St. Helena and is extinct in the wild. After settlers arrived, the species was extensively exploited for its excellent timber and bark which was used for tanning hides. By 1718, the species was already extremely rare. Further losses occurred when flax plantations began in the late 1800s. By the mid of the twentieth century, only one redwood individual survived and this single tree is the source of all the St. Helena's Redwoods known in cultivation today. Inbreeding depression and a depauperate gene pool are the most serious threat to the future survival of this species.

Trochetiopsis ebenus, called Saint Helena Ebony, is a critically endangered shrub found only on this island. The species declined sharply in the eighteenth century, mainly due to overgrazing by goats, and was once thought to be extinct. In 1980, two shrubs were rediscovered on the island. All existing material in cultivation is derived from these two individuals. This species was previously burned to produce mortar. The wood was also used in the nineteenth century for turnery and ornament making and was introduced to British gardens around 1800.

The hybrid of cultivated plants of *Trochetiopsis erythroxylon* × *ebenus* may provide the only chance of survival for this part of the gene pool (Cairns-Wicks 2003).

Also *Mellissia begoniifolia* is a critically endangered shrub found on St. Helena Island. The wild population currently numbers some 16 individuals. The size of this population fluctuates year by year, largely depending on weather conditions, but also on predation pressure. Currently only one plant in the population can be considered mature and it is from this that the majority of seeds has been collected to establish plants in cultivation. Threats include attacks from aphids and caterpillars, mice and rabbits. Growing in such a dry environment, the plants are also prone to drought (www.iucnredlist.org).

Nesiota elliptica was a small endemic tree on St. Helena that grew on the highest parts of the island's eastern central ridge. It became very rare in the nineteenth century and by 1875 only 12–15 trees were recorded. The species had been thought to become extinct until a single tree was discovered in 1977. This tree was found to suffer from fungal infections which might have been exacerbated by damage sustained during attempts to conserve it. Because cuttings were difficult to root, because the species very rarely set good seed as it was almost completely self-incompatible, and because of fungal infections the species died in the wild in 1994 and became globally extinct in 2003. No other live material, plants, seeds or tissues, remain in local or international collections. The extinction of this plant has been attributed to habitat loss through felling for timber and to make way for plantations.

2.1.10 Germany

Oenanthe conioides is an annual or biannual pioneer herb living at the Elbe river in and nearby Hamburg. The number of individuals is fluctuating from year to year; normally some hundred individuals are counted. A serious problem for the survival of this species is the destruction of habitat including alteration of water depth, currents and tides, caused by governmental authorities of the harbour of Hamburg and river traffic. Different neophytes inhabit the remaining natural habitats which in this case obviously do not cause any problem to *Oenanthe conioides*. Fortunately, the species is building a soil seed bank in muddy substrates (Below et al. 1996).

2.1.11 Alborán, Spain

The annual herb *Diplotaxis siettiana* was last seen in 1974 in the wild. At that time seeds were fortunately collected on the island of Alborán, south of Spain's mainland, where the plant was growing in a tiny area around a helicopter platform. The seeds were multiplied at the University of Madrid, and distributed to some botanical gardens. Under cultivation conditions high germination rates can fortunately be achieved. A re-introduction programme has been started and has become more and more successful since 1999 (Montmollin and Strahm 2005).

2.1.12 Sicily Archipelago, Italy

Some 30 trees of *Abies nebrodensis* grow at an altitude of 1,500 m altitude on limestone soil. The Madonie Mountains, Sicily, rising to 1,980 m, were once covered by *Abies nebrodensis*. Degraded natural habitat, the poor health of specimens propagated in tree nurseries, the limited population size, and threat from fire represent the biggest threats to the species. Hybridization with non-native firs results in genetic contamination.

Foresters immediately initiated conservation measures. However, soil degradation in the natural habitat has made re-introduction difficult. Researchers from Palermo University are currently investigating the species' ideal growth conditions. The species has grown well in several European botanic gardens. An EU LIFE-financed project is being carried out to conserve the existing population. The project includes implementing an action plan which would include forest management, conservation, and the gradual elimination of non-indigenous fir species. The goal is to stabilize the current population and improve the survival rate based on natural reproduction. Their location within the Madonie Regional Park guarantees some level of protection. In 1978, following seed collection, the forestry service cultivated 110,000 young trees in a nursery. Since the survival rate in nature is extremely low, an adoption programme was set up in parallel. 40,000 young plants have been planted in the Botanical Garden of Palermo, Sicily, as well as in gardens and second homes in the Madonie Mountains, slightly away from their natural area of distribution. Several mature trees also grow in botanic gardens and arboreta elsewhere in Europe. For *ex situ* cultivation of *A. nebrodensis*, areas should be selected that are not home to other fir trees to prevent genetic contamination (Farjon et al. 2006; Montmollin and Strahm 2005; Virgilio et al. 2000; Ducci et al. 1999).

Bupleurum dianthifolium, a small shrub, is endemic to the island of Marettimo, part of the Egadi archipelago, west of Sicily, Italy. It grows in only a few locations on the northern side of the island in an area of 5 km^2. It is estimated that approximately 300–500 individuals remain. The small, cushion-shaped perennial shrub grows on calcareous cliffs at an altitude of 20–600 m, preferring north-facing slopes and growing in the cracks of limestone rock faces.

This species is considered to be an old, paleoendemic taxon, which means that it was once much more widely distributed than today, and probably grew throughout the mountains of the Mediterranean when the region had a tropical climate. The plant reproduces from seeds only, a common characteristic of plants growing in such habitats. Mist is probably its main source of water (Gianguzzi and La Mantia 2006; Gianguzzi et al. 2003; Fabbri 1969). In this case global warming might help this species to survive.

2.1.13 Malta Archipelago

Cheirolophus crassifolius has a patchy distribution along the northwestern and southern cliffs of the islands of Malta. The total wild population is estimated at a thousand individuals. This perennial shrub is confined to coralline limestone of seaside cliffs and scree, growing in full sun. The so-called Maltese Rock-centaury is the National Plant of Malta. The species displays some ancient traits in its habitat preference and flower morphology, and is considered to be paleoendemic, which means that it speciated in the distant past and may have been much more widely distributed than today.

The species is threatened by a number of factors. First, it is rare to find juvenile plants of this long-lived species, possibly due to the larvae of an unidentified moth observed attacking the developing fruits. Second, the habitat is under threat from quarrying, as fragile boulder cliffs collapse from the pressure wave of nearby dynamite explosions. Third, a number of sites have been affected by human disturbance, especially those most easily accessible (Stevens and Lanfranco 2006; Lanfranco 1996, 1995).

2.1.14 Cyprus

Scilla morrisii is found only in three locations on Cyprus. Less than 600 individuals are known to exist and the survival of the species depends on conservation of the remaining oak forests. Logging for timber, road construction and expansion of farmland has considerably reduced these forests and large, old oak trees have become rare and scattered where there used to be a closed forest cover. Currently the *S. morrisii* population does not seem to be declining. The species entered the 2006 Red List as critically endangered (Kadis and Christodoulou 2006; Montmolin and Strahm 2005; Meikle 1977, 1985).

Centaurea akamantis is only found on the Akamas peninsula in the northwestern part of Cyprus. There are only two small fragmented subpopulations covering an area of less than 1 km^2, and the extent and quality of its habitat is declining. One subpopulation has only 50 individuals, and the other approximately 500. The subpopulations are geographically isolated from each other and if one of them disappears it is unlikely to be recolonized from the other.

This semi-woody herbaceous plant colonizes steep and humid limestone cliffs in the Avakas and Argaki ton Koufon ("Stream of Snakes") Gorges. *Centaurea akamantis* is characterized by an extremely long flowering and fruiting period.

The number of mature individuals has remained stable since 1993, when the species was first described, although the increasing number of visitors to the Akamas peninsula is contributing to a decline in habitat quality. Grazing poses a serious threat, even though it is not permitted in these areas (Kadis and Christodoulou 2006).

2.1.15 Other Extremely Rare Plants on Mediterranean Islands

Montmollin and Strahm (2005) published a booklet about the so-called "Top 50 Mediterranean Island Plants: wild plants at the brink of extinction, and what is needed to save them". Many of the plants listed are restricted to a few or single populations with countable numbers of individuals. Examples are *Silene hecisiae* (Aeolian Islands, c. 430 individuals), *Apium bermejoi* (Minorca, <100 ind.), *Arenaria bolosii* (Majorca, <200 ind.), *Euphorbia margalidiana* (Ses Margalides, Balearic Islands, <200 ind.), *Ligusticum huteri* (Majorca, <100 ind.), *Bupleurum kakiskalae* (Crete, c. 100 ind.), *Convolvulus argyrothamnos* (Crete, c. 30–35 ind.), *Horstrissea dolinicola* (Crete, a few dozen ind.), *Allium calamarophilon* (Euboea, a few ind.), *Minuartia dirphya* (Euboea, <250 ind.), *Ribes sardoum* (Sardinia, c. 100 ind.), *Bupleurum elatum* (Sicily, c. 400–600 ind.), *Zelkovia sicula* (Sicily, c. 200–250 ind.).

2.1.16 Algeria and Tunisia

The orchid *Serapias stenopetala* is found in small, scattered populations of less than 250 individuals in total in Algeria and Tunisia (www.iucnredlist.org). This critically endangered species is threatened by the destruction of its habitats through building of new roads, trampling and grazing by livestock, and the creation of an Animal Park in Brabtia, Algeria. Trade of all orchids is regulated under Annex B of the Convention on International Trade in Endangered Species of Wild Fauna and Flora (CITES), and a specific policy of protection and conservation is urgently needed (De Bélair 2009; De Bélair et al. 2005; De Bélair and Boussouak 2002).

2.1.17 South Africa, Namibia

Aloe pillansii is a critically endangered tree aloe occurring primarily in the mountainous area of the Northern Cape, South Africa and southern Namibia. A decline in the population has reduced the numbers to less than 200 mature individuals. No recruitment has been recorded at any of the main sub-populations probably due to the impacts of grazing by goats and donkeys, and the older plants are dying. Today, the species is the focus of a new survey and possible reintroduction programme by members of the IUCN/SSC Southern African Plant Specialist Group (Loots and Mannheimer 2003; Hilton-Taylor 1998; Williamson 1998; Menne 1992).

Two collections were made of *Kniphofia leucocephala* in 1970 near Lake Msingazi in northern KwaZulu-Natal, South Africa. The plant grows in coastal wetlands (Baijnaith 1992). No further plants were collected until 20 years later when a new population was discovered in a wetland that was being planted to timber trees. By that time the habitat had almost completely been transformed into commercial forestry plantations and urban expansion. Hence it is impossible to guess what the past range and population size have been. The population discovered in 1990 is still the only known living population, and the wetland where it grows is owned by a forestry company and is completely surrounded by plantations (cf. Scott-Shaw 1999). In 1991 there were about 70 plants, a number declining to about 21 individuals in 1998. Fortunately, the forestry company developed a strong conservation focus, the wetland was declared a natural heritage site, considerable effort has been put in since 2000 to rehabilitate the wetland and the forestry company implemented a conservation management plan. The result is that the *K. leucocephala* population is now thriving and numbers have increased to around 350 individuals (www.platzafrica.com and www.iucnredlist.org, downloaded 21st of September 2010).

Only a single individual of the Cycad *Encephalartos woodii* – now extinct in the wild – was ever found in South Africa. Its extinction in the wild may have been hastened by over-exploitation for medicinal purposes by local people. There is no likelihood of ever reintroducing the species back into the wild as there are only male plants in existence, and the risk of theft would be too great (Donaldson 2009; Baillie et al. 2004). Unless a female plant is found, *E. woodii* will never reproduce naturally. However, the next best thing has been accomplished. *Encephalartos woodii* forms fertile hybrids with *E. natalensis*. If each offspring is subsequently crossed with *E. woodii* and the process is then repeated, after several generations, the female offspring will be close to a female of *Encephalartos woodii*. Genetic analysis of chloroplast DNA of F1 hybrids between *E. woodii* and *E. natalensis* showed that all chloroplasts are inherited from the female *E. natalensis*, indicating that multigenerational hybrid offspring would have *E. natalensis* chloroplasts and could never be pure *E. woodii* (Cafasso et al. 2001; Osborne 1986; Giddy 1984).

2.1.18 Tanzania

Toussaintia patriciae is an endangered species known from less than 30 trees in the Udzwunga Mountains National Park and West Kilombero Nature Reserve (Deroin and Luke 2005). The plant occurs in very low numbers where found, though it is cryptic when not flowering and hopefully may be more common than is currently known. It is considered relatively secure at present, as the population is present in protected areas and occurs above the altitude at which firewood collectors are allowed to operate. However, this species could become threatened very quickly if human activities, especially wood collection, increase (Eastern Arc Mountains & Coastal Forests CEPF Plant Assessment Project Participants 2006).

Aloe pembana is another example of an endemic plant of Tanzania with few individuals. The species is confined to a small area on Pemba Island (Misali Island Conservation Area). The total population numbers between 50 and 250 mature individuals which grow in scrub vegetation right on the beach. It is threatened by fishermen through trampling and collecting for medicinal purposes. There is also increasing tourism on the island which may have an impact (Eastern Arc Mountains & Coastal Forests CEPF Plant Assessment Project Participants 2006; Carter 1994).

2.1.19 Kenya

Afrothismia baerae is confined to Shimba Hills National Reserve, Kenya. A single population of seven plants (Cheek 2003) was recorded only from a tiny area less than 0.5 m in diameter. The parasitic plant is growing in an evergreen coastal forest remnant on the roots of *Zanha golungensis* (Sapindaceae). Species in this parasitic genus are often very hard to find because of their scarcity, diminutive stature and the fact that unless flowering the plants are invisible above ground. Flowering only lasts for a month or so and plants may not flower every year. The area was logged in the past by the Forestry Department but meanwhile that has been stopped and the impact of elephants was being reduced (Eastern Arc Mountains & Coastal Forests CEPF Plant Assessment Project Participants 2006; Cheek 2003).

2.1.20 Madagascar

Recent surveys revealed only a few individuals of the orchid *Grammangis spectabilis* in its habitat in SE Madagascar. The species lives on large trees, in seasonally dry, deciduous forest or woodland (Cribb and Hermans 2009).

A few individuals of two other orchids have been recorded in the plateau region of the same island. *Erasanthe henrici* var. *isaloensis* inhabits humid forests at 800–1,000 m. *Angraecum longicalcar* is living in xerophytic vegetation and in gallery forest at 1,000–2,000 m on trachyt rock (Cribb and Hermans 2009). Pressure on the relating vegetation types means extinction risk for these orchids.

Only few adult specimens have been recorded in a couple of other plant taxa in Madagascar, including e.g. *Dalbergia bathiei*, *Delonix velutina*, *Phylloxylon xiphoclada* (cf. Labat & Moat in Goodman and Benstead 2003). Fortunately, in some cases there is hope to find other populations of these and other rare plants in regions where botanists or biogeographers have not been before.

2.1.21 Mauritius

On the island of Mauritius most native plant communities – almost all forest – have been destroyed and converted to sugar-cane plantations or else they are badly degraded. The remaining forests are strongly influenced by different neobiota such as rat, deer, pig, tenrec, macaque monkey, mongoose, man and different neophytes. The decline of some tree species may be due to habitat destruction, timber harvesting, introduced pigs and macaques which disturb and eat the fruits, and introduced plants. As a result, in 1973 there were only 13 very old *Sideroxylon grandiflorum* specimens recorded, a tree endemic to the island. Fortunately, to aid the seed in germination, botanists and foresters now know to use turkeys and gem polishers to erode the pericarp of the fruits to allow germination (Hershey 2004; Rouillard and Guého 1999; Walter and Gillett 1998; Davis et al. 1994; Staub 1993).

Hibiscus fragilis is another example of the many critically endangered plant species unique to the island of Mauritius. There are only c. 40–60 mature plants left at the two known localities and they are not regenerating because of introduced alien species. Although the species is easy to propagate from cuttings, long-term maintenance in botanic gardens is problematic because the species hybridizes easily with the introduced garden plant *Hibiscus rosa-sinensis*. The only hope for the continued survival of the species is management of the wild populations, clearance of the alien invaders and restocking from known cultivated sources (Bachraz and Strahm 2000; Bosser et al. 1987).

2.1.22 Sri Lanka

Cryptocoryne bogneri is an aquatic plant still living on Sri Lanka. The species was believed to be extinct as it had not been recorded since 1900. However, in 1999 a researcher discovered a new population of this species (www.iucn.org, downloaded 21 September 2010). More than 250 individual plants were recorded in a very small area of a 75 ha patch of swamp forest on the edge of an extensive rubber estate. This species is not commercially exploited and the present owner of the site location is conserving the land as the area contains many other endemic species. The water level at the site fluctuates naturally but it is not known that there is regular fluctuation in the numbers of mature individuals. Expanding agricultural activities and human settlements have impacted the other sites where this species was previously recorded (cf. Jacobsen 1987; De Wit 1975).

2.1.23 China

Abies beshanzuensis and *Abies yuanbaoshanensis* are two critically endangered species of China, the first living in Baishanzu Mountain, Zheijang, the second in

Yuanbao Mountain, Guanxi. Both species occur outside the general range of fir species. Only five living specimens of *Abies beshanzuensis* have been recorded in a mixed forest in the wild. Major threats are expansive agriculture and fires, coupled with poor regeneration. These are thought to have largely been responsible for the decline of the species (Conifer Specialist Group 1998; Fu and Chin 1992). The number of individuals of *Abies yuanbaoshanensis* is estimated at ~100. Young trees of this plant are very rare and regeneration is hampered by long coning intervals, seed predation by squirrels and competition with *Sinarundinaria* species. Fortunately, effective protection of the population is given by the Forest Department (Hilton-Taylor 2000; Fu and Chin 1992).

2.1.24 Philippines

Rafflesia magnifica is among the group of plants that produce the largest single flowers in the world. Only a few individuals of *R. magnifica* have been recorded, all of them male. The species is listed as critically endangered because of its very small population size and restricted range. Construction of a national highway in the area has facilitated easier human access and disturbance poses a threat to this rare plant as its unusual flowers are often treated as visitor attractions. Parts of the forest are also being converted into banana plantations (Madulid et al. 2008, 2005).

2.1.25 New Caledonia

First discovered in 1988, *Pittosporum tannianum* from New Caledonia was thought to have gone extinct in the 1990s. But, in 2002 the species was rediscovered. Three individuals are known to exist in the wild today giving this species the opportunity to avoid extinction. Loss and degradation of its sclerophyllous forest habitat is the main threat to the species (Baillie et al. 2004; Tirel and Veillon 2002).

2.1.26 Vanuatu

Carpoxylon macrospermum occurs on the islands Aneityum, Tanna and Futuna of the Vanuatu Archipelago. Approximately 40 individuals exist in lowland rainforest and another 120 mature trees are cultivated around villages. Regeneration is moderate. This palm tree is of ornamental interest (Dowl 1998).

2.1.27 New Zealand

There are only two species in the genus *Chordospartium*, both endemic to New Zealand. *Chordospartium muritai* is confined to a site in Clifford Bay in Marlborough. Only 12 wild plants have been counted in a remnant of coastal forest (Oates and Lange 1998; Hunt 1996; Purdie 1985).

2.1.28 Fiji Islands

Acmopyle sahniana is a critically endangered species of Central Viti Levu, Fiji. Very few individuals are still living on forested mountain ridges. There is only little evidence of regeneration (Conifer Specialist Group 2000; Oldfield et al. 1998; Smith 1979).

2.1.29 Hawaii Islands

Listed as critically endangered, the forest tree *Caesalpinia kavaiensis* is found only on the islands of Hawaii and O'ahu. Only some 60 individuals are known, however, many of them are old and probably non-reproductive. Threats include pigs, cattle, deer, goats, introduced plants, rats, fire, volcanic eruptions, sheep, black twig borer and collection by humans (Bruegmann and Caraway 2003; Wagner et al. 1999).

Hesperomannia arbuscula, listed as critically endangered, is a small shrubby tree known only from the Hawaiian islands of Maui and O'ahu. There has been an observed population decline of 25–50 % over the last years and the number of known individuals is less than 25. Main threats to the species are habitat degradation by pigs, predation by rats, and trampling or collecting by humans (World Conservation Monitoring Centre 1998; Oldfield et al. 1998).

The whole genus *Hibiscadelphus* is endemic to the Hawaiian Islands. Seven species have been described. Three of them have become globally extinct, the other four species are listed as critically endangered. One of them is *Hibiscadelphus distans.* Occurring as a shrub or small tree, the species is known from a single population on Kauai. Fewer than 20 individuals are recorded in a patch of dry forest (World Conservation Monitoring Centre 1998; Wagner et al. 1990, 2005; Gentry 1986).

Gentry (1986) reported on a couple of species in the tropics which are known from a single to a few individuals only. At least some 100 vascular plant taxa worldwide are known to have become extinct in historical times and more than 1,500 are classified as extinct in the wild or critically endangered. Unfortunately, and in contrary to the goals of CBD, the number of extinctions is still increasing (Baillie et al. 2004; www.iucnredlist.org).

2.2 Number of Endemic Taxa (E)

The species is regarded as a fundamental unit of biodiversity, species richness as the fundamental dimension of biodiversity, and richness of endemic species in a region as one of the fundamental characteristics of Biodiversity Hotspots and value to nature conservation policies. A severe but frequently ignored complication when counting the number of species or measuring the richness of endemic species is the lack of a universally applied definition of what a species is. For any given assemblage, the numbers of species may potentially differ, dependent on the species concept (cf. Gaston 1996a; Mallet 1996). In practice, one can find counts of endemic species and subspecies or varieties for many regions in the world on the internet, in books and scientific journals. What we normally do not know is if these various numbers are comparable and if the taxa have been split into many, or lumped to fewer, species or subspecies. As an interim result we have to accept that comparing numbers of endemic taxa in different geographical regions might result in biased conclusions; the number of endemic subspecies in one geographical region may be comparable with the number of endemic species in another one.

If we ignore such systematic biases and if we use the number of endemic vascular plant taxa – including very often endemic species and subspecies – then China (15,000–18,000 endemics), Indonesia (17,500), Colombia (15,000), Australia (14,000), Mexico (12,500), Venezuela (8,000), Madagascar (6,500), India (5,000), Peru (5,400), Ecuador (4,000), Bolivia (4,000), and the USA (4,000) are the countries with the highest numbers of endemic plant species (numbers rounded, see Huang et al. 2011 for China, and Groombridge and Jenkins 2002 for all these countries).

2.3 Proportion of Endemics (E/S)

Regional endemism is often characterised by the proportion of endemics: the number of endemic taxa (E) divided by the total number of taxa (S) [as a percentage: (E $\times$ 100 %)/S]. Many islands have higher proportions of endemics than any mainland region, e.g. Madagascar, Hawaii Islands, New Caledonia, New Guinea, New Zealand, St. Helena, etc. (cf. overview in Hobohm 2000).

In the biogeographical context, one has to be very careful when comparing proportions (Magurran and McGill 2011; Ungricht 2004; Hobohm 2000, 2008). For example, 15 % of the whole vascular flora of Vanuatu Island is endemic whereas the rate for Corsica is 6 %. But on their own, these numbers do not allow any conclusion concerning the richness of endemic taxa in the two regions. Both islands are inhabited by nearly the same number of 150 endemic taxa, but the size of Vanuatu (14,763 km^2) is almost twice the size of Corsica (8,723 km^2). This means that the density of endemic taxa (the number of endemic species per unit area) is higher on Corsica than on Vanuatu. The densities of endemic plant species on the Madeira Islands (11 % of the vascular flora is endemic), Balearic Islands (7 %) or Crete (7 %)

are also higher than or similar to concentrations on Hawaii (47 %), St. Helena (80 %) or New Zealand (82 %; cf. Hobohm 2000, 2008). Therefore, when a high percentage of the vascular flora of an area is endemic, this does not necessarily indicate a high density of endemics or a centre of endemism. If a single species is living on an island and if this taxon is endemic to the island then rate of endemism is 100 % and the density also is maximal. It is thus important to clearly distinguish between total number of endemics, rate (or proportion) of endemism and endemic density.

2.4 Bykow's Index of Endemism (BI)

Based on available empirical data about the proportions of endemics found in different regions, as well as the consideration that on the scale of the entire globe, all taxa are endemic, Bykow (1979) developed a regression equation describing the relationship between area and the expected proportion of endemics in that area. The formula for this equation is:

$$\log e_n = 0.373 \times \log a - 1.043,$$

where $a =$ size of the area in km^2, $e_n =$ expected endemism (as percentage value). The residuals can be calculated as $i = |e_f/e_n|$; with $i =$ index of endemism, $e_f =$ proportion of endemics as a percentage of the whole species composition (empirical value for a region). According Bykow's Index, 1 % endemics can be expected in an area of ~625 km^2, 2.8 % endemics in an area of 10,000 km^2, 6.6 % in 100,000 km^2, and 15.7 % endemics in a region of 1,000,000 km^2.

This measure is used as a simple and quite good predictor of endemism that results from a combination of both isolation and ecological continuity of a region (cf. Bruchmann 2011; Georghiou and Delipetrou 2010). The authors calculated high Bykow-values for the Canary Islands, Madagascar, Chile, South Africa and California. Hobohm (1999) compared different regions within Europe and found an extraordinarily high value for a small Balearic islet called Ses Margalides. *Euphorbia margalideana*, discovered in 1978 by Heinrich Kuhbier (1978) and endemic to this islet, is living together with only seven other vascular plant species.

2.5 Density, Endemic Species Diversity in Space (E/A)

The density of endemics in a region (number of endemics over area, E/A) describes how many endemic taxa per unit area occur. It is a measurement of the concentration of endemics. In general, the density of endemics is used as indicator for the duration of uninterrupted evolution and/or evolutionary speed in a region.

The number of endemic species divided by area gives a very simple measurement that roughly indicates high or low degrees of endemism. This method is valid for areas of similar sizes, but it is extremely difficult to compare density values calculated for regions with different area sizes in a mathematically correct way (Magurran and McGill 2011). For example, say we want to compare the concentration of endemics on the endemic-rich island Madagascar with that of the Cape Verde Island Brava (55 km^2), which with a single endemic species seems to be relatively endemic-poor. 7,750–8,160 endemic vascular plant species inhabit Madagascar (585,000–587000 km^2; Gillespie and Clague 2009; Burga and Zanola 2007; Hobohm 2000; Davis et al. 1994). If we divide these numbers by the size of the area – which is not in fact a valid method (see literature cited for EARs) – to get the number of endemics per km^2 or 10,000 km^2 – we would find higher values for Brava (0.018 endemic species per km^2 or 182 endemics per 10,000 km^2) compared to Madagascar (0.032–0.039 endemics per km^2 or 132–139 endemics per 10,000 km^2). This is the consequence of attempting to directly compare areas of different sizes.

If we are only taking areas of the same size into account we can compare absolute numbers of endemics (E) directly (Table 2.1), and the interpretation of E then is similar to the interpretation of density (concentration) measures of endemic taxa. For regions of different sizes, a different method is necessary (EARs, Sect. 2.6).

According to Table 2.1, Mauritius has a density of endemic vascular plant species three times higher than Tenerife, Canary Islands, while endemism of Tenerife is slightly higher than endemism of Sierra Nevada, Spain mainland, but much higher than endemism of Cuatro Cienagas, Mexico mainland. The numbers of Tenerife and Sierra Nevada are quite well comparable because they share important parts of the flora and biogeographers (Izquierdo et al. 2004; Castroviejo et al. 1986 ff.). The tables also give some geographical numbers which could be relevant. Distance to mainland means also reduction of dispersal opportunities. Altitude is used as indicator for the diversity of climate zones and habitats (environmental heterogeneity). Latitude is related to albedo, warmth and influence of former glaciation periodes.

We can also compare regions with the same E and different A directly (Table 2.2). In this case the locality with the smaller size of area represents the higher concentration (density) of endemics.

In Table 2.2 we list areas with approximately the same number of endemic species in their vascular floras. But according to the numbers in Table 2.2, the highest density of endemics goes to Amistad which is higher than Polynesia-Micronesia and Cuba, much higher than SW Australia, and very much higher than the Himalaya Mountains. How can we explain these differences? If we take the environmental parameters of Table 2.2 into account we easily can come to the conclusion that these numbers (and area) alone cannot properly explain the differences in endemism between the regions. Other factors must be taken into consideration including the geographical position of Amistad which is located between two oceans (W-E) sandwiched by climate-stabilising currents, and two continents with dispersal abilities from two different floristic subcontinents to the North and South. A complete explanation must take three types of processes and conditions into account: global and continental evolutionary processes (due to climate constancy or change and geological activity or quiescense) which generates the continental species pool (i), regional dispersal opportunities (ii) and the history

Table 2.1 Regions with a similar size of area in declining order of concentration in endemic vascular plant species

Region (km^2)	No. of endemic species	Latitudes (degrees)	Climates/annual precipitation/Pleistocene	Altitudes (m asl.)	Distance to mainland (km)	No. of vascular plant species in total
Mauritius, Indian Ocean (1,940)	300	20S	Wet tropical/1,000–4,000 mm/not glaciated	826	1,800	900
Tenerife, Canary Islands, Atlantic Ocean (2,050)	106	28N	Dry Mediterranean to orogenic dry/100–1,000 mm/not glaciated	3,718	280	1,367
Sierra Nevada, Spain, Europe (2,000)	80	36/37N	Mediterranean to orogenic cool/358–2,500 mm/glaciated above 2,500 m	600–3,479	–	2,100
Cuatro Cienagas, Mexico (2,000)	23	27N	Dry subtropical to orogenic/c. 200-?/not glaciated	740–2,500 (−3,000)	–	860

Data from Burga et al. (2004), Nau (2003), Hobohm (2000), Davis et al. (1997)

Table 2.2 Regions with a similar number of endemic vascular plant species in declining order of concentration

Region	Size of area (km^2)	Latitudes (degrees)	Recent climate annual precipitation rate Pleistocene	Altitudes (m a.s.l.)	Minimum distance to mainland (km)	Total no. of vascular plant species
La Amistad Costa Rica, Panama, Middle America (c. 3,000 endemics)	c. 10,200	8–10N	Wet tropical to tropical alpine, 2,800–6,840 mm, not glaciated during Pleistocene	3,819	–	c. 10,000
Polynesia-Micronesia (3,074 endemics)	47,000	25S–30N	Dry to wet tropical, 400–15,000 mm, not glaciated	0–4,205	c. 2,000	c. 6,600
Cuba Caribbean Islands (c. 3,100 endemics)	108,722	19–22N	Warm tropical with a winter-dry period, 800–3,000 mm, not glaciated during Pleistocene	1,994	500	6,375
SW Australia (2,948 endemics)	356,717	26–36S	Mediterranean, 300–1,500 mm, not glaciated during Pleistocene	1,110	–	5,571
Himalaya Asia (c. 3,100 endemics)	741,706	20–35N	Warm wet tropical to extremely cold and dry, winterdry or summer-rain in almost all altitudes, 150–12,000 mm, large parts glaciated during Pleistocene	100–8,848	–	c. 10,000

Data from Mittermeier et al. (2005), Burga et al. (2004), Nau (2003), Beard et al. (2000); Hobohm (2000); Davis et al. (1997, 1995)

of the landscape ecology (iii) including all the local biotic and abiotic interactions, processes controlling vegetation structures, biomass, productivity, destruction of vegetation and reduction of biomass.

2.6 Endemics-Area Relationships (EARs)

One of the best fit laws in ecology and biogeography is the *power law* or *Arrhenius-equation* (see e.g. Magurran and McGill 2011; Dengler 2009; Arrhenius 1921; and many other ecologists and biogeographers). This equation describes Species-Area Relationships (SARs) in many cases better than other equations.

$$\log S = z \times \log A + \log c \text{ or } S = cA^{z}$$

where S = total species number of an area, A = area size, z and c = empirical constants. This equation describes the increase in the number of species with increasing size of area surveyed. It can reformed for the analysis of the endemics, in terms of E instead of S.

$$\log E = z \times \log A + \log c \text{ or } E = cA^{z}$$

The curve and mathematical description of the relationship between area and richness depends on the dispersion of the species in space and the design of analysis, e.g. on the spatial extent and size and position of the plots surveyed (Fig. 2.2).

From the definition of the term *endemic*, the z-value in a nested plot design – every small plot is included in a larger one (Fig. 2.2) – should be as high or higher than 1 because the number of endemics should increase at the same rate or more rapidly than the total number of species (i.e. when area doubles, on average the number of endemics will at least double as well). Recently Storch et al. (2012) found that different curves (for amphibians, birds and mammals) collapse into one universal power law with a slope (z-value) close to 1 after the area is rescaled by using the range sizes of the taxa.

If we assume that the terrestrial area on earth is ~147,930,000 km^2 (logA = 8.17) and the number of endemics is 300,000 (logE = 5.477; cf. Baillie et al. 2004) and then extrapolate beyond this point, we predict that the average sized area in which there is a single endemic species (logE = 0) should be greater than or equal to 493 km^2. Hence Brava (1 endemic species, 55 km^2) represents an area with higher than average endemism.

We can compare the values of logE and logA – 2.7 as a simple heuristic to determine if observed values of density (E/A) are below or above the trendline (regression) for the whole world. The formula results from the assumption that the total number of vascular plant species on earth is 300,000, that the terrestrial area is c. 147,930,000 km^2, that z in the power law is 1 (nested plot design), and that the power law describes the relationship between species number and area in an adequate way. Because of these complex assumptions the formula can only be used as a very rough short-cut analysis.

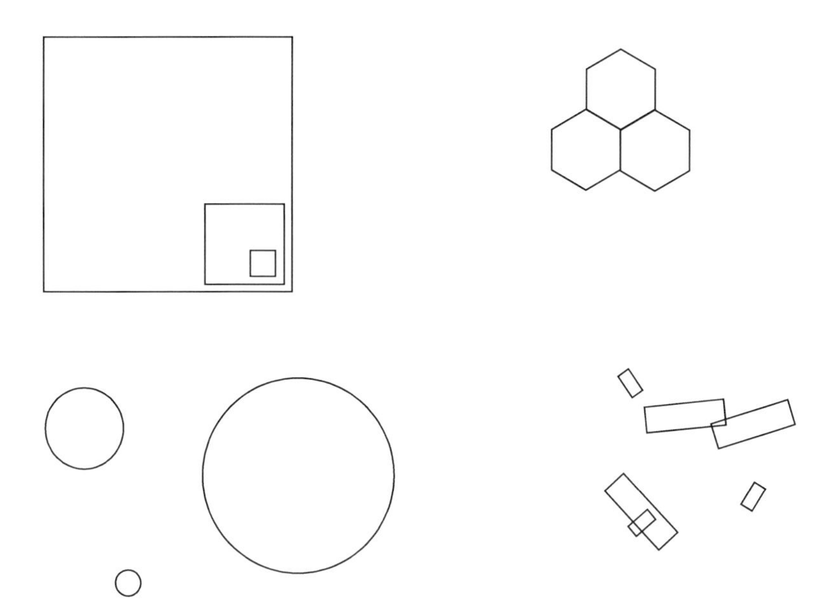

Fig. 2.2 Nested plots (*quadrats*), split plots with shared flanks (*hexagons*), dispersed, independent plots (*circles*), and randomly dispersed, particularly overlapping plots (*rectangles*); for discussion of different designs of multiscale sampling and how the design is controlling the result see Dengler (2009), Stohlgren (2007)

If the value of logE is higher than logA – 2.7, then the density (concentration) of endemics is above average (for example, Brava, 55 km^2, 1 endemic plant species: logA = 1.74, logE = 0; 0 is higher than 1.74–2.7 = – 1.04, Madagascar, 585,000 km^2, 8,000 endemics: logA = 5.77, logE = 3.9; 3.9 is higher than 5.77–2.7 = 3.07).

If the z-value for the regression is higher than 1 then the slope becomes steeper and the distance to the trendline from each residual above the trendline is increasing. Recall that the only fixed point of the regression in this nested plot design is the residual for the whole earth (300,000 endemics, 147,930,000 km^2) and the reason for z > or = 1 is the definition of endemism, the logic of EARs in a nested design and few empirical data (cf. Storch et al. 2012).

If we wish to analyse empirical data for dispersed and independent plots such as islands or archipelagos rather than nested plots, and try to find a regression for this design in logE-logA space (referred to as either type IV according to Scheiner 2003, 2004, ISARs according to Whittaker and Fernandez-Palacios 2007, and dispersed, independent plots in Stohlgren 2007) then the z-values are often lower than 1 (cf. Georghiou and Delipetrou 2010; Werner and Buszko 2005; Hobohm 2000). This is not in conflict with the assumptions above because those assumptions are valid for nested plot designs only (Fig. 2.2).

Regardless of the assumptions, with different types of regressions it is possible to find whether a residual is located close to or distant from the trendline, and if the concentration of endemism in a region is higher or lower than average. The numerical procedure has been developed independently by different authors (cf. Groombridge and Jenkins 2002; Hobohm 2000; Thomasson 1999). In every case the regression equation has to be checked not only mathematically but also with regards to biogeographical plausibility. The z-value can even become negative (!) if too many residuals represent large and cold or arid regions with little endemism and many small islands in warmer or wetter regions with higher endemism (which is the case e.g. in Europe). We found a negative relationship between the number of endemic vascular plant taxa and the size of regions in Europe in a log-log space (unpublished). This example shows that it can be difficult to find a meaningful regression equation, especially for the islands design (dispersed plots). Further, there are many theoretical discussions concerning the correct empirical design or relevance of a regression equation (cf. Dengler 2009; Stohlgren 2007; Scheiner 2003, 2004), and it is difficult at the moment to conclusively determine the biological or biogeographical meaning of the residuals, of z or c, or which processes they reflect, respectively.

EARs have been used to estimate the number of species that will go extinct after suitable habitats in a region are destroyed (cf. Kinzig and Harte 2000). Recently, He and Hubbell (2011) published a paper titled "*Species-area relationships always overestimate extinction rates from habitat loss*". Recent controversy about calculating extinction rates using Species-Area Relationships (SARs) versus EARs or via comparison of SARs and EARs is surprising, given the lack of data about real extinction events (i) and the discussion about the "correct" numerical procedures, designs and curves when calculating such relationships (cf. Dengler 2009; Whittaker and Fernandez-Palacios 2007; Scheiner 2003, 2004) (ii). It seems to be more useful at the moment to focus less on calculating former extinctions or the risk of extinction by using SARs or EARs or both, given the scientific discussion that is ongoing, and better to concentrate on real threats faced by concrete rare and endemic organisms.

2.7 Range-Size Rarity (RSR) and Weighted Endemism (WE, CWE)

The range size of a taxon is often calculated as the sum of geographical grid cells or unit areas in which a taxon occurs. Growing attention is being paid to latitudinal and altitudinal gradients in the mean (or median) geographic range size of species in an area. The mean range size has been found to increase towards higher latitudes in a sufficient variety of higher taxa (Rapoport's Rule, cf. Rapoport 1982; Gaston 1996a, b) although one can find exceptions of the rule.

Range-Size Rarity can be measured as the sum of the inverse range sizes per grid cell or area unit (cf. Lovett et al. 2000). Slightly different modifications of this methodology have also been published (cf. Crisp et al. 2001; Linder 2001; Laffan and Crisp 2003).

$$\mathrm{RSR} = \sum (1/\mathrm{C_j})$$

with C_j = number of grid cells in which the taxon j occurs. Thus, a species restricted to a single grid cell would be scored as 1 for the cell where it occurs and with 0 for the other grid cells where it doesn't occur. A species occurring in two grid cells would be scored as 0.5 for each of the two cells. RSR is the sum of all inverse range sizes (= 1/number of grid cells) for all plant taxa that occur in a grid cell. This method was also applied for geographical units of different size which is statistically questionable.

RSR can be used as a simple measure of the distribution of species endemism in space. For the measurement of RSR for endemic species it is not necessary to analyse the whole vascular plant species composition of every site. This method makes it possible to identify centres of endemism over large areas. In many cases the RSR values are calculated only for a single genus, family or order (cf. Kessler 2002). The value of RSR for each grid cell depends on the extent of the species composition and total number of grid cells. This means that any modification in the database will change the values of many or all grid cells. Similar to the measure of the proportion of endemics, a high value can result for different reasons. In general, two explanations are possible for all species groups which confirm Rapoport's Rule: a decreasing (1) or increasing (2) number of species with large ranges towards higher latitudes, in combination with an increasing (1) or decreasing (2) number of species with small ranges in relation to latitude. If we find a great number of species with small ranges (endemics) in the tropics and only a few at higher latitudes then the absolute number of species with large ranges can theoretically also be higher in the tropics; for a confirmation of Rapoport's Rule only an increasing mean (or median) of the range sizes with latitude is important.

Weighted endemism (WE) is calculated as the sum of the reciprocals of the total number of cells each species is found in. Weighted endemism is correlated with richness, and so a better measure, Corrected Weighted Endemism (CWE) (Crisp et al. 2001) removes this richness effect. To do so, WE is divided by the total number of species in each cell. For example, let us presume the existence of three species in a single grid cell, which has four immediate neighbouring cells to the North, East, South and West. One of the species is highly range-restricted and occurs in three of the five grids in our sample area and in four grids in total (one outside our sample area). The second species is weakly range-restricted with two of 30 and the third species is relatively widespread with four of 3,100. The calculation of the CWE for this grid cell is

$$\mathrm{CWE} = (3/4 + 2/30 + 4/3,100)/3 = 0.473$$

This differs from RSR in that the values can be computed for different clusters (packages) of grid cells (so called "sample windows"; Laffan and Crisp 2003). In our example the sample window is five grid cells.

2.8 Parsimony Analysis of Endemism (PAE)

The possibility of applying a parsimony method to delimit areas of endemism was developed by Rosen (1988; see also Giokas and Sfenthourakis 2008; Santos 2005; Garcia-Barros et al. 2002). Comparable to RSR analyses, PAE basically unites areas based on their shared taxa.

As described in Morrone (1994) the method comprises the following five steps: Draw quadrats or other unit areas of the same size on a map of the region to be analysed (1), construct a data matrix where columns represent the species and rows represent the sample units (quadrats) with 1 if a species is present and 0 if it is absent (2), perform a parsimony analysis (3), e.g. as described in Rovito et al. (2004), delimit the groups of quadrats defined by at least two or more species (4), and finally, superimpose the groups delimited in the cladogram onto the quadrats and map the areas of endemism (5).

To identify areas of endemism or conservation priorities, the method often was modified or combined with different more or less complex modeling procedures (cf. Rovito et al. 2004).

The method often was criticized, especially when it was used to detect biogeographic histories involving dispersal (Brooks and Veller 2003; Peterson 2008). However, similar to analyses of RSR the method can be useful to identify and describe centres of endemism.

2.9 Phylogenetic Measurements of Endemism (PE)

With the rapid development of DNA-sequencing technologies, there has been a phylogenetic revolution in biology. There has been great diversification in phylogenetic metrics meant to measure diversity, as well as large advances in the amount and quality of information encoded by phylogenies. Phylogenetic trees quantify the evolutionary information represented in groups of taxa. Their structure – both degree of relatedness and age (branch lengths) of taxa – can be used as measures of taxonomic distinctiveness and time since evolution of biological features or genetic isolation (Huang et al. 2012; Nipperess et al. 2010; Rosauer et al. 2009; Sechrest et al. 2002; Vane-Wright et al. 1991). Moreover, with genetic information and information about recent distribution patterns of a taxon, it is possible to reconstruct historical colonization routes (Dixon et al. 2009; Erkens et al. 2007). For example, based on geographical distribution data and information about the genetic diversity within populations of *Saponaria bellidifolia*, Csergö et al. (2009) suggested that the Romanian populations of this plant – which have relatively low genetic diversity – have most likely originated from a Holocene range expansion from the glacial refugia in Bulgaria or Italy where the populations are genetically more diverse.

Combining phylogenetic analyses with biogeographical parameters such as endemism is of interest for nature conservation policies as well. The term *phylogenetic diversity* originates with Faith (1992), and represents the sum of the branch lengths of a phylogenetic tree representing a certain taxon (e.g. genus, family) in a region (Faith and Baker 2006). This value is high if there are many old taxa (e.g. species) represented by the cladogram. It is lower if the taxa are younger and/or if the number of taxa is smaller. This also means that the measurement has to be standardised to capture differences in area or other variables of interest. Recently Rosauer et al. (2009, cf. also Sechrest et al. 2002) introduced a new metric, *phylogenetic endemism* (PE). This value is high if a small region captures a large fraction of a clade's evolutionary diversity. The value is smaller if the region is larger or if the fraction is smaller (Cadotte and Davies 2010; Rosauer et al. 2009). Using another metric of phylogenetic endemism, Tucker et al. (2012) showed that the distribution of high phylogenetic endemism was different for the Proteaceae of the South African fynbos as compared to the distribution of overall species diversity. This has important implications, since current reserves capture species richness well, but endemic richness poorly.

Calculating these or other values which combine evolutionary distinctiveness with geographical diversity of endemism measurements can require empirical data from the field and laboratory that is not always readily available (Huang et al. 2012). Because this scientific approach is rather young it is clear that for many regions in the world we are – at the moment – far away from having the necessary information.

2.10 Habitat Preferences (E/H)

Most likely Rikli (1943, 1946) was the first biogeographer to include chapters in his books about endemism in vascular plants in relation to habitat and ecological conditions. For Mediterranean endemics he emphasizes the meaning of rocky habitat, garigue, steppe, coastal and desert habitats. The publications of Pawlowski (1969; Alps and Balcanic Mountains), Gamisans and Marzocchi (1996; Corse), Tan and Iatrou (2001; Peloponnese, Greece), Petrova (2006; Bulgaria), and Rabitsch and Essl (2009; Austria) are further important examples focusing on endemism and characterizing habitat and environmental conditions.

Independent of the distributional range, every taxon is more or less strongly related to one or more habitat types. *Rhizphora mangle* is an example of a widely distributed plant which almost only occurs in mangroves (cf. Bailey 1998; Huston 1994); compared to other habitats such as *grassland* or *shrubland* this habitat type is relatively well defined (influence of tides, salt or brackish water, sand or muddy soil, tropical/subtropical climate, coastal and estuarine areas). Other plant taxa are less strongly connected to a single habitat type or narrow habitat conditions. In contrary to *Rhizophora mangle*, e.g. *Cirsium hypopsilum* represents a narrow geographical range in combination with a wider habitat specificity. This species is endemic only to the Mountains of Peloponnesus, Stera Ellas, S Pindhos and Evvia in Greece, where the main habitats and ecological conditions are deforested rocky slopes and screes, and ravines in open *Pinus nigra* and *Abies cephalonica* forest, *Pinus halepensis* woodland, and spiny *Astragalus* thorncushion communities, on calcareous marl, limestone or schist, between 300 and 2,350 m (Tan and Iatrou 2001). Another

Photo 2.1 *Saussurea pseudotilesii*, endemic to Kamchatka Peninsula and Commander Islands in the Russian Far East, NE Asia, in pioneer vegetation on tephra (Photographed by Dietbert Thannheiser, Kamchatka Peninsula)

example of a geographically restricted species which occurs at different altitudes between sea level and up to c. 1,000 m, on different soil types and in different habitat types such as ruderal vegetation, coastal dunes, megaforb communities, or tundra is *Saussurea pseudotilesii* on Kamtchatka peninsula. We assume that the dispersal mode of hydrochory in the marine environment and ecological conditions promote the wide geographical distribution in the case of *Rhizophora mangle* whereas dispersal or pollination mode and/or the marine environment limits the ranges of *Cirsium hypopsilum* and *Saussurea pseudotilesii* (Photo 2.1).

These examples also show that it might become complicated to quantify a strong or weak habitat preference. If we want to explore which and how many endemics in a region are related to a certain habitat type or vegetation structure we first have to do it in a descriptive or qualitative manner.

The second step can be to count endemics in a certain region which are recorded in predefined habitat categories (E/H, cf. Bruchmann 2011). Normally endemics in a region are not absolutely restricted to a single habitat type. For a numerical procedure this means that the same endemic taxon can be listed in different habitat types. However, this metric underlines the meaning of certain habitats for conservation priorities.

According to the IUCN Red List (www.iucn.redlist.org; downloaded 9/2011) 421 vascular plant species of the categories EW (extinct in the wild) and CR (critically endangered) are living or should live in forests. The second largest group (145 species) is associated with shrublands, followed by plants of rocky areas (109), wetlands (61), grassland (46), marine coastal, supratidal (43), desert (21), savanna (16) and other habitat types (21). Most of these species are restricted to very small areas. In this case the habitat types are not strictly defined. E.g. the term *desert* is related to obviously different habitat types with different precipitation regimes and vegetation structures.

2.11 Perception of Endemics

Since the term *endemic* was defined by De Candolle in (1820), endemic species have received growing interest. However, we assume that the general perception of different endemic species by the public varies, depending on the range size, threats, and also on many other circumstances. The perception might also be dependent on questions such as if the endemic plant is beautiful or charismatic, where it occurs, or if the species is economically valuable.

The IUCN Red List, first published in 1966 as Red Data Book, was one of the first Red Lists worldwide. Conceived in 1963, this system set standards for species listing and nature conservation assessment efforts. Today many thousands of rare and threatened species are named in a huge number of Red Lists, appendices of national or international laws, or they are labeled as flagship species, keystone species, umbrella species or target species to underline their meaning for nature conservation activities. Red Lists, the concept of target species (cf. Ozinga and Schaminée 2005) and laws are the result of scientific work and political processes including legislation. The resulting attributes – Red List species, protected species, target species – can be used as indicators for the political, conservational, and social perception of rarity or threats.

To test this we first posed a null hypothesis and alternative hypothesis (i). Second, we searched regions of comparable size (ii): a single region on the one hand and a region of the same size which is composed by two or more regions as defined in Flora Europaea (cf. Tutin et al. 1996a, b, c, d, e), on the other. And then (iii) we compared the lists of endemic vascular plants of these regions (EvaplantE; last updated version) with lists of vascular plants which are named in the IUCN Red List (version 2011.1, downloaded 1/2012), which are protected by the Bern Convention (Appendix I; revised 2002) and European Habitats Directive 92/43 (2003, Annex II), and finally with the list of vascular plant species which are characterised as target species in Europe (Ozinga and Schaminée 2005).

We assume that the real ranges of the endemics within these pairs of regions are normally much smaller than the regions themselves, that they on average are comparable in size and that they together represent a similar frequency distribution of range sizes at local to regional scales. However, the discussion what range size of a taxon exactly means and how it can be measured is actually a long and ongoing discussion (cf. Gaston 1996a, b, c).

Table 2.3 Numbers of endemic vascular plant species in two regions of comparable range size (Italy vs. Greece plus Albania and Bulgaria), number of the relating protected species (Bern Convention and/or European Habitats Directive), of species named in the IUCN Red List, and of the species characterised as target species (Ozinga and Schaminée 2005). $^{**}p<0.01$

Region (km^2)	No. of endemics (vascular plant species)	No. of protected species	No. of species named in IUCN Red List	No. of target species
Italy (250,631)	145	19 (13.1 %)	31 (21.4 %)	127 (87.6 %)
Greece, Albania, Bulgaria (261,245)	91	5 (5.5 %)	7 (8 %)	22 (24.2 %)
Significance of differences (chi-square)		n.s.	**	**

Our null hypothesis in this context is that the perception of endemics results from national/political delimitations and global threats in equal measure. We assume that the opposite is more likely and that the perception of endemics is geographically biased (alternative hypothesis). We guess that the perception of endemics restricted to a single country is stronger than the perception of endemics which occur in two or more countries even if these countries together cover the same range. Furthermore, the perception and appreciation of island endemics is probably stronger than the perception of mainland endemics in regions of the same size.

For Europe we examined a few pairs of regions which are of similar area size. France is comparable with Portugal plus Spain, Italy with Greece plus Albania and Bulgaria, France with Poland plus Czech Republic, Slovakia and Hungary, European Northern Russia with Scandinavia plus Baltic Countries. Sicily is a little bit smaller than the Krim Peninsula which again is only a little bit smaller than Albania. The area of Great Britain is comparable to the Czech Republic plus Slovak Republik or to Bulgaria, respectively. Unfortunately, the northern regions and small countries with low numbers of endemic vascular plants do not meet the criteria of the relating statistics (Chi-Square Test for Homogeneity).

Thus, only two pairs of regions could numerically be analysed. The following two tables show the results.

All differences of the perception indicators for endemics in a single country are higher than those for combined countries (i.e. endemics which occur in two or more regions). However, only the differences for the numbers of species listed in the IUCN Red List and in the list of target species (in Ozinga and Schaminée 2005) are highly significant (Tables 2.3 and 2.4).

These results clearly are not enough to understand whether there is a global trend. Further studies should be undertaken with a higher number of comparable regions. At the very least we can conclude that the perception of endemics in different regions may be biased and we assume that these differences depend on the impact of geographically different scientific, economic and political efforts.

The photographs 2.2, 2.3, 2.4, 2.5, and 2.6 show images of very attractive plant species from Australia.

Table 2.4 Numbers of endemic vascular plant species in two regions of comparable range size (France vs. Portugal plus Spain), number of the relating protected species (Bern Convention and/or European Habitats Directive), of species named in the IUCN Red List, and of the species characterised as target Species (Ozinga and Schaminée 2005). $^{**}p<0.01$

Region (km^2)	No. of endemics (vascular plant species)	No. of protected species	No. of species named in IUCN Red List	No. of target species
France (539,527)	77	10 (13 %)	18 (23.4 %)	63 (81.8 %)
Spain, Portugal (582,626)	191	12 (6.3 %)	15 (7.9 %)	28 (14.7 %)
Significance of differences (chi-square)		n.s.	**	**

Photo 2.2 *Anigozanthus rufus* in species-rich kwongan (Fitzgerald River National Park, some 600 km SE of Perth, near the south coast of Australia; photographed by Marinus Werger). Kwongan is the general Australian term for Mediterranean type woody scrub vegetation. This perennial plant is endemic to dry sandy, silicious regions. The species and other members of the genus are grown commercially in many parts of the world

Photo 2.3 *Helipterum splendens* in a dense spring carpet community together with a few other annuals, near Cue, some 450 km NE of Perth (photographed by Marinus Werger). This plant is also commercially grown for gardening

Photo 2.4 *Swainsona formosa* (*Clianthus formosus*) at Coral Bay, Western Australia (photographed by Marinus Werger). The annual plant species is native to the arid regions of central and north-western Australia

Photo 2.5 *Banksia hookeriana* and *Xylomelum occidentale* on sandplain shrubland near Enneaba, Western Australia (photographed by Marinus Werger). Both species belong to the Proteaceae; this family is very species-rich and mainly distributed in Australia, South Africa and southern South America

Photo 2.6 *Banksia menziesii* in Kwongan near Cataby, Western Australia (photographed by Marinus Werger). All but one *Banksia* species are endemic to Australia. Western Australia is the centre of the diversity of banksias

References

Alves THS (2004) Biometria de frutos e sementes e germinação de *Dimorphandra mollis* e *Dimorphandra wilsonii* Rizz. In: 55th Congresso Nacional de Botânica, Viçosa, MG Brasil

Anderson EF (2001) The cactus family. Timber Press, Portland

Anderson EF, Fitz Maurice WA, Fitz Maurice B (2002) *Acharagma aguirreanum*. In: IUCN 2010. IUCN Red List of threatened species. Version 2010.3. www.iucnredlist.org. Downloaded on 18 Sept 2010

Arrhenius O (1921) Species and area. J Ecol 9:95–99

Bachraz V, Strahm W (2000) *Hibiscus fragilis*. In: IUCN 2010. IUCN Red List of threatened species. Version 2010.3. www.iucnredlist.org. Downloaded on 21 Sept 2010

Baijnaith H (1992) *Kniphofia leucocephala* (Asphodelaceae) a new white-flowered red-hot poker from South Africa. South Afr J Bot 58(6):482–485

Bailey RG (1998) Ecoregions: the ecosystem biogeography of the oceans and continents. Springer, New York

Baillie JEM, Hilton-Taylor C, Stuart N (2004) 2004 IUCN Red List of threatened species: a global species assessment. IUCN, Gland/Cambridge

Beard JS, Chapman AR, Gioia P (2000) Species richness and endemism in the Western Australian flora. J Biogeogr 27:1257–1268

Below H, Poppendieck H-H, Hobohm C (1996) Verbreitung und Vergesellschaftung von *Oenanthe conioides* (Nolte) Lange im Tidegebiet der Elbe. Tuexenia 16:299–310

Bosser J, Cadet Th, Julien HR, Marais W (1987) Flore des Mascareignes, 151 Malvacees to 12 Oxalidacees. Sugar Industry Research Institute (Mauritius), ORSTOM (Paris) and RBG, Kew

Brooks DR, van Veller GP (2003) Critique of parsimony analysis of endemicity as a method of historical biogeography. J Biogeogr 30(6):819–825

Bruchmann I (2011) Plant endemism in Europe: spatial distribution and habitat affinities of endemic vascular plants. Dissertation University of Flensburg, Flensburg. URL: www.zhb-flensburg.de/dissert/bruchmann

Bruegmann MM, Caraway V (2003) *Caesalpinia kavaiensis*. In: IUCN 2010. IUCN Red List of threatened species. Version 2010.3. www.iucnredlist.org. Downloaded on 21 Sept 2010

Burga CA, Zanola S (2007) Madagaskar – Hot Spot der Biodiversität. Exkursionsbericht und Landeskunde. Schriftenreihe Physische Geographie Bodenkunde und Biogeographie 55:1–201

Burga CA, Klötzli F, Grabherr G (eds) (2004) Gebirge der Erde. Ulmer, Stuttgart

Cadotte MW, Davies TJ (2010) Rarest of the rare: advances in combining evolutionary distictiveness and scarcity to inform conservation at biogeographical scales. Divers Distrib 16:376–385

Cafasso D, Cozzolino S, Caputo P, De Luca P (2001) Maternal inheritance of plastids in Encephalartos Lehm. (Zamiaceae, Cycadales). Genome 44:239–241

Cairns-Wicks R (2003) *Trochetiopsis erythroxylon*. In: IUCN 2010. IUCN Red List of threatened species. Version 2010.3. www.iucnredlist.org. Downloaded on 21 Sept 2012

Carine MA, Humphries CJ, Guma IR, Reyes-Betancort JA, Guerra AS (2009) Areas and algorithms: evaluating numerical approaches for the delimitation of areas of endemism in the Canary Islands archipelago. J Biogeogr 36:593–611

Carter S (1994) Aloaceae. In: Polhill RM (ed) Flora of tropical East Africa. A.A. Balkema, Rotterdam

Castroviejo S (1998) Flora Iberica Vol. 6: Rosaceae. C.S.I.C., Madrid

Castroviejo S (2001) Flora Iberica Vol. 14: Myoporaceae – Campanulaceae. C.S.I.C., Madrid

Castroviejo S (2003) Flora Iberica Vol. 10: Araliaceae – Umbelliferae. C.S.I.C., Madrid

Castroviejo S (2005) Flora Iberica Vol. 21: Smilacaceae – Orchidaceae. C.S.I.C., Madrid

Castroviejo S, Aedo C, Cirujano S, Lainz M, Montserra P, Morales R, Munoz Garmendia F, Navarro C, Paiva J, Soriano C (1993a) Flora Iberica Vol. 3: Plumbaginaceae (partim) – Capparaceae. C.S.I.C., Madrid

Castroviejo S, Aedo C, Gomez Campo C, Lainz M, Montserra P, Morales R, Munoz Garmendia F, Nieto Feliner G, Rico E, Talavera S et al (1993b) Flora Iberica Vol. 4: Cruciferae – Monotropaceae. C.S.I.C., Madrid
Castroviejo S, Aedo C, Benedi C, Lainz M, Munoz Garmendia P, Nieto Feliner G, Paiva J (1997a) Flora Iberica Vol. 8: Haloragaceae-Euphorbiaceae. C.S.I.C., Madrid
Castroviejo S, Aedo C, Lainz M, Morales R, Munoz Garmendia F, Nieto Feliner G, Paiva J (1997b) Flora Iberica Vol. 5: Ebenaceae – Saxifragaceae. C.S.I.C., Madrid
Castroviejo S, Benedi C, Rico E, Güemes J, Herrero A (2009) Flora Iberica Vol. 13: Plantaginaceae-Scrophulariaceae. C.S.I.C., Madrid
Castroviejo S, Lainz M, Lopez Gonzalez G, Montserra P, Munoz Garmendia F, Paiva J, Villar L (1986) Flora Iberica Vol. 1: Lycopodiaceae – Papaveraceae. C.S.I.C., Madrid
Castroviejo S, Lainz M, Lopez Gonzalez G, Montserra P, Munoz Garmendia F, Paiva J, Villar L (1990) Flora Iberica Vol. 2: Platanaceae – Plumbaginaceae (partim). C.S.I.C., Madrid
Castroviejo S, Luceno M, Galän A, Jimenez Mejfas P, Cabezas F, Medina L (2007) Flora Iberica Vol. 18: Cyperaceae-Pontederiaceae. C.S.I.C., Madrid
Castroviejo S, Talavera S (2006) Flora Iberica Vol. 7(2): Leguminosae. C.S.I.C., Madrid
Cheek M (2003) A new species of *Afrothismia* (Burmanniaceae) from Kenya. Kew Bull 58:951–955
Clubbe C, Pollard B, Smith-Abbott J, Walker R, Woodfield N (2003) *Acacia anegadensis*. In: IUCN 2010. IUCN Red List of threatened species. Version 2010.3. www.iucnredlist.org. Downloaded on 21 Sept 2010
Conifer Specialist Group (1998) Abies beshanzuensis. In: IUCN 2012. IUCN Red List of threatened species. Version 2012.1. www.iucnredlist.org. Downloaded on 29 Aug 2012
Conifer Specialist Group (2000) *Abies beshanzuensis*, *Abies yuanbaoshanensis*, *Acmopyle sahniana*. In: IUCN 2010. IUCN Red List of threatened species. Version 2010.3. www.iucnredlist.org. Downloaded on 18 Sept 2010
Cribb P, Hermans J (2009) Field guide to the orchids of Madagascar. Kew Publishing, Kew
Crisp MD, Laffan S, Linder HP, Monro A (2001) Endemism in the Australian flora. J Biogeogr 28:183–198
Csergö A-M, Schönswetter P, Mara G, Deak T, Boscaiu N, Höhn M (2009) Genetic structure of peripheral, island-like populations: a case study of *Saponaria bellidifolia* Sm. (Caryophyllaceae) from the Southeastern Carpathians. Plant Syst Evol 278:33–41
D'Arcy WG (1971) The island of Anegada and its flora. Atoll Res Bull 139:1–21
Davis SD, Heywood VH, Hamilton AC (eds) (1994) Centres of plant diversity, vol 1, Europe, Africa, South West Asia and the Middle East. IUCN Publications, Unit, Cambridge
Davis SD, Heywood VH, Hamilton AC (eds) (1995) Centres of plant diversity, vol 2, Asia, Australasia and the Pacific. IUCN Publications, Unit, Cambridge
Davis SD, Heywood VH, Herrera-MacBryde O, Villa-Lobos J, Hamilton AC (eds) (1997) Centres of plant diversity, vol 3, The Americas. IUCN Publications, Unit, Cambridge
De Bélair G (2009) *Serapias stenopetala*. In: IUCN 2010. IUCN Red List of threatened species. Version 2010.3. www.iucnredlist.org. Downloaded on 21 Sept 2010
De Bélair G, Boussouak R (2002) Une orchidée endémique de Numidie, oubliée: *Serapias stenopetala* Maire & Stephenson 1930. L Orchidophile 153:189–196
De Bélair G, Véla E, Boussouak R (2005) Inventaire des Orchidées de Numidie (N-E Algérie) sur vingt années. J Eur Orchids 37(2):291–401
De Candolle AB (1820) Essai elementaire de geographie botanique. In: Dictionnaire des sciences naturelles 18. Flevrault, Strasbourg, pp 1–64
de Montmollin B, Strahm W (eds) (2005) The top 50 Mediterranean island plants: wild plants at the brink of extinction, and what is needed to save them. IUCN/SSC Mediterranean Islands Plant Specialist Group, Gland/Cambridge
De Wit HCD (1975) *Cryptocoryne alba* de Wit (nov. sp.) en *Cryptocoryne bogneri* de Wit (nov. sp.). Het Aquar 45(12):326–327
Dengler J (2009) Which function describes the species-area relationship best? A review and empirical evaluation. J Biogeogr 36(4):728–744

Deroin T, Luke Q (2005) A new Toussaintia (Annonaceae) from Tanzania. J East Afr Nat History 94(1):165–174
Dixon CJ, Schönswetter P, Vargas P, Ertl S, Schneeweiss GM (2009) Bayesian hypothesis testing supports long-distance Pleistocene migrations in a European high mountain plant (Androsace vitaliana, Primulaceae). Mol Phylogenet Evol 53:580–591
Donaldson JS (2009) *Encephalartos woodii*. In: IUCN 2010. IUCN Red List of threatened species. Version 2010.3. www.iucnredlist.org. Downloaded on 21 Sept 2010
Dowl JL (1998) *Carpoxylon macrospermum*. In: IUCN 2010. IUCN Red List of threatened species. Version 2010.3. www.iucnredlist.org. Downloaded on 27 Sept 2010
Ducci F, Proietti R, Favre J-M (1999) Allozyme assessment of genetic diversity within the relic Sicilian fir *Abies nebrodensis* (Lojac.). *Mattei*. Ann For Sci 56:345–355
Eastern Arc Mountains & Coastal Forests CEPF Plant Assessment Project Participants (2006) *Afrothismia baerae*, *Aloe pembana*, *Toussaintia patriciae*. In: IUCN 2010. IUCN Red List of threatened species. Version 2010.3. www.iucnredlist.org. Downloaded on 18, 21 and 25 Sept 2010
Erkens RHJ, Chatrou LW, Maas JW, van der Niet T, Savolainen V (2007) A rapid diversification of rainforest trees (Guatteria; Annonaceae) following dispersal from Central into South America. Mol Phylogenet Evol 44:399–411
Evans R 1998. *Cryosophila cookii*. In: IUCN 2010. IUCN Red List of threatened species. Version 2010.3. www.iucnredlist.org. Downloaded on 27 Sept 2011
Fabbri F (1969) Il numero cromosomico di "*Bupleurum dianthifolium*" Guss. Endemismo di Marittimo (Isole Egadi). Inform Bot Ital 1:164–167
Faith DP (1992) Conservation evaluation and phylogenetic diversity. Biol Conserv 61:1–10
Faith DP, Baker AM (2006) Phylogentic diversity (pd) and biodiversity conservation: some bioinformatics challenges. Evolut Bioinform Online 2:121–128
Farjon A, Pasta S, Troìa A (2006) *Abies nebrodensis*. In: IUCN 2010. IUCN Red List of threatened species. Version 2010.3. www.iucnredlist.org. Downloaded on 18 Sept 2010
Fu L-K, Chin C-M (eds) (1992) China plant red data book – rare and endangered plants, vol 1. Science Press, Beijing
Gamisans J, Marzocchi J-F (1996) La flore endémique de la Corse. Aix-en-Provence, Edisud
Garcia-Barros E, Gurrea P, Lucianez MJ, Cano JM, Munuira ML, Moreno JC, Sainz H, Sanz MJ, Simon JC (2002) Parsimony analysis of endemicity and its application to animal and plant geographical distributions in the Ibero-Balearic region (Western Mediterranean). J Biogeogr 29:109–124
Gaston KJ (ed) (1996a) Biodiversity: a biology of numbers and difference. Blackwell Science, Oxford
Gaston KJ (1996b) Species richness: measure and measurement. In: Gaston KJ (ed) Biodiversity: a biology of numbers and difference. Blackwell Science, Oxford, pp 77–113
Gaston KJ (1996c) Species-range-size distributions: patterns, mechanisms and implications. Trends Ecol Evol 11(5):197–201
Gentry AH (1986) Endemism in tropical versus temperate plant communities. In: Soulé ME (ed) Conservation biology: the science of scarcity and diversity. Sinauer Associates, Inc.-Publisher, Sunderland, pp 153–182
Georghiou K, Delipetrou P (2010) Patterns & traits of the endemic plants of Greece. Bot J Linn Soc 162:130–422
Gianguzzi L, La Mantia A (2006) *Bupleurum dianthifolium*. In: IUCN 2010. IUCN Red List of threatened species. Version 2010.3. www.iucnredlist.org. Downloaded on 25 Sept 2010
Gianguzzi L, Scuderi L, La Mantia A (2003) Dati preliminari per una caratterizzazione sinfitosociologica e cartografica del paesaggio vegetale dell'Isola di Marettimo (Arcipelago delle Egadi, Canale di Sicilia). In: Proceedings of the 12th Società Italiana di Fitosociologia Congress, Venezia, 12–14 Feb 2003
Giddy C (1984) Cycads of South Africa, 2nd edn. C. Struik Publishers, Cape Town
Gillespie RG, Clague DA (eds) (2009) Encyclopedia of islands. University Press of California, Berkeley

Giokas S, Sfenthourakis S (2008) An improved method for the identification of areas of endemism using species co-occurrences. J Biogeogr 35:893–902
Glass C (1998) Identification guide to threatened Cacti of México. CANTE
Goodman SM, Benstead JP (eds) (2003) The natural history of Madagascar. The University of Chicago Press, Chicago/London
Groombridge B, Jenkins MD (2002) World atlas of biodiversity: earth's living resources in the 21st century. University of California Press, Berkeley
He F, Hubbell S (2011) Species-area relationships always overestimate extinction rates from habitat loss. Nature 473:368–371
Hershey DR (2004) The widespread misconception that the tambalocoque absolutely required the dodo for its seed to germinate. Plant Sci Bull 50:105–108
Hilton-Taylor C (1998) *Aloe pillansii*. In: IUCN 2010. IUCN Red List of threatened species. Version 2010.3. www.iucnredlist.org. Downloaded on 21 Sept 2010
Hilton-Taylor C (2000) 2000 IUCN Red List of threatened species. IUCN, Gland/Cambridge
Hobohm C (1999) *Euphorbia margalidiana* und Bykow's Index of Endemicity – ein Beitrag zur Biogeographie ausgewählter Inseln und Archipele. Abhandlungen Naturwissenschaftlicher Verein zu Bremen 44:367–375
Hobohm C (2000) Plant species diversity and endemism on islands and archipelagos, with special reference to the Macaronesian Islands. Flora 195:9–24
Hobohm C (2008) Ökologie und Verbreitung endemischer Gefäßpflanzen in Europa. Tuexenia 28:7–22
Huang J-H, Chen J-H, Ying J-S, Ma K-P (2011) Features and distribution patterns of Chinese endemic seed plant species. J Syst Evol 49(2):81–94
Huang J, Chen B, Liu C, Lai J, Zhang J, Ma K (2012) Identifying hotspots of endemic woody seed plant diversity in China. Divers Distributions 18:673–688
Hunt DR (ed) (1996) Temperate trees under threat. In: Proceedings of an international dendrological society symposium on the conservation status of temperate trees. Royal Botanical Gardens, Kew, U.K. 30 Sept–1 Oct 1994
Hunt D (1999). CITES cactaceae checklist, 2nd ed. Royal Botanic Gardens, Kew and International Organization for Succulent Plant Study (IOS)
Huston MA (1994) Biological diversity. Cambridge University Press, Cambridge
Izquierdo I, Martin JL, Zurita N, Arechavaleta M (eds) (2004) Lista de especies silvestres de Canarias (hongos, plantas y animales terrestres) 2004. – Consejeria de Medio Ambiente y Ordenacion Territorial, Gobierno de Canarias
Jacobsen N (1987) *Cryptocoryne*. In: Dassanayake MD, Fosberg FR (eds) A revised handbook to the flora of Ceylon Amerind Publishing Co. Pvt. Ltd., New Delhi, vol 6, pp 85–99
Kadis C, Christodoulou S (2006) *Centaurea akamantis*, *Scilla morrisii*. In: IUCN 2010. IUCN Red List of threatened species. Version 2010.3. www.iucnredlist.org. Downloaded on 21 and 27 Sept 2010
Kessler M (2002) Environmental patterns and ecological correlates of range size among bromeliad communities of Andean forests in Bolivia. Bot Rev 68(1):100–127
Kinzig AP, Harte J (2000) Implications of endemics-area relationships for estimates of species extinctions. Ecology 81(12):3305–3311
Kraft NJB, Baldwin BG, Ackerly DD (2010) Range size, taxon age and hotspots of neoendemism in the California flora. Divers Distrib 16:403–413
Kuhbier H (1978) *Euphorbia margalidiana*–eine neue Wolfsmilch-Art von den Pityusen (Balearen/Spanien). Veröff Überseemuseum Bremen A 5:25–37
Kühn I (2007) Incorporating spatial autocorrelation may invert observed patterns. Divers Distrib 13:66–69
Laffan SW, Crisp M (2003) Assessing endemism at multiple spatial scales, with an example from the Australian vascular flora. J Biogeogr 30:511–520
Lanfranco E (1995) The Maltese flora and conservation. Ecologia Mediterranea 21(1–2):165–168
Lanfranco E (1996) The flora and vegetation of Gozo. In: Farrugia J, Briguglio L (eds) A focus on Gozo. University of Malta, Malta

Linder HP (2001) Plant diversity and endemisms in sub-Saharan tropical Africa. J Biogeogr 28:169–182
Loots S, Mannheimer C (2003) The status of *Aloe pillansii* L. Guthrie (Asphodelaceae) in Namibia. Bradleya 21:57–62
Lovett JC, Rudd S, Taplin J, Frimodt-Moeller C (2000) Patterns of plant diversity in Africa south of the Sahara and their implications for conservation management. Biodivers Conserv 9:37–46
Madrigal Sánchez X, Guridi Gómez L (2002) Los árboles silvestres del Municipio de Morelia, Michoacán. México. Rev Ciencia Nicolaita 33:29–57
Madrigal Sánchez X, Rzedowski J (1988) Una especie nueva de *Diospyros* (Ebenaceae) del municipio de Morelia, estado de Michoacán (México). Acta Botánica Mexicana 1:3–6
Madulid DA, Tandang DN, Agoo EMG (2005) *Rafflesia magnifica* (Rafflesiaceae), a new species from Mindanao, Philippines. Acta Manilana 50:1–6
Madulid DA, Tandang DN, Agoo EMG (2008) *Rafflesia magnifica*. In: IUCN 2010. IUCN Red List of threatened species. Version 2010.3. www.iucnredlist.org. Downloaded on 21 September 2010
Magurran AE, McGill BJ (eds) (2011) Biological diversity: frontiers in measurement and assessment. Oxford University Press, Oxford
Mallet J (1996) The genetics of biological diversity: from varieties to species. In: Gaston KJ (ed) Biodiversity: a biology of numbers and difference. Blackwell Science, Oxford, pp 13–47
Meikle RD (1977, 1985) Flora of Cyprus. The Bentham-Moxon trust. Royal Botanic Gardens, Kew
Menne W (1992) Future of new *Kniphofia* in jeopardy. Plant Life 7:1
Mittermeier RA, Gil PR, Hoffman M, Pilgrim J, Brooks T, Mittermeier CG, Lamoreux J, da Fonseconda GAB (2005) Hotspots revisited: earth's biologically richest and most endangered terrestrial ecoregions. Cemex, Mexico City
Moreira Fernandez F (2006) Dimorphandra wilsonii. In: IUCN 2012. IUCN Red List of threatened species. Version 2012.1. www.iucnredlist.org. Downloaded on 29 Aug 2012
Moreno P, Pitman N (2003) *Centropogon cazaletii*. In: IUCN 2010. IUCN Red List of threatened species. Version 2010.3. www.iucnredlist.org. Downloaded on 27 Sept 2010
Morrone JJ (1994) On the identification of areas of endemism. Syst Biol 43:438–441
Nau C (2003) Das Insel-Lexikon. Heel Verlag, Barcelona
Nipperess DA, Faith DP, Barton K (2010) Resemblance in phylogenetic diversity among ecological assemblages. J Veg Sci 21(5):809–820
Oates MR, De Lange PJ (1998) *Chordospartium muritai*. In: IUCN 2010. IUCN Red List of threatened species. Version 2010.3. www.iucnredlist.org. Downloaded on 27 Sept 2010
Oldfield S, Lusty C, MacKinven A (1998) The world list of threatened trees. World Conservation Press, Cambridge
Osborne R (1986) Encephalartos woodii. Encephalartos 5:4–10
Ozinga WA, Schaminée JHJ (2005) Target species – species of European concern. A data-base driven selection of plant and animal species for the implementation of the Pan European Ecological Network. Wageningen, Alterra, Alterra-report 1119
Pawlowski B (1969) Der Endemismus in der Flora der Alpen, der Karpaten und der Balkanischen Gebirge im Verhältnis zu den Pflanzengesellschaften. Mitteilungen der ostalpin-dinarischen pflanzensoziologischen Arbeitsgemeinschaft 9:167–178
Peterson AT (2008) Parsimony analysis of endemism (PAE) and studies of Mexican biogeography. Revista Mexicana de Biodiversidad 79:541–542
Petrova A (2006) Atlas of Bulgarian endemic plants. Gea Libris Publishing House, Sofia
Proctor D, Fleming V (eds) (1999) Biodiversity: the UK overseas territories. Joint Nature Conservation Committee, Peterborough
Purdie AW (1985) *Chordospartium muritai* (Papilionaceae) – a rare new species of New Zealand tree broom. New Zeal J Bot 23:157–161
Rabitsch W, Essl F (eds) (2009) Endemiten – Kostbarkeiten in Österreichs Pflanzen- und Tierwelt. Umweltbundesamt, Klagenfurth
Rapoport EH (1982) Areography: geographical strategies of species. Bergamon Press, New York
Rikli M (1943) Das Pflanzenkleid der Mittelmeerländer 1. Hans Huber, Bern

Rikli M (1946) Das Pflanzenkleid der Mittelmeerländer 2. Hans Huber, Bern
Rosauer D, Laffan SW, Crisp MD, Donnellan SC, Cool LG (2009) Phylogenetic endemism: a new approach for identifying geographical concentrations of evolutionary history. Mol Ecol 18:4061–4072
Rosen BR (1988) From fossils to earth history: applied historical biogeography. In: Myers A, Giller P (eds) Analytical biogeography: an integrated approach to the study of animal and plant distribution. Chapman and Hall, London, pp 437–481
Rouillard G, Guého J (1999) Les Plantes et Leur Histoire a l' Ile Maurice. Louis Bouton, Morne Brabant
Rovito SM, Arroyo MTK, Pliscoff P (2004) Distributional modelling and parsimony analysis of endemicity of senecio in the Mediterranean-type climate area of central Chile. J Biogeogr 31:1623–1636
Santos CMD (2005) Parsimony analysis of endemicity: time for an epitaph? J Biogeogr 32:1284–1286
Scheiner SM (2003) Six types of species-area curves. Glob Ecol Biogeogr 12:441–447
Scheiner SM (2004) A melange of curves – further dialogue about species-area relationships. Glob Ecol Biogeogr 13:479–484
Scott-Shaw CR (1999) Rare and threatened plants of KwaZulu-Natal and neighbouring regions. KwaZulu-Natal Nature Conservation Service, Pietermaritzburg
Sechrest W, Brooks TM, Da Fonseca GAB, Konstant WR, Mittermeier RA, Purvis A, Rylands AB, Gittleman JL (2002) Hotspots and the conservation of evolutionary history. Proc Natl Acad Sci USA 99:2067–2071
Skottsberg C (1911) The wilds of Patagonia: a narrative of the Swedish expedition to Patagonia Tierra Del Fuego and the Falkland islands in 1907–1909. Edward Arnolds, London
Smith AC (1979) Flora vitiensis nova: a new flora of Fiji. Pacific Tropical Botanic Garden, Hawaii
Smith-Abbott J, Walker R, Clubbe C (2002) Integrating National Parks, Education and Community Development (British Virgin Islands). Unpublished final report to the Darwin Initiative Secretariat, London, UK
Staub F (1993) Fauna of Mauritius and associated flora. Precigraph Limited, Mauritius
Stevens D, Lanfranco E (2006) *Cheirolophus crassifolius*. In: IUCN 2010. IUCN Red List of threatened species. International Union for Conservation of Nature. Version 2010.3. www.iucnredlist.org. Downloaded on 27 Aug 2012
Stohlgren TJ (2007) Measuring plant diversity. Oxford University Press, Oxford
Storch D, Keil P, Jetz W (2012) Universal species-area and endemics-area relationships at continental scales. Nature 488:78–81
Tan K, Iatrou G (2001) Endemic plants of Greece: the Peloponnese. Gad Publishers Ltd., Copenhagen
Thomasson M (1999) Réflexions sur la biodiversité: richesse, originalité et endémicité floristiques. Acta Botanica Gallica 146(4):403–419
Tirel C, Veillon J-M (2002) Pittosporaceae. Flore de la Nouvelle-Calédonie t. 24. MNHN, Paris, 178 p
Tucker CM, Rebelo AG, Davies JD, Cadotte MW (2012) The distribution of biodiversity: linking richness to geographical and evolutionary rarity in a biodiversity hotspot. Conserv Biol 26(4):593–601
Tutin TG, Burges NA, Chater AO, Edmondson JR, Heywood VH, Moore DM, Valentine DH, Walters SM, Webb DA (1996a) Flora Europaea, 2nd edn, vol 1, Psilotaceae-Platanaceae. Cambridge University Press, Cambridge
Tutin TG, Heywood VH, Burges NA, Moore DM, Valentine DH, Walters SM, Webb DA (1996b) Flora Europaea, vol 2, Rosaceae-Umbelliferae. Cambridge University Press, Cambridge
Tutin TG, Heywood VH, Burges NA, Valentine DH, Walters SM, Webb DA (1996c) Flora Europaea, vol 3, Diapensiaceae-Myoporaceae. Cambridge University Press, Cambridge
Tutin TG, Heywood VH, Burges NA, Valentine DH, Walters SM, Webb DA (1996d) Flora Europaea, vol 4, Plantaginaceae-Compositae (and Rubiaceae). Cambridge University Press, Cambridge

Tutin TG, Heywood VH, Burges NA, Valentine DH, Walters SM, Webb DA (1996e) Flora Europaea, vol 5, Alismataceae-Orchidaceae. Cambridge University Press, Cambridge

Ungricht S (2004) How many plant species are there? And how many are threatened with extinction? Endemic species in global biodiversity and conservation assessments. Taxon 53:481–484

Valencia R, Pitman N, León-Yánez S, Jørgensen PM (eds) (2000) Libro Rojo de las Plantas Endémicas del Ecuador. Publicaciones del Herbario QCA, Ponticicia Universidad Católica del Ecuador, Quito

Vane-Wright RI, Humphries CJ, Williams PH (1991) What to protect – systematics and the agony of choice. Biol Conserv 55:235–254

Villaseñor Gómez L (2005) La biodiversidad en Michoacán. Estudio de Estado. CONABIO-SUMA-UMSNH, México

Virgilio F, Schicchi R, La Mela VD (2000) Aggiornamento dell'inventario della popolazione relitta di *Abies nebrodensis* (Lojac.). Naturalista Siciliano 24(1–2):13–54

Wagner WL, Herbst DR, Sohmer S (1990) Manual of the flowering plants of Hawaii. University of Hawaii Press, Bishop Museum Press, Honolulu

Wagner WL, Herbst DR, Sohmer S (1999) Manual of the flowering plants of Hawaii. Bernice Pauahi Bishop Mus Spec Pub 91:1–1918

Wagner WL, Herbst DR, Lorence DH (2005) Flora of the Hawaiian islands – online. Smithsonia National Museum of History. http://ravenel.si.edu/botany/pacificislandbiodiversity/hawaiianflora/index.htm

Walter KS, Gillett HJ (eds) (1998) 1997 IUCN Red List of threatened plants. Compiled by the World Conservation Monitoring Centre IUCN, Gland/Cambridge

Werner U, Buszko J (2005) Detecting biodiversity hotspots using species-area and endemics-area relationships: the case of butterflies. Biodivers Conserv 14:1977–1988

Whittaker RJ, Fernandez-Palacios JM (2007) Island biogeography: ecology, evolution, and conservation, 2nd edn. Oxford University Press, Oxford

Williamson G (1998) The ecological status of *Aloe pillansii* (Aloaceae) in the Richtersveld with particular reference to Cornellskop. Bradleya 16(1998):1–8

World Conservation Monitoring Centre (1998) *Auerodendron pauciflorum*, *Cercocarpus traskiae*, *Hesperomannia arborescens*. In: IUCN 2010. IUCN Red List of threatened species. Version 2010.3. www.iucnredlist.org. Downloaded between 21 Sept 2010 and 10 Oct 2012

Part II
Endemic Vascular Plants Over Time

This part focuses on conditions and processes related to evolutionary and historical times, to long-term changes and constancy, to the growth of pedigrees, migration and dispersal processes.

What are the historical and evolutionary reasons for the uneven distribution of endemic species across the world? What mechanisms favour or reduce endemism in evolutionary terms? What environmental conditions may raise or reduce numbers of endemics within an evolved biotic assemblage in a given region? Does constancy or change – in terms of climate, landscapes, substrates - more effectively stimulate or promote the composition of endemics in a region? Evolution is necessary for speciation. Habitat destruction is one of the main factors that threaten the existence of taxa today. Thus, the factors which determine increases in the number of taxa in a region are different from those which control decreases.

This simple fact leads also to the conclusion that conservation management aimed at increasing the number of individuals in a small population must differ from that aimed at stopping decrease and extinction risk of populations.

Chapter 3
Factors That Create and Increase Endemism

Ines Bruchmann and Carsten Hobohm

3.1 Introduction

Today's species inventory is only a snapshot in time of the ever progressing evolution of being. Different theories on the evolution and changes in species compositions of endemics have been discussed over the last 150 years, but to date these theories give no conclusive indication of the evolutionary mechanisms that make a taxon become a geographically restricted – i.e. an endemic – taxon.

It would be difficult to find a single endemic vascular plant species in an arctic lowland region that was covered by glaciers a few thousand years ago (cf. Talbot et al. 1999). In contrast, many regionally restricted vascular plants inhabit subtropical or tropical regions with high environmental heterogeneity (e.g. Mittermeier et al. 2005). Why is this so? What are the historical and evolutionary processes behind the patterns that favour speciation and endemism and reduce the extinction risk of entire systematic entities?

An endemic vascular plant, in general, is the result of speciation or simply of downsizing the range (Fig. 3.1). During allopatric speciation, a taxon splits into two or more geographically separated and, due to different selective pressures, genetically different taxa. Sympatric speciation is the term for different biological processes such as mutation, hybrid speciation or speciation via polyploidization which result in two or more descendant taxa which occupy the same geographical region. Disturbance of habitats is one of the processes that leads to a smaller range of a former more wide-spread taxon. All these processes most likely result in alterations of the biology of the endemic taxa compared to the ancestors, e.g. in lower heterozygosity, genetic drift, lower adaptability, higher risk of inbreeding depression, and so on.

I. Bruchmann • C. Hobohm (✉)
Ecology and Environmental Education Working Group, Interdisciplinary Institute of Environmental, Social and Human Studies, University of Flensburg, Flensburg, Germany
e-mail: hobohm@uni-flensburg.de

C. Hobohm (ed.), *Endemism in Vascular Plants*, Plant and Vegetation 9,
DOI 10.1007/978-94-007-6913-7_3, © Springer Science+Business Media Dordrecht 2014

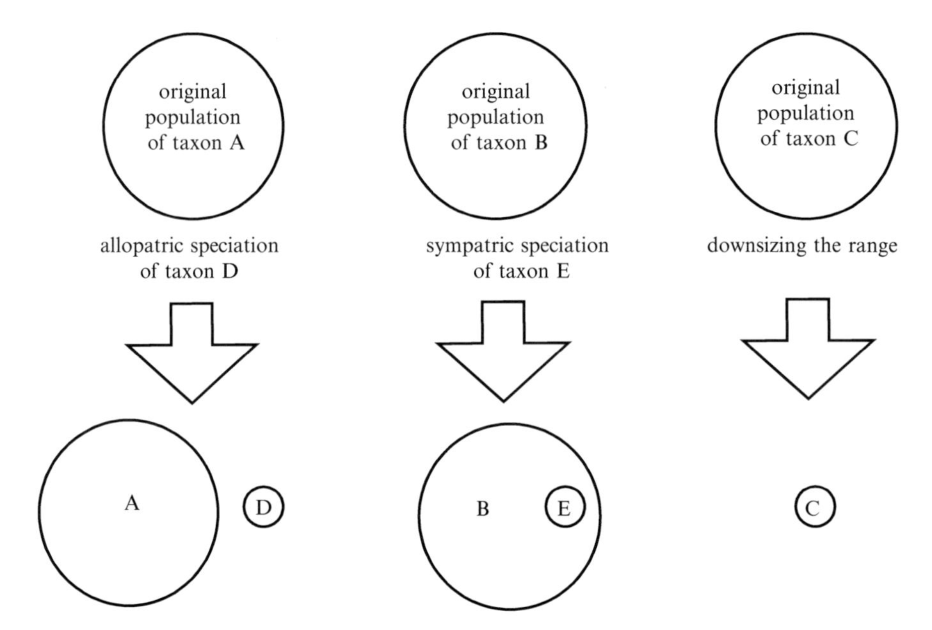

Fig. 3.1 Development and evolution of an endemic taxon by allopatric speciation, sympatric speciation, or downsizing the range

Factors and processes that increase endemism, in general, are related to biological traits and environmental conditions. Biological traits are, for example, life-form and life-cycle, genome and genetical processes, pollination, production of propagules and dispersal mode. Environmental conditions that allow a continuous, uninterrupted evolution, and that lower the extinction risk (e.g. high environmental heterogeneity), or favour high evolutionary speed (warmth, length of vegetation periode, water?) promote speciation processes (e.g. Bruchmann 2011; Graham et al. 2006; Hobohm 2003; Cowling and Lombard 2002; Givnish 2000; Rosenzweig 1995; Huston 1994; Hendrych 1982).

Especially paleoclimate data are considered relevant for the understanding of biogeography and ecology of endemism but the relationship is not thoroughly understood (Parmentier et al. 2007; Jansson 2003). One reason is that historical data which might have been relevant for explaining regional diversity of endemics are still lacking for many ecoregions.

Two sets of time-related hypotheses concerning environmental conditions that increase the diversity of (endemic) vascular plant taxa in space have been proposed, one about short-term processes (such as water-energy supply, productivity, and other ecological processes), the other focusing on long-term processes and historical events such as (the absence of) volcanic eruptions, plate tectonic movements, long-term climatic stability, Pleistocene glaciations, or recent warming. Kreft and Jetz (2007), O'Brien (1998, 2006), Cowling and Lombard (2002), Vetaas and Grytnes (2002), Whittaker et al. (2001), Körner (2000), Hobohm (2000a, b), Huston (1994),

Werger et al. (1987), Wright (1983), Brown (1981), e.g., give an overview of diverse hypotheses. Additionally, altitudinal gradients, environmental heterogeneity (Orme et al. 2005; Whittaker et al. 2001; Qian and Ricklefs 2000), or geographical positions of mainland regions and islands (theory of island biogeography, cf. Lomolino 2001; Simberloff and Wilson 1969; MacArthur and Wilson 1967) seem to play an important role in speciation and stabilization of species compositions.

3.2 Neo- and Paleoendemism

One of the oldest endemism concepts is that of neo- or palaeoendemism, introduced by Engler (1879–1882), who classified endemic taxa according to their inferred evolutionary age. The basic assumption is that taxa which are isolated at a high taxonomical level e.g. the monotypic gymnosperms *Welwitschia mirabilis* or *Ginkgo biloba*, are also very old in evolutionary terms (Khoshoo and Ahuja 1963). Engler hypothesised that palaeoendemics are evolutionary leftovers that have survived long evolutionary periods while various factors had led to the extinction of other congeners; e.g., *Ginkgo biloba* is the last remaining member of the taxonomic division Ginkgophyta (Royer et al. 2003). Neoendemics, on the other hand, are described as recently evolved, phylogenetically young; they rank most often at low taxonomical levels and their relatively recent speciation process is reflected in the existence of several closely related congeners – whether widespread or restricted.

Following this basic idea of neo- and palaeoendemism Favarger and Contandriopoulos (1961) attempted to systematise endemics on the basis of genome analyses and made some effort to order endemic plants according to their ploidy level. They argued that the formation of new species is often provoked by the multiplication of the taxon's chromosome set; thus, taxa with a low ploidy-level should be older in evolutionary terms and of higher taxonomic ranking than taxa with high ploidy-levels. The authors distinguished three categories of neoendemics: (1) 'apoendemics', which they defined as taxa with a higher ploidy level than their closest relatives; (2) 'patroendemics', which have a low ploidy level and have presumably spawned younger taxa with higher ploidy levels; (3) 'schizoendemics', if the ploidy levels of the endemic taxon and its close relatives are equal (vicariant species).

Normally, we assume that endemics with a lower ploidy level originated earlier than those with a higher level. In fact, conclusive evaluations of the age of taxa and the time of speciation processes often point towards plausibilities rather than proven evidence (cf. Trigas and Iatrou 2006).

Several studies indicate that polyploidism, mainly allopolyploidy, is quite a common evolutionary strategy, especially in plant evolution (Wood et al. 2009). However, this finding gave no clear indication for the status of endemism itself; different ploidy levels in plants will not answer the question as to whether a plant will be geographically restricted (endemic) or widely distributed (non-endemic) after it has evolved. It has, however, been shown that polyploidism brings some evolutionary advantages which promote speciation (Wood et al. 2009). Thus, the Favarger and

Contandriopoulos concept of systematising endemics according to ploidy levels is contradicted by many findings on aberrant ploidy levels in endemic plants. Furthermore, the idea of close relatives based on morphological similarities is often questioned (e.g. Trigas and Iatrou 2006; Pinter et al. 2002; Trewick et al. 2002).

3.3 Constancy and Change

Today's patterns of endemism do not only mirror the impact of contemporary climate on species ranges. Past climate regimes strongly influenced the conditions of survival, the migration of species and therefore recent species compositions (Bruchmann 2011; Stewart et al. 2010; Lesica et al. 2006; Cowling and Lombard 2002). Past climate changes (e.g. glaciation cycles of the Pleistocene) have led to significant changes in today's distribution patterns of species. Stewart et al. (2010) stated that '*geographical ranges of species expand and contract in cyclical manner according to glacial cycles*'. Cowling and Lombard (2002) recognised today's diversity patterns as temporary *steady-state-diversities* which are results of past processes in speciation and extinction. They hypothesise that regions with high proportions of range-restricted and rare species have '*experienced higher rates of speciation and/or lower rates of extinction*' (Cowling and Lombard 2002).

Cain (1944: 216) wrote: " . . . *the lands of the northern hemisphere which were covered by the Pleistocene ice sheets seem to be conspicuously low in endemics.*" The meaning of ecological constancy through time is important at all regional habitat scales (cf. Jansson 2003; Cowling and Lombard 2002; Pärtel 2002). However, many factors can destroy or change ecoregions, landscapes and habitats and the compositions of inhabitants in a region, for example climate change, volcanic eruptions, fire, and man (Fig. 3.2).

Climate stability was identified as an important evolutionary factor. The longer the periods of environmental stability in time the higher the chance of persistence of species pools and the survival of highly range-restricted, specialised species with low dispersal ability (Jansson 2003). This hypothesis is called *Stability-Time-Hypothesis* or *Climate-Stability-Hypothesis* (cf. Cowling and Lombard 2002; Dynesius and Jansson 2000; Latham and Ricklefs 1993).

Fjeldså and Lovett (1997) found that the richness of Afrotropical neo- and palaeoendemic plants was highest in regions of high climatic stability and suggested the factor long-term climatic stability as an important premise for the high number of endemic species in the Eastern Arc Biodiversity Hotspot. Cowling and Lombard (2002) found much higher overall species richness and higher numbers in endemic plants in the western compared to the eastern Cape Floristic Region which could not be sufficiently explained by patterns of habitat heterogeneity or contemporary climatic regimes. They found that the greater stability of historical climate in the western Cape compared to the eastern Cape had favoured speciation events and lowered rates of extinction.

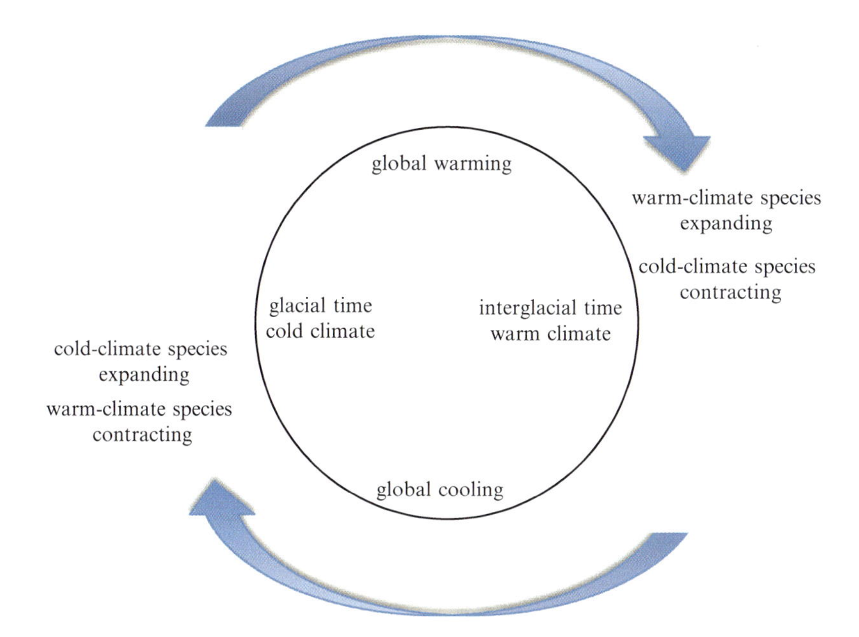

Fig. 3.2 Glaciation cycles as drivers of area expansion and contradiction

Indeed, absolutely stable ecological conditions do not exist. Hence, the stability of conditions should also be measured on a relative scale. Environmental conditions can be relatively constant, as in many regions of the humid tropics, or they are characterised by more or less severe changes such as glacial cycles, volcanic eruptions, or other destruction e.g. caused by man.

Furthermore, the ecological effects of global warming or cooling, of wet and dry periods can be diminished by the buffering capacities of the environment. In this context, waterbodies and environmental heterogeneity seem to play important roles.

The Milankovitch climate oscillations which are the periodic changes of the Earth's orbit and the cause of periodic warming and cooling of the Earth's atmosphere (e.g. Webb and Bartlein 1992; Imbrie et al. 1989) were suggested to be the major factor influencing today's geographical patterns of species distribution at regional to global scales. According to Dynesius and Jansson (2000) the Milankovitch changes of species ranges induced by oscillation were termed 'orbitally forced range dynamics'. The authors hypothesise that the buffering capacities of oceanic climates or topographically caused associations of microclimates can lower the impact of climate change on regional species pools.

Most of the regions with a high diversity of endemic vascular plants are found close to the coast, where major oceanic surface currents are not far away (Konstant et al. 2005; Mittermeier et al. 2005; Linder 2001; Davis et al. 1994, 1995, 1997; Shmida and Wilson 1985; Stebbins and Major 1965). One clear result of the analyses of Crisp et al. (2001) is that all the major centres of Australian vascular plant endemism are near-coastal. Thus, strong marine currents effectively stabilize climate conditions of different neighbouring terrestrial regions. The influences in certain terrestrial areas seem to be very old and constant phenomena dependent on the size and position of land masses and on the positions of the poles. The existence of evolutionary old palaeoendemic species in isolated island floras was also traced back to the effects of the regional buffering of the surrounding oceans against long-term climatic extremes (Jansson 2003; Cronk 1997).

We assume that the related climates are normally also relatively constant over time. This is the case in many wet tropical ecoregions with high rates of water-energy-supply. But Mediterranean climate regions and the Succulent Karoo in South Africa also seem to be relatively stable in terms of climate conditions through evolutionary time.

In these endemic-rich biomes – and this also applies to areas with low rainfall – the annual season with rain or mist is highly predictable and totally dry periods of several years do not occur (Mittermeier et al. 2005; Dean and Milton 1999; Bailey 1998; Davis and Richardson 1995).

In this biogeographical context of endemism related to ecological constancy it is also of interest that annual temperature amplitudes of many mountain tops are smaller than those of mid-altitude or foothill regions. In the Alps, for example, the highest zones and mountain-top regions represent annual temperature amplitudes of less than 15 °C (mean temperature warmest month minus mean temperature coldest month; unpublished data from Deutscher Wetterdienst 2008), whereas mid and low regions have higher amplitudes. In this sense, the climates of high mountain areas tend to be more stable than those of lower regions, and this favours endemism. The alpine flora (above the treeline), which is rich in local and regional endemics in many mountain ranges of the world, covers 3 % of the vegetated land area, but includes some 4 % of all known plant species (Körner 2000). The climate in high-mountain areas may also have been relatively stable in evolutionary time, thus favouring endemism.

Most of the regions with high diversity are also characterised by high environmental heterogeneity and habitat diversity. We assume that environmental heterogeneity (different expositions, inclinations, substrates, water and nutrient conditions, etc.) can stimulate and promote speciation. Furthermore, especially under changing conditions, environmental heterogeneity can secure the survival of (endemic) species.

It is undisputed that of two plant individuals of the same taxon with the same competitive vigour neither would be able to displace the other. Therefore, we do not find any argument to support the opinion that two different species could not permanently coexist in the same environment. Furthermore, we assume that this is quite normal (cf. Kreft et al. 2008; Pronk et al. 2007; Huston 1994; Remmert 1991; Stebbins and Major 1965). Different taxa can coexist in the same environment, under the same ecological constraints and conditions, for a long time. Habitat diversity or environmental heterogeneity would not be necessary for the coexistence of different (endemic) species under stable ecological conditions. However, we assume that the meaning of the environmental heterogeneity for the species pool increases if the conditions, for example climate, change.

Survival of the biota in large regions under changing conditions results from having a sufficient number of patches to ensure that changes in one patch will not affect other patches, and that there is only little or no net change over a large area. For example, almost all the mid- and high altitudes of the Alps were covered by Pleistocene glaciers. However, the elevated zones of the Alps harbour many endemic vascular plants which survived in the valleys or at lower altitudes between the Alps and the Mediterranean Basin. Stability at this scale is a stochastic phenomenon which involves local and regional processes like catastrophes, gap dynamics and successions, and it is different from the deterministic mathematical equilibrium of the competitive exclusion theory.

The effects of diverse microclimates which provided refuge areas with relative environmental stability are seen in many mountainous regions. The high endemic diversity of the flora of the Mediterranean Basin with several palaeoendemic plants, for example, is largely explained by the persistence of Tertiary (relict) species in refugial areas during glacial times. Due to the general favourable geographical position of the Mediterranean and its high topographical variability (habitat diversity), species evolved and persist in several small climatically stable areas which mitigate the effects of climatic extremes (Tzedakis 2009). Médail and Diadema (2009) reviewed phylogeographical studies and identified more than 50 refugial areas of plants distributed throughout the circum-Mediterranean today. High congruence was found between the geographical position of the former refugial areas and today's regional biodiversity hotspots with high rates of endemism (Médail and Diadema 2009; Médail and Quézel 1999). A similar picture was sketched for New Zealand's alpine plants where the climatically stable refuge areas are largely congruent with today's areas of high endemism (McGlone et al. 2001). The authors state that "*Pleistocene species selection appears [. . .] important, especially in the selection of species that can tolerate cyclical appearance and disappearance of environments, and thus tolerate long periods of small population sizes in suboptimal habitat followed by massive range expansion during short periods of optimal environments*" (Mcglone et al. 2001).

For a long time the notion of climate stability lowering extinction rates and favouring speciation was mainly discussed from the perspective of the classical 'glacial refugium concept' (see summary in (Stewart et al. 2010). The classical refugia idea suggests that re-immigration of species in interglacial times started from only a few macro-refugia at lower latitudes. Calculations of the theoretical speed of the recolonisation process compared with historical data indicated that recolonisation took place much faster than interpolated from species' real dispersal speed. This phenomenon (Reid's Paradox) led to the recognition of micro-refugials (e.g. Rull 2004, 2009; Rull et al. 1988) also called cryptic refugia (Stewart et al. 2010; Clark et al. 2003), i.e. very small disjunct refugial areas at higher latitudes that can only harbour very small refuge populations of species. Stewart et al. (2010) distinguished between the different spatial and temporal categories of the term 'refugia' and introduced the concept of 'interglacial refugia' which describes the survival of cold-climate adapted species in interglacial times: While in interglacial periods populations of warm-climate adapted species expand from their refugials, populations of cold-climate adapted species contract into their cooler refugial areas (e.g. *Dryas integrifolia*).

Regardless of glacial or interglacial macro- or micro-refugia, the existence of refugia is important for the survival of species during unfavourable time periods.

3.4 Age of the Geological Substrate and Landscape

The age of the geological substrate and the time scales of prolonged landscape genesis have a strong influence on evolution and thus on contemporary diversity patterns (speciation and extinction rates, species diversity, endemism, etc.). The geological substrate forms the basis for the landscape morphology (landscape heterogeneity), determines the subsequent processes of erosion (e.g. hard vs. soft bedrock) and soil genesis. The major geological substrates determine soil properties such as pH, salinity, water-holding capacities, nutrient-availability, etc. which in turn determine to some degree the species composition of sites. How long it takes before pioneer species invade, the first stable populations become established and speciation leads to the establishment of vegetation with endemics depends on the degree of isolation of a newly evolved landscape.

The island Reunion, for example, is larger, more elevated, obviously more habitat-rich, and is situated closer to the most important species pools Madagascar and Africa, yet it is hardly richer in vascular plant species in total than Mauritius. Both Reunion and Mauritius are tropical islands which lie in the same climate zone and are of volcanic origin; however, Mauritius is much richer in endemic plant taxa (and endemic birds, Hobohm 2000b; Staub 1993). One obvious explanation for the higher endemism on Mauritius is that the island is some five million years older than Reunion i.e. landscape genesis and the establishment of the biotic inventory have had longer to develop on Mauritius than on Reunion.

However, in addressing species richness or endemism the age of landscapes or geological substrates is not the only dimension affecting the evolution of biota. The principal question with regard to the factor landscape age is: How long can evolution run without major disturbances?

It is hypothesised that long, undisturbed evolutionary histories in floras increase rates of highly adapted specialists and also of range-restricted endemic species. Hopper (2009) identified three ancient, geologically stable landscapes, with oceanically buffered climates which have not been glaciated since the early Cretaceous (140 Ma): These are (a) the Pantepui of the Guyana Highlands, (b) South Africa's Greater Cape (comprising the Cape Floristic Region and the Succulent Karoo Biome), and (c) the Southwest Australian Floristic Region, (Hopper 2009). All these ancient landscapes are characterised by impoverished soils and inhabited by extremely diverse floras with endemism rates of between 40 and 69 %. In order to explain evolution and the ecology of the biotic inventory in these old, climatically buffered, infertile landscapes (OCBILs) Hopper categorised landscapes according to (1) landscape age, (2) climate buffering, and (3) soil nutrient status, and contrasted OCBILs to young, often disturbed, fertile landscapes (YODFELs). Young landscapes can be landscapes which originated from volcanic eruptions or landscapes resulting from rapid tectonic events (e.g. uplifts of alpine regions) or landscapes transformed by Quarternary glacial regimes.

Mucina and Wardell-Johnson (2011) incorporated and broadened Hopper's framework and presented the concept of old stable landscape (OSL). Evolutionary development of biodiversity in different landscapes is related to the three dimensions landscape age (including soil nutrients), climate stability (whether stable or highly dynamic with predictable climatic dynamics) and fire regime predictability. The authors found old stable landscapes all over the globe where areas of old surfaces are congruent with areas of long-lasting climates (see comprehensive study by Mucina and Wardell-Johnson 2011).

3.5 Dispersal, Isolation, Hard and Soft Boundaries

The importance of geographical migration- or dispersal barriers was identified at the latest by MacArthur and Wilson's theory on Island Biogeography (Macarthur and Wilson 1967). This theory states that the closer an island is situated towards the continental species pool the higher the chance of species colonisation and the better the chance of a stabilising gene-flow between continent and island population. Long distances to the species pool, however, limit gene flow and promote speciation of daughter populations. Thus, dispersal barriers, such as long distances across the sea, reduce gene-flow and increase the degree of isolation of local populations, a precondition which promotes speciation events and local endemism.

Since this theory was developed various studies have correlated the degree of endemism with the degree of geographical isolation, which is measured as the distance from an island to the nearest continent. The main result is that the level of endemism (in % of the whole composition or indigenous taxa) is often highly correlated with isolation. In contrast, the relationship between isolation and density values (concentration of endemics) is often not significantly correlated (Bruchmann 2011; Panitsa et al. 2010; Hannus and Von Numers 2008; Nikolic et al. 2008; Reyes-Betancort et al. 2008; Kreft et al. 2007; Hobohm 2000b; Kruckeberg and Rabinowitz 1985).

Dispersal barriers may be distinguished by their geographical distinctiveness. Geographically clearly defined dispersal barriers such as mountain ranges and other topographic contrasts, as well as distances across the sea or other water barriers (e.g. streams) are called *hard boundaries* (Vetaas and Grytnes 2002). Dispersal barriers that are defined by ecological properties such as edaphic or climatic gradients can be defined as *soft boundaries*. While hard boundaries are easily recognisable and can be mapped, the extent of soft boundaries is less distinct and the effects more diffuse, making it more difficult to identify them: Examples of *soft boundaries* are habitats that differ markedly ecologically from those typical of the regional environments, e.g. heavy metal outcrops, and are comparable to *ecological islands* where reduced gene-flow and isolation is caused by ecological constraints/conditions.

Another good example of a *soft boundary* is the transition zone along the rainfall gradient between the seasonal winter rainfall zone (west) and the non-seasonal rainfall zone (east) of the Cape Floristic Region (SW Africa, cf. Cowling and Lombard 2002). Overall species diversity and strong species-turnover is visible there. Sprouting plants survive the less predictable conditions of the eastern zone, while the more stable ecological conditions of the winter rainfall zone favour non-sprouting plants. The authors hypothesise that the low species diversity of the eastern zone is determined by lower speciation rates and elevated extinction rates caused by contemporary and historical climate regimes.

However, hard and soft boundaries work in the same direction as they both reduce dispersal opportunities and successful colonisation.

Cardona and Contandriopoulos (1979) examined the distribution of endemic plants in the Mediterranean area. They found that (geographical) separation is an important factor in the evolution of endemics: "*The geographical distribution of endemics and corresponding taxa indicates these areas which have had species in common, isolation being the principal factor in speciation*" (Cardona and Contandriopoulos 1979: 140; see also Haffer 1969).

According to this hypothesis geographical separation is important for genetic isolation. It is commonly suggested that the stronger the degree of separation the more endemic taxa are found. Although several examples show such a relationship, the general effect of separation on endemism appears highly overestimated. The isolation factor means that islands far from the continents should show higher endemism than mainland areas. However, the situation is clearly more complicated (cf. e.g. Rull 2004; Melendo et al. 2003). In fact, genetic isolation is the precondition for speciation and geographic separation favours genetic isolation. But, many other sympatric isolation processes are known (Bell 2008; Cockburn 1991).

Starting with a large species pool in continuous mainland areas or continental islands tends to promote speciation and endemism more effectively than a small species pool in separated areas such as oceanic islands far away from the continents. The separation/dispersal/colonisation complex is the reason why islands tend to have fewer species in total than mainland areas, and why oceanic islands are less species-rich than continental ones (Kreft et al. 2008; Hobohm 2000b; MacArthur and Wilson 1967). Separation on the one hand reduces successful dispersal and total species richness, which is the basis for evolutionary processes, speciation and endemism. On the other hand, geographical separation clearly favours genetic isolation, which is the precondition for speciation. Thus, the separation/isolation factor both promotes and restricts endemism.

The uniqueness of the flora from the remote tableaux summits of the Tepuis Ecoregion in northern South America has been explained both as the result of a long history of evolution in isolation (lost world-hypothesis) and by alternating upward and downward displacements during glacial-interglacial Quarternary cycles (vertical displacement-hypothesis; see Rull 2004 and the discussion on Haffer's refugial hypothesis in Colinvaux 1998).

Topographic barriers may contribute to the restriction of gene-flow and thus allow speciation to occur. Nonetheless, the existence of many regions with high endemism and no apparent topographic barriers demonstrates that such barriers are not essential to the maintenance of high endemism or speciation (cf. Kreft et al. 2008; Hobohm 2008, 2000b; Huston 1994).

Several studies discuss the role of species dispersal for endemism. Lesica et al. (2006) hypothesised that many recently evolved, and thus young, populations of neoendemic species simply have not had enough time yet to spread to their full potential range.

Uninterrupted gene-flow between individuals or populations of a species is the basis for allele diversity (heterozygosity) which in turn defines the adaptive abilities of the species in its local existence. If there is a fundamental change in the selective pressures (e.g. habitat changes, immigration of new predators etc.), populations with high allele diversity may buffer and resist ecological changes better than populations with a higher degree of homozygosity (Jamieson 2007).

Thus, species range is determined by dispersal ability in two ways: (i) survival and persistence in an already occupied region and (ii) migration success i.e. invading new favourable habitats in the case of changing environmental pressures. Gene-flow in plants is mainly determined by their success in sexual reproduction and their seed-dispersal abilities. Limited gene-flow in plant populations can be induced by several factors: external abiotic factors (dispersal barriers, distance between suitable habitats, edaphic constraints), species immanent, trait-induced factors (e.g. generation length, propagule investment), biotic interactions in sexual reproduction and dispersal (e.g. pollination/dispersal via biotic agents).

Beside abiotic dispersal boundaries several causes that are species immanent and are reflected in life-history traits are of interest. This could be, for example, the plant species' (a) generation length (e.g. annual therophytes vs. perennial phanerophytes, also age of sexual maturity); (b) pollination mode (abiotic pollination vs. complex specialised biotic pollination); (c) mode of seed dispersal (short-distance vs. long-distance), and (d) propagule investment (r-strategy vs. K-strategy) as was quantified by Shmida and Werger (1992) for plants of the Canary Islands.

Many regional endemic vascular plant taxa belong to the most species-rich plant families Asteraceae (many regions outside the tropics) and Orchidaceae (e.g. wet tropics). In both families long-distance dispersal by wind is quite normal. Very often, total regional species-richness is correlated with high endemism occurring in a small number of the same plant families where insect-pollination is common (Lavergne et al. 2004; Melendo et al. 2003; Tutin et al. 1996a, b, c, d, e; cf. many other floras; Gaston and Williams 1996; Shmida and Werger 1992). Other species-rich families such as Poaceae, Chenopodiaceae, or Cyperaceae tend to be relatively poor in endemic species at regional scales. These are anemochorous as well but wind-pollinated. From studies of endemics in isolated island floras, it is known that many endemic plants show long-distance dispersal modes (e.g. anemochory or endoornithochory). Crossing long distances of unsuitable habitats and arriving in new, unsettled regions such as isolated islands is the pre-condition for founder effects and events of adaptive radiation (evolving of neoendemics). Plant species that disperse via slow or small terrestrial animals would never reach an isolated island.

Photo 3.1 *Dendroseris litoralis*, Asteraceae, normally pollinated by humming birds whereas seeds are dispersed by wind, near San Juan Bautista on Robinson Crusoe Island (photographed by Marinus Werger). The species is endemic to this island. The natural population is now reduced to a few individuals

On the other hand, the occurrence of plants that disperse via biotic agents (e.g. ants; myrmecochorous) should normally be more restricted than that of wind-dispersed congeners (Holmgren and Poorter 2007; Helme and Trinder-Smith 2006; GiméNez et al. 2004; Renner 1990; Gentry 1986). *Centaurea corymbosa* (Asteraceae), for example, spreads exclusively with the help of ants. The range of the biotic agents dispersing the seeds also confines *Centaurea corymbosa*'s range (about 3 km^2 (Imbert 2006; Colas et al. 1997). Hopper (2009) argued that plants occurring in young, often disturbed, fertile landscapes (YODFELs) are good colonisers with high dispersal ability and tend to be biological or nutritional generalists, while inhabitants of old climatically buffered, infertile landscapes (OCBILs) tend to be well adapted, ecological specialists with reduced dispersibility and therefore show increased endemism (see comprehensive explanations in Hopper 2009).

However, we cannot conclude that most endemics are dispersed by slow animals such as ants or snails, whereas more widespread taxa are dispersed by wind, bats or birds.

Very specialised mutual dependencies between plants and animals, other than dispersal mode, might have a partially strong impact on the range expansion of plant species. We assume that the pollination mode (by insects or birds) in general is another important factor for speciation/radiation processes and the amount of endemism at regional to continental scales (Photo 3.1).

3.6 Growth of Phylogenetic Trees and Evolutionary Speed

The evolution and the evolutionary speed of the species in a region depend on environmental conditions (temperature, precipitation) and biological traits such as duration of the lifecycle, fertility, mode of pollination and dispersal.

Rensch (1954) was most likely the first scientist to state that the speed of evolution might depend on temperature. According to the hypothesis of "*higher speciation rate*" or "*evolutionary speed*" speciation rates are greater at lower latitudes and towards warmer climates than at high latitudes with lower temperatures (Ricklefs 1973). Goldie et al. (2010) found out that evolutionary speed in arid Australia is limited by water. We hypothesise that the availability of energy (warmth, sunlight) and water (precipitation) over the year is related to evolutionary speed. This idea is consistent with the so-called "water-energy hypothesis" (Svenning et al. 2009; Moser et al. 2005; Hutchinson 1959) which relates climate with species diversity.

The interpretation of evolutionary speed in phylogenetic branches may vary depending on different assumptions. Estimates on the divergence times resulting from values of gene-divergence are a function of mutation rates those vary largely between different DNA-types (Tzedakis 2009), depending on whether it is nuclear DNA, localised in the nucleolus, or extrachromosomal DNA, localised in mitochondria or chloroplasts. In fact, the branch lengths of the clade can only mirror the degree of divergence of the sampled alleles but the degree of divergence in alleles or genes is not congruent with evolutionary time (Rosauer et al. 2009, see also Heads 2010). According to Silvertown (2004) nuclear DNA may be subject to recombination events when different lineages hybridise, and therefore results must be interpreted cautiously.

The processes and speed of speciation also vary between species. Some taxa have undergone a much faster evolution than others. Thus, mutation rates in genes also depend on external promoting factors. For example, new niches lead to monophyletic radiation or selective pressures (e.g. environmental stress). Several ancient lineages of palaeoendemic species persist over long evolutionary time periods without fundamental changes (e.g. monotypic *Welwitschia mirabilis*) while other lineages show high rates of speciation. Hopper recently found that newly evolved species (e.g. neoendemics) tend to occur in young landscapes (YODFELs), while old lineages persist mostly in old, climatically stabilised landscapes (OCBILs). Mucina and Wardell-Johnson (2011) discuss the "*emergence of evolutionary innovations*" induced by environmental stress (e.g. nutrient stress, fire disturbance). Events of higher evolutionary speed e.g. fast adaptive radiation were reported from several island regions i.e. young, newly established landscapes. Many island endemics were identified as originating from monophyletic speciation. Thus, several species evolve in short time periods from one parental species (e.g. Rosauer et al. 2009; Silvertown 2004; Pinter et al. 2002).

References

Bailey RG (1998) Ecoregions: the ecosystem biogeography of the oceans and continents. Springer, New York

Bell G (2008) Selection: the mechanism of evolution. 2. Aufl. Oxford University Press, Oxford, p 553 S

Brown JH (1981) Two decades of homage to Santa Rosalia: toward a general theory of diversity. Am Zool 21:877–888

Bruchmann I (2011) Plant endemism in Europe: spatial distribution and habitat affinities of endemic vascular plants. Dissertation, University of Flensburg, Flensburg. URL: www.zhb-flensburg.de/dissert/bruchmann

Cain SA (1944) Foundations of plant geography. Harper and Brothers, New York

Cardona MA, Contandriopoulos J (1979) Endemism and evolution in the islands of the Western Mediterranean. In: Bramwell D (ed) Plants and islands. Academic, London/New York/Toronto, pp 133–170

Clark JS, Fastie C, Hurtt G, Jackson ST, Johnson C, King C, Lewis M, Lynch J, Pacala S, Prentice IC, Schupp EW, Webb T, Wyckoff P (2003) Reid's Paradox of rapid plant migration. Bioscience 48:13–24

Cockburn A (1991) An introduction to evolutionary ecology. Blackwell Publications, Oxford

Colas B, Olivieri I, Riba M (1997) *Centaurea corymbosa*, a cliff-dwelling species tottering on the brink of extinction: a demographic and genetic study. Proc Natl Acad Sci USA 94:3471–3476

Colinvaux PA (1998) A new vicariance model for Amazonian endemics. Glob Ecol Biogeogr Lett 7:95–96

Cowling RM, Lombard AT (2002) Heterogeneity, speciation/extinction history and climate: explaining regional plant diversity patterns in the Cape Floristic Region. Divers Distrib 8:163–179

Crisp MD, Laffan S, Linder HP, Monro A (2001) Endemism in the Australian flora. J Biogeogr 28:183–198

Cronk QCB (1997) Islands: stability, diversity, conservation. Biodivers Conserv 6:477–493

Davis GW, Richardson DM (eds) (1995). Mediterranean-type ecosystems. Ecol Stud 109:1–42. Springer, Berlin/Heidelberg/New York

Davis SD, Heywood VH, Hamilton AC (eds) (1994) Centres of plant diversity, vol 1, Europe, Africa, South West Asia and the Middle East. IUCN Publications, Unit, Cambridge

Davis SD, Heywood VH, Hamilton AC (eds) (1995) Centres of plant diversity, vol 2, Asia, Australasia and the Pacific. IUCN Publications, Unit, Cambridge

Davis SD, Heywood VH, Herrera-MacBryde O, Villa-Lobos J, Hamilton AC (eds) (1997) Centres of plant diversity, vol 3, the Americas. IUCN Publications, Unit, Cambridge

Dean WRJ, Milton SJ (eds) (1999) The Karoo. Ecological patterns and processes. Cambridge University Press, Cambridge

Deutscher Wetterdienst (ed) (2008) Climate data of high mountain areas. Unpubl. CD, Hamburg

Dynesius M, Jansson R (2000) Evolutionary consequences of changes in species' geographical distributions driven by milankovitch climate oscillations. Proc Natl Acad Sci USA 16:9115–9120

Engler A (1879–1882) Versuch einer Entwicklungsgeschichte der extratropischen Florengebiete der nördlichen Hemisphäre. Wilhelm Engelmann, Leipzig

Favarger C, Contandriopoulos (1961) Essai sur lèndimism. Berichte der Schweizerischen Botanischen Gesellschaft 71:384–406

Fjeldså J, Lovett JC (1997) Geographcal patterns of old and young species in African forest biota: the significance of specific montane areas as evolutionary centres. Biodivers Conserv 6:hbox325–346

Gaston KJ, Williams PH (1996) Spatial patterns in taxonomic diversity. In: Gaston KJ (ed) Biodiversity. A biology of numbers and difference. Blackwell Science, Cambridge

Gentry AH (1986) Endemism in tropical versus temperate plant communities. In: Soulé ME (ed) Conservation biology: the science of scarcity and diversity. Sinauer Associates, Inc.-Publisher, Sunderland, pp 153–182

Giménez E, Melendo M, Valle F, Gómez-Mercado F, Cano E (2004) Endemic flora biodiversity in the south of the Iberian peninsula: altitudinal distribution, life forms and dispersal modes. Biodivers Conserv 13:2641–2660
Givnish TJ (2000) Adaptive radiation, dispersal, and diversification of the Hawaiian Lobeliads. In: Kato M (ed) The biology of biodiversity. Springer, Tokyo, pp 67–90
Goldie X, Gillman L, Crisp M, Wright S (2010) Evolutionary speed limited by water in arid Australia. Proc Biol Sci 277:2645–2653
Graham CH, Moritz C, Williams SE (2006) Habitat history improves prediction of biodiversity in rainforest fauna. PNAS 103(3):632–636
Haffer J (1969) Speciation in Amazonian forest birds. Science 165:131–137
Hannus J-J, Von Numers M (2008) Vascular plant species richness in relation to habitat diversity and island area in the Finnish archipelago. J Biogeogr 35:1077–1086
Heads M (2010) Old taxa on young islands: a critique of the use of island age to date island-endemic clades and calibrate phylogenies. Syst Biol Adv Access 60:1–15
Helme NA, Trinder-Smith TH (2006) The endemic flora of the cape peninsula, South Africa. South Afr J Bot 72:205–210
Hendrych R (1982) Material and notes about the geography of the highly stenochoric to monotopic endemic species of the European flora. Acta Universitatis Carolinae-Biologica, 335–372
Hobohm C (2000a) Biodiversität. Quelle & Meyer, Wiebelsheim
Hobohm C (2000b) Plant species diversity and endemism on islands and archipelagos, with special reference to the Macaronesian Islands. Flora 195:9–24
Hobohm C (2003) Characterization and ranking of biodiversity hotspots: centres of species richness and endemism. Biodivers Conserv 12:279–287
Hobohm C (2008) Ökologie und Verbreitung endemischer Gefäßpflanzen in Europa. Tuexenia 28:7–22
Holmgren M, Poorter L (2007) Does a ruderal strategy dominate the endemic flora of the West African forests? J Biogeogr 34:1100–1111
Hopper SD (2009) OCBIL theory: towards an integrated understanding of the evolution, ecology and conservation of biodiversity on old, climatically buffered, infertile landscapes. Plant Soil 322:49–86
Huston MA (1994) Biological diversity. Cambridge University Press, Cambridge
Hutchinson GE (1959) Homage to Santa Rosalia or why are there so many kinds of animals. Am Nat 93:145–159
Imbert E (2006) Dispersal by ants in *Centaurea Corymbosa* (Asteraceae): what is the elaiosome for. Plant Spec Biol 21:109–117
Imbrie J, McIntyre A, Mix A (1989) Oceanic response to orbital forcing in the late quaternary: observational and experimental strategies. In: Berger A, Schneider A, Duplessy JC (eds) Climate and geo-sciences. Kluwer, Dordrecht, pp 121–164
Jamieson IG (2007) Has the debate over genetics and extinction of island endemics truly been resolved? Anim Conserv 10:139–144
Jansson R (2003) Global patterns in endemism explained by past climatic change. Proc R Soc Lond 270:583–590
Khoshoo N, Ahuja MR (1963) The chromosomes and relationships of *Welwitschia mirabilis*. Chromosoma 14:522–533
Konstant WR, Taylor D, Wake DA, Loarie SR, Bittman R, Ertter B (2005) California floristic province. In: Mittermeier RA, Gil PR, Hoffmann M, Pilgrim J, Brooks T, Mittermeier CG, Lamoreux J, Da Fonseca GAB (eds) Hotspots revisited: earth's biologically richest and most endangered terrestrial ecoregions. Cemex, Sierra Madre
Körner C (2000) Why are there global gradients in species richness? Mountains might hold the answer. Trends Ecol Evol 15(12):513f
Kreft H, Jetz W (2007) Global patterns and determinants of vascular plant diversity. Proc Natl Acad Sci USA 104:5925–5930
Kreft H, Jetz W, Mutke J, Kier G, Barthlott W (2007) Global diversity of island floras from a macroecological perspective. Ecol Lett 11:116–127

Kreft H, Jetz W, Mutke J, Kier G, Barthlott W (2008) Global diversity of island floras from a macroecological perspective. Ecol Lett 11:116–127
Kruckeberg AR, Rabinowitz D (1985) Biological aspects of rarity in higher plants. Annu Rev Ecol Syst 16:447–479
Latham RE, Ricklefs RE (1993) Global patterns of tree species richness in moist forests: energy-diversity theory does not account for variation in species richness. Oikos 67:325–333
Lavergne S, Thompson JD, Garnier E, Debussche M (2004) The biology and ecology of narrow endemic and widespread plants: a comparative study of trait variation in 20 congeneric pairs. Oikos 107:505–518
Lesica P, Yurkewycz R, Crone EE (2006) Rare plants are common where you find them. Am J Bot 93:454–459
Linder HP (2001) Plant diversity and endemisms in sub-Saharan tropical Africa. J Biogeogr 28:169–182
Lomolino MV (2001) Elevational gradients of species-density: historical and prospective views. Glob Ecol Biogeogr 10:3–13
MacArthur RH, Wilson EO (1967) The theory of island biogeography. Princetown University Press, Princetown
McGlone MS, Duncan RP, Heenan PB (2001) Endemism, species selection and the origin and distribution of the vascular plant flora of New Zealand. J Biogeogr 28:199–216
Médail F, Diadema K (2009) Glacial refugia influence plant diversity patterns in the Mediterranean Basin. J Biogeogr 36:1333–1345
Médail F, Quezel P (1999) Biodiversity hotspots in the Mediterranean Basin: setting global conservation priorities. Conserv Biol 13:1510–1513
Melendo M, Gimenez E, Cano E, Gomez-Mercado F, Valle F (2003) The endemic flora in the south of the Iberian Peninsula: taxonomic composition, biological spectrum, pollination, reproductive mode and dispersal. Flora 198:260–276
Mittermeier RA, Gil PR, Hoffman M, Pilgrim J, Brooks T, Mittermeier CG, Lamoreux J, da Fonseconda GAB (2005) Hotspots revisited: earth's biologically richest and most endangered terrestrial ecoregions. Cemex, Mexico City
Moser D, Dullinger S, Englisch T, Niklfeld H, Plutzar C, Sauberer N, Zechmeister HG, Grabherr G (2005) Environmental determinants of vascular plant species richness in the Austrian Alps. J Biogeogr 32:1117–1127
Mucina L, Wardell-Johnson G (2011) Landscape age and soil fertility, climatic stability, and fire regime predictability: beyond the OCBIL framework. Plant Soil 341:1–23
Nikolic T, Antonic O, Alegro AL, Dobrovic I, Bogdanovic S, Libera Z, Reetnika I (2008) Plant species diversity of Adriatic islands: an introductory survey. Plant Biosyst 142:435–445
O'Brien EM (1998) Water-energy dynamics, climate, and prediction of woody plant species richness: an interim general model. J Biogeogr 25:379–398
O'Brien EM (2006) Biological relativity to water–energy dynamics. J Biogeogr 33:1868–1888
Orme CDL, Davis RG, Burgess M, Eigenbrod F, Pickup N, Olson VA, Webster AJ, Ding TS, Rasmussen PC, Rigely RS, Stattersfield AJ, Bennett PM, Blackburn TM, Gaston KJ, Owens IPF (2005) Global hospots of species richness are not congruent with endemism or threat. Nature 436:1016–1019
Panitsa M, Trigas P, Iatrou G, Sfenthourakis S (2010) Factors affecting plant species richness and endemism on land-bridge islands – an example from the east Aegean archipelago. Acta Oecologica 36:431–437
Parmentier I, Malhi Y, Senterre B, Whittaker RJ, Alonso A, Balinga MPB, Bakayoko A, Bongers FJJM, Chatelein C, Comiskey J et al (2007) The odd man out? Might climate explain the lower tree α-diversity of African rain forests relative to Amazonian rain forests? J Ecol 95:1058–1071
Pärtel M (2002) Local plant diversity patterns and evolutionary history at the regional scale. Ecology 83(9):2361–2366

Pinter I, Bakker F, Barrett J, Cox C, Gibby M, Henderson S, Morgan-Richards M, Rumsey F, Russell S, Trewick S, Schneider H, Vogel J (2002) Phylogenetic and biosystematic relationships in four highly disjunct polyploid complexes in the subgenera Ceterach and Phyllitis in Asplenium (Aspleniaceae). Org Divers Evol 2:299–311
Pronk TE, Schieving F, Anten NPR, Werger MJA (2007) Plants that differ in height investment can coexist if they are distributing non-uniformerly within an area. Ecol Complex 4:182–191
Qian H, Ricklefs RE (2000) Large-scale processes and the Asian bias in species diversity of temperate plants. Nature 407:180–182
Remmert H (ed) (1991) The mosaic-cycle concept of ecosystems, vol 122, Ecological studies. Springer, New York, 168 pp
Renner SS (1990) Reproduction and evolution in some genera of neotropical Melastomataceae. Mem New York Bot Gard 55:143–152
Rensch B (1954) Neuere Problem der Abstammungslehre. Die transspezifische Evolution. Encke, Stuttgart
Reyes-Betancort JA, Santos Guerra A, Guma IR, Humphries CJ, Carine MA (2008) Diversity, rarity and the evolution and conservation of the Canary islands endemic flora. Anales del Jardín Botánico de Madrid 65:25–45
Ricklefs RE (1973) Ecology. Nelson & Sons, London
Rosauer D, Laffan SW, Crisp MD, Donnellan SC, Cool LG (2009) Phylogenetic endemism: a new approach for identifying geographical concentrations of evolutionary history. Mol Ecol 18:4061–4072
Rosenzweig ML (1995) Species diversity in space and time. Cambridge University Press, Cambridge
Royer DL, Hickey LJ, Wing S (2003) Ecological conservatism in the 'living fossil' ginkgo. Paleobiology 29:84–104
Rull V (2004) Biogeography of the 'lost world': a palaeoecological perspective. Earth-Sci Rev 67:125–137
Rull V (2009) Microrefugia. J Biogeogr 36:481–484
Rull V, Schubert C, Aravena R (1988) Palynological studies in the Venezuelan Guayana Shield: preliminary results. Curr Res Pleistocene 5:54–56
Shmida A, Wilson M (1985) Biological determinants of species diversity. J Biogeogr 12:1–20
Shmida A, Werger MJA (1992) Growth form diversity on the Canary Islands. Vegetatio 102:183–199
Silvertown J (2004) The ghost of competition past in the phylogeny of island endemic plants. J Ecol 92:168–173
Simberloff D, Wilson EO (1969) Experimental zoogeography of islands: the colonization of empty islands. Ecology 50:278–296
Staub F (1993) Fauna of Mauritius and associated flora. Precigraph Limited, Mauritius
Stebbins GL, Major J (1965) Endemism and speciation in the California flora. Ecol Monogr 35:2–35
Stewart JR, Lister AM, Barnes I, DaléN L (2010) Refugia revisited: individualistic responses of species in space and time. Proc R Soc Lond Ser B Biol Sci 277:661–671
Svenning J-C, Normand S, Skov F (2009) Plio-Pleistocene climate change and geographic heterogeneity in plant diversity-environment relationships. Ecography 32:13–21
Talbot SS, Yurtsev BA, Murray DF, Argus GW, Bay C, Elvebakk A (1999) Atlas of rare endemic vascular plants in the Arctic. Conservation of Arctic flora and fauna (CAFF) technical report 3:73 pp
Trewick SA, Morgan-Richards M, Russell SJ, Henderson S, Rumsey FJ, Pintér I, Barrett JA, Gibby M, Ogel JC (2002) Polyploidy, phylogeography and pleistocene refugia of the rockfern asplenium ceterach: Evidence from chloroplast DNA. Mol Ecol 11:2003–2012
Trigas P, Iatrou G (2006) The local endemic flora of Evvia (W Aegean, Greece). Willdenowia 36:257–270

Tutin TG, Burges NA, Chater AO, Edmondson JR, Heywood VH, Moore DM, Valentine DH, Walters SM, Webb DA (1996a) Flora Europaea, 2nd edn, vol 1, Psilotaceae-Platanaceae. Cambridge University Press, Cambridge

Tutin TG, Heywood VH, Burges NA, Moore DM, Valentine DH, Walters SM, Webb DA (1996b) lora Europaea, vol 2, Rosaceae-Umbelliferae. Cambridge University Press, Cambridge

Tutin TG, Heywood VH, Burges NA, Valentine DH, Walters SM, Webb DA (1996c) Flora Europaea, vol 3, Diapensiaceae-Myoporaceae. Cambridge University Press, Cambridge

Tutin TG, Heywood VH, Burges NA, Valentine DH, Walters SM, Webb DA (1996d) Flora Europaea, vol 4, Plantaginaceae-Compositae (and Rubiaceae). Cambridge University Press, Cambridge

Tutin TG, Heywood VH, Burges NA, Valentine DH, Walters SM, Webb DA (1996e) Flora Europaea, vol 5, Alismataceae-Orchidaceae. Cambridge University Press, Cambridge

Tzedakis PC (2009) Museums and cradles of Mediterranean biodiversity. J Biogeogr 36:1033–1034

Vetaas OR, Grytnes J-A (2002) Distribution of vascular species richness and endemic richness along the Himalayan elevation gradient in Nepal. Glob Ecol Biogeogr 11:291–301

Webb T, Bartlein PJ (1992) Global changes during the last 3 million years – climatic controls and biotic responses. Annu Rev Ecol Syst 23:141–173

Werger MJA, During HJ, van Rijnberk H (1987) Leaf diversity of three vegetation types of Tenerife, Canary Islands. In: Huiskes AHL, Blom CWPM, Rozema J (eds) Vegetation between land and sea. Dr. W. Junk Publishers, Dordrecht, pp 107–118

Whittaker RJ, Willis KJ, Field R (2001) Scale and species richness: towards a general, hierarchical theory of species diversity. J Biogeogr 28:453–470

Wood T, Takebayashi N, Barker MS, Mayrose I, Greenspoon PB, Rieseberg LH (2009) The frequency of polyploid speciation in vascular plants. PNAS 106:13875–13879

Wright DH (1983) Species-energy theory: an extension of species-area theory. Oikos 41:496–506

Chapter 4
Factors That Threaten and Reduce Endemism

Carsten Hobohm and Ines Bruchmann

4.1 Introduction

Species can become extinct for a number of natural reasons, such as the extinction of a key prey species or pollinator, diseases, being out-competed by invasive species, because of changes in environmental conditions, geological events (e.g. volcanic eruption), or simply by evolution. In the case of radiation, the parental species may disappear as well if it evolves into two or more new species: perhaps without any loss of genes.

In general, the risk of extinction increases if the size of a taxon's range decreases (Payne and Finnegan 2007). Most of the critically endangered plant taxa on the IUCN Red List (cf. www.iucnredlist.org) are restricted to small ranges. Fontaine et al. (2007) found that small-range restricted endemics 'are by far the most at risk of extinction'.

Factors that threaten local populations or endemic species at regional scales are related to alterations of environmental conditions or species compositions on the one hand, and to biological traits such as reproductive wastage on the other hand.

Random events such as catastrophes hit endemic species as hard as widespread taxa, but thanks to their higher abundance or the wider ranges of occurrence non-endemics have better survival opportunities than geographically restricted taxa. For example, a single bank slippage can be a catastrophe to the extremely rare endemic fern *Anogramma ascensionis* living on Ascension Island as this event could destroy most of the world s population (Işik 2011; Lambdon et al. 2010).

A single species such as the killer fungus *Batrachochytrium dendrobatidis* can cause dramatic changes in populations and increase the extinction risk of species, as the global amphibian decline shows (Mendelson et al. 2006). As a further example,

C. Hobohm (✉) • I. Bruchmann
Ecology and Environmental Education Working Group, Interdisciplinary Institute of Environmental, Social and Human Studies, University of Flensburg, Flensburg, Germany
e-mail: hobohm@uni-flensburg.de

C. Hobohm (ed.), *Endemism in Vascular Plants*, Plant and Vegetation 9,
DOI 10.1007/978-94-007-6913-7_4, © Springer Science+Business Media Dordrecht 2014

the root pathogen *Phytophthora cinnamomi* is threatening rare and endemic plants of the Stirling Range National Park in Australia (Barrett et al. 2007).

Species consisting of small or fragmented populations are also more affected by the consequences of reduced genetic diversity, genetic drift, effects of inbreeding and the accompanying loss of fitness, in comparison with widespread taxa. Thus, the ability to adapt in changing environments is also reduced (Jamieson 2007; Frankham 2005).

Sixty-three animal and plant species have recently become extinct in the wild (31 vascular plants) and 3,947 species (1,723 vascular plants) are critically endangered (www.iucnredlist.org; downloaded 10/2012). According to the IUCN Red List, the main causes of high extinction risk are habitat disturbance and biological resource use. Both are the result of the increasing human population and the consequent demand for water, food, and other resources.

In most cases, the *critically endangered* vascular plant species on the IUCN Red List (CE; www.iucnredlist.org) are endemic to small regions.

According to this list (downloaded 10/2012) the critically endangered vascular plant species (1,703) are threatened by *biological resource use* (354), *invasive and other problematic species and genes* (317), *agriculture and aquaculture* (310), *residential and commercial development* (241), *natural system modifications* (212), *human intrusions and disturbance* (91), *climate change and severe weather* (68), *geological events* (58), *energy production and mining* (47), *transportation and service corridors* (32), *pollution* (29), and *other options* (4). Some species are listed in more than one threat category.

It becomes evident that threats caused by man seem to have more and worse impacts on rare vascular plant species than threats caused by natural processes.

4.2 Extinctions in the Past

A mass extinction period is a strong decrease in species diversity on earth. Normally, five mass extinctions – *the big five* – are distinguished (cf. Alroy 2008; Arens and West 2008; Butterfield 2007; Wignal 2001, also for the following). However, based on the oscillating curves for species numbers over time it is also possible to distinguish more than five mass extinctions. Each of these mass extinction periods lasted many thousands, if not millions, of years. Therefore, the term *extinction event* seems to be a little misleading. The hit of a meteorite also becomes implausible as a cause of a mass extinction considering the duration of such periods.

In the past 300 million years there have been three major mass extinctions (the last three of the big five) when a large proportion of animal and plant species died (Permian-Triassic mass extinction 260–250 Ma ago, Triassic-Jurassic mass extinction 210–200 Ma ago, Cretaceous-Paleogene mass extinction 70–65 Ma ago).

There is still discussion about the reasons for the great mass extinctions. In general, mass extinctions might result when the biosphere undergoes a short-term shock (e.g. volcanic activity, asteroid impact) or – more plausibly – when the biosphere undergoes stress from long-term changes (e.g. due to climate change and/or continental drift).

During the Pleistocene and Holocene – around 60,000–3,000 years ago – numerous predominantly large animal species became extinct in continental regions. This phenomenon is not called a mass extinction. However, all over the world c. 100 genera with far more species disappeared during this period. The main reasons suggested for this are climate change and overkill by man (Louys et al. 2007; Martin and Klein 1989). Recently Burney and Flannery (2005: 395) highlighted the evidence of the hunting or overkill hypothesis: … "*global pattern of human arrival to such landmasses, followed by faunal collapse and other ecological changes, appears without known exception.*"

Another and more recent wave of extinctions is mainly related to island floras and faunas. This period began some hundred years ago and is still in progress. The native animals and plants were affected by alien species and the diseases they brought with them. Hundreds of years ships have brought new and often invasive species across the sea. In most cases, the extinctions of indigenous animals on islands are the result of the immigration of a new guild, predator or disease. Often aggressive rats, pigs, dogs, goats or other alien animals killed the native animals such as ground-breeding and flightless birds. Goats, in particular, opened and converted many habitats on islands and devasted indigenous plants. Competition between invasive and indigenous species is also mentioned as a cause of extinction of indigenous island-species. However, we are not aware of any case in which competition has been conclusively shown to be the reason for extinction (cf. Sax et al. 2005).

We can summarise extinction waves as follows:

The mass extinctions might have resulted from geological processes, climate change or changes following asteroid impact. Each lasted thousands or millions of years.

Historical megafaunal extinctions mainly on continents over the last 60,000–3,000 years were most likely caused by hunting. Climate change may have accelerated the extinction events in several cases.

Extinctions – of many bird species, but also reptiles, amphibians, less often mammals and vascular plants – during the last centuries on islands and archipelagos were caused by alien invaders, including alien species and hunting pirates or other voyagers who crossed the ocean by ship.

4.3 Disturbance of Habitats

Habitat loss, change of habitat quality (degradation), habitat fragmentation and disturbance of habitat dynamics are identified to be the major drivers of species decline not only for vascular plants (cf. www.iucnredlist.org). Endemic species in particular suffer from these habitat disturbances, as their range is limited. The European assessment shows that almost half the endemic plant species in Europe are threatened (Bilz et al. 2011; Ozinga and Schaminee 2005).

In a regional assessment of the IUCN Red List in Europe 1,826 vascular plant species were mainly '*affected by loss, degradation and/or increased fragmentation of their habitats that result from unsustainable human mismanagement of the environment*' (Bilz et al. 2011). The same is true for European terrestrial mammals (Temple and Terry 2007), reptiles (impacts of habitat loss/degradation; Cox and Temple 2009), amphibians (impacts of habitat loss/degradation, also pollution; Temple and Cox 2009), non-marine molluscs (impacts of urbanisation and land-use intensification, also water quality; Cuttelod et al. 2011), butterflies (impacts of land-use intensification and abandonment; Van Swaay et al. 2010), dragonflies (impacts of water-pollution and eutrophication; Kalkman et al. 2010), and saproxylic beetles (impact of logging and wood harvesting; Nieto and Alexander 2010).

Habitat loss and degradation is mostly caused by intensive livestock farming, infrastructure development and urbanisation but also by mining activities. Today, human recreation activities and tourism have a strong impact, especially in coastal and mountain areas. The loss of habitat quality is caused by pollution/contamination or fertilisation, changing of management regimes (including changes in ecological dynamics, e.g. land abandonment, regulation of water systems) or fragmentation of habitats that lead to isolation of populations (IUCN threats classification scheme: www.iucnredlist.org; cf. Bilz et al. 2011).

Some examples:

The aquatic fern *Marsilea batardae* (Marsileaceae, which lives in temporarily flooded areas of the Iberian Peninsula) suffers from the general destruction and degradation of water bodies such as the construction of dams, embankment of streams or pollution (Lansdown and Medina Domingo 2011).

Rising pressure from tourism development is threatening the endemic plant *Chaenorhinum serpyllifolium* ssp. *lusitanicum* (Scrophulariaceae) that inhabits rocks, cliffs and shores on the coast of southwestern Portugal.

On a regional scale, Bruchmann and Hobohm (2010) raised concerns about the creeping loss of extent and quality of Europe's grassland ecosystems that host plenty of plants endemic to Europe. The first composite report on the conservation status of habitat types (as required under the European habitats directive) states that all over Europe strong pressure is also exerted on wetland and coastal habitat types (Commission of the European Communities 2009).

On a global scale, the third Global Biodiversity Outlook reports ongoing loss of natural habitats in most parts of the world (affected habitats are tropical forests, mangroves, freshwater wetlands, sea ice habitats, salt marshes, coral reefs, and some others; Secretary of the Convention on Biological Diversity 2010).

4.4 Biological Resource Use

Biological resource use means the use of wild plants and animals and includes fishing, logging, medical or cultural uses of wild species, but also food collecting. Biological resource use can result in severe disturbance of habitats.

The large tree species *Pouteria polysepala*, for example, is only known from non-flooded places near the mouth of Rio Javari in Brazil (www.iucnredlist.org; also in the following). Another example of a valuable timber species is the up to 40 m tall *Microberlinia bisulcata* in Cameroon. Because of large-scale habitat decline, forest clearance for agriculture, and logging in the local populations of these tree species, they are rapidly declining.

Medical or cultural uses can be further reasons for species decline: *Alocasia* species have been collected for ornamental and also for medicinal use (e.g. *A. atropurpurea* which is found only in a rather small area at Mt. Polis on Luzon Island, or *A. sanderiana* which is confined to primary and secondary forest on Mindanao Island, both Philippines).

These are only few of many examples of overexploitation. All over the world, many other plants are collected for ornamental and/or medicinal use and as raw material for construction, e.g. rattans in South East Asia and hemiepiphytes in the Amazon region. Often the economic value of these plants is estimated to be very high, which in many cases leads to unsustainable harvesting and illegal trading (Caldecott et al. 1994).

4.5 Invasive and Other Problematic Species

Neophytes and invasive plants – also native species can become invasive (!) – can have several different effects on habitats. Possible effects are competition, changes in environmental conditions, the food web or vegetation structure.

> The transportation and successful introduction of non-native species to new locations and habitats is a pervasive component of human-induced global change, one that presents distinct challenges for the conservation and management of biological diversity (Cassey 2005: 9, in Sax et al. 2005).

Imbert et al. (2011) concluded that endemic species often occur in stressful habitats such as rocky habitats, steep slopes, heavy metal outcrops or soils of low fertility. However, to date only few experiments have verified a low competitive ability of endemic species, while other studies describe, in contrast, a greater competitive ability of endemic species compared to widespread species or show no clear difference in the competitive abilities of locally restricted and widespread species (see summary in Imbert et al. 2011; Lavergne et al. 2004; Murray et al. 2002; Gaston and Kunin 1997).

If the range of an invasive species is increasing, while at the same time and in the same area the range of an endemic is decreasing, then there are theoretically three different relationships which can explain this: (i) a direct cause-effect relationship, e.g. due to competition, (ii) an indirect cause-effect relationship, e.g. due to human influence: positive for the one, negative for the other, or (iii) independent processes.

In the case of large mainland regions, there is only weak scientific evidence for interspecific competition that threatens the existence of species. In Europe, for example, not one native vascular plant species or animal species has become extinct as a result of competing alien or invasive species. It seems that local communities in large continuous terrestrial areas can survive the arrival and the establishment of a neophyte even if this neophyte behaves invasively (Sax et al. 2005; Kowarik 2003).

This picture has to be painted more pessimistically for species living at the edges of their potential distribution ranges or living on small islands, as these species there do not have the chance to migrate if conditions change. The described threats caused by alien species result from alterations in the habitat structure, ecosystem functioning, introduced diseases or the introduction of a member of a new ecological guild (niche). In the case of some island biota certain ecological guilds have not evolved, e.g. many of the endemic island birds which have become extinct were simply eaten up by dogs, rats, pigs and other introduced vertebrates, while before their introduction adequate predators did not occur on the island. So, ground birds were not adapted to predators, were trusting and did not fight or flee and became easy prey. *Homo sapiens* as a planning and acting hunter, farmer or (de)forester also functions as an agent of a new guild.

Many critically endangered plant taxa are more severely affected by man and introduced vertebrates and the accompanying changes in ecological conditions than by direct competition with invasive neophytes.

However, the impact of invasive pressures also depends on the biotic resistance of native biota. This depends on how drastic the invasion events from abroad are in terms of proportions of endemic, non-endemic native and introduced species. Olesen et al. (2002) and Gray et al. (2005) reported from Ascension Island where a native flora of 25 species (mostly ferns, including 10 endemic taxa) is confronted with about 280 introduced neophytes. For Mauritius, 680 native angiospermous species (including 312 endemic) face 736 introduced ones (Strahm 1993).

Some invasive species are able to dominate areas inhabited by small-range endemics and – depending on habitat type – may also displace them by altering ecological conditions (e.g. light, nutrients and moisture availability).

Several studies indicate that species interactions may impede and sometimes also facilitate the invasion of alien species (whether invading plant or invading pollinating animal (Olesen et al. 2002). If the invading plant has a specialised and complex pollination biology (e.g. complex mutualistic plant insect-pollination) then it will have a lower probability of establishing stable populations than an invading plant with a more generalised pollination biology (e.g. wind pollination). However, invading species can also disrupt existing plant-pollinator systems (see summary in Traveset and Richardson 2006). The honeybee, for example, is a generalised pollinator that visits and exploits many plant species very effectively. Qualitatively, however, honeybees are often less effective pollinators than native co-evolved pollinators (e.g. because of non-selective plant visits). The high frequencies of visiting flowers by honey bees may reduce the visitation by native pollinators, with the result that the latter might be outcompeted (Traveset and Richardson 2006). On the other hand, however, invading pollinators might also have neutral or positive effects on threatened native island floras. *Nesocodon mauritianus* (Campanulaceae), for example, a rare endemic plant of Mauritius, is today more frequently visited and pollinated by the introduced bird *Pycnonotus jocosus* (red-whiskered bulbul) than by a native bird-pollinator (*Hypsipetes olivaceus* Olesen et al. 1998). In New Zealand the decline of some native bird-pollinators is largely buffered by the introduced *Zosterops lateralis* (white-chested silvereye; Traveset and Richardson 2006).

4.6 Climate Change

Species with restricted ranges are said to be most endangered with respect to climate change (Jansson 2003; Thuiller et al. 2005; Pompe et al. 2008). The IUCN Red List (www.iucnredlist.org) describes 67 *critically endangered* vascular plant species that are threatened by *climate change and severe weather*. Almost all of these species are endemic to very small areas of the world, and in many cases populations are reported to already be very small and fragile, i.e. they have been reduced to only a few individuals. Interestingly, though, for none of the listed critically endangered species *climate change and severe weather* is the exclusive threat. In all listed cases other reasons also affect or have even more impact on population trends than *climate change*. Hence, in many cases climate change is an additional threat that weakens the resistance of the already endangered plant species. The prominence of the threats *climate change* and *severe weather* is probably based on the assumption that single climatic catastrophes such as drought, heavy rain or thunderstorm (e.g. leading to slope slumping or other destructions) can be able to erase the remaining populations of some species.

During the last decade much attention has been paid to describing potential impacts of anthropogenically caused climate change, to interpolating species range shifts or to predicting extinction risks of several species (Broennimann et al. 2006; Thuiller et al. 2006; Keith et al. 2008; Ohlemüller et al. 2008).

On a macro-scale level, Thuiller et al. (2005), for example, found for projected distributions of 1,350 European plants species under different climate change scenarios that even under moderate climate scenarios many European plant species could become threatened. Species loss was mainly attributed to climate variables describing temperature and moisture conditions. Depending on the scenario, species occurring in mountainous regions are said to be most affected (up to 60 % species loss), while Europe's boreal regions were even projected to gain species that immigrate from other regions. The authors argue that high proportions of the highly specialised mountain flora, narrowly bound to a few marginal habitat niches, are not able to manage range shifts and therefore will become extinct, while species of the boreal zones are already well adapted to hot and dry summers and therefore may better tolerate future climatic conditions. However, the authors themselves point to the large potential of uncertainty in predictions, as the coarse macro-scale may hide potential (micro-)refugial areas for species, especially in heterogeneous mountain areas. On a local scale, under different scenarios, including predictors of local soil characteristics and land use, Pompe et al. (2008) projected range shifts of 845 plant species occurring in Germany. They predicted high rates of species turnover and species loss in southwestern and eastern parts of Germany and projected severe changes in Germany's species pool as they expected that central and southern parts of the country would be affected by species gains, i.e. immigrating species from neighbouring regions. Over all scenarios, it appears that in the future species ranges in some regions of Germany will become smaller under climate change, which might weaken the resistance of the species to additional stresses, e.g. habitat fragmentation.

Ohlemüller et al. (2008), for example, found that species "*with small range sizes tend to occur in areas with rare, i.e. localized, climatic conditions*" and also found a disproportionate shrinkage of these areas under different scenarios of climate change.

However, all these studies have a high degree of uncertainty, whether inherent to the modelling technique or related to the non-inclusion of the reproductive potential of plants, effects of biotic interactions (e.g. plant-pollinator systems, impacts of immigrating species, etc.), species evolution (landscape history, climatic buffering systems, adaption) or often simply because they ignore the environmental heterogeneity at local scales (Pompe et al. 2008).

It was shown that impacts of climate change may vary substantially depending on the region under consideration: An assessment of the potential impacts of anthropogenic climate change on Namibia's endemic flora projected substantial changes in vegetation structure and ecosystem functioning, on the one hand, but anticipated a surprisingly high potential of resistance of endemic plants. Depending on life-forms endemics are projected to be either negatively affected (geophytes, perennials, trees), or to persist in a relatively stable state (annuals or succulents), or even to be able to extend their range (some annuals; Thuiller et al. 2006). The low sensitivity of endemic species to recent climate change might be explained by the theories of old stabilised landscapes or local buffering capacities (see chapters Climate Stability and Age of Landscapes; Mucina and Wardell-Johnson 2011).

According to Jansson (2003) it is not surprising that the effects of present climate change are not as severe in some areas of high endemism as in many other regions because areas of high endemism result from historically stable climates and uninterrupted long-term evolution. However, he raised concerns that worst case scenarios of 3–5 °C of global warming would exceed the tolerance of many species as the response of species to climatic change (e.g. habitat tracking) is additionally impeded by the rapid transformation of ecosystems.

4.7 Problems of Reproduction

The success of producing fruits and seeds depends on both environmental conditions and biological aspects such as genetic constitution. Under adequate conditions plants normally produce more and better germinating seeds than under unfavourable conditions (Ellenberg 1996; Marshall and Grace 1992). The plants which occur at the highest elevations in the Alps are not able to produce any seeds simply because the length of the season is too short (Körner; lecture at the meeting of IAVS in

Lyon 2011). The success of producing fruits and seeds also depends on the biology of the plant species e.g. on the age of maturity and on interspecific relationships such as pollination mode. If the appropriate pollinator is absent and autogamy is impossible then production of fertile seeds is impossible as well.

A decreasing population size generally results in lower heterozygosity. Thus, endemics will normally have a lower heterozygosity in comparison with more widespread taxa, which leads to a lower fitness of a population. The result might be lower adaptivity under changing conditions or inbreeding depression.

Several studies showed that the reproductive output of endemics was lower compared to their widespread congeners (Murray et al. 2002). However, reduced reproductive output does not automatically mean reduced dispersal distance. Lavergne et al. (2004) raised concerns over the hypothesis that endemics are generally poor dispersers as there are strong dependencies on the different environments and phylogenetic contexts of the examined floras.

For example, *Zelkova sicula*, a critically endangered species of the family Ulmaceae, occurs on the island of Sicily and today spreads largely vegetatively. Only one decreasing population of 200–250 individuals exists. During recent years only a few flowering individuals have been seen and the fruits produced appeared to be sterile. To date it is unclear why reproduction in this taxon is reduced. Perhaps the hydrological conditions are inadequate for seed production of this species under recent climate conditions (Garfi 2006).

4.8 Biotic Crisis and Conservation of Endemic Vascular Plants

The recent *biotic crisis* is often described as the *sixth extinction crisis* (Ceballos et al. 2010; Leakey and Lewin 2008) and one which is mainly caused by habitat destruction and biological resource use. However, other causes such as diseases (amphibian decline) are important as well. This new wave of extinctions can become a mass extinction which most likely will last less than a few hundred years. This extinction wave is related to the growing human population of five (1987), six (1999), seven (2012), eight (2027?) and nine (2046?) billion people on earth.

The Convention on Biological Diversity (CBD) is the most important legally binding treaty with the conservation of biological diversity as its main goal. Almost all countries in the world have now signed this convention. Thus, there is huge agreement that it is necessary to halt the loss of biodiversity and to protect the species diversity on earth. On the other hand, the diversity of ideas regarding ways of achieving this is rich as well.

These include economic, psychological, religious, social, cultural and other aspects of affecting or sparing nature. Different indicator, monitoring and assessment systems for evaluating the progress in halting the loss of biodiversity have been achieved. These are, for example, all the Red Lists at regional to global scales, the biodiversity hotspot analyses (Myers 1988, 1990), the 2005 Millennium Ecosystem

Assessment (Millennium Ecosystem Assessment Board 2005), and a set of 26 indicators as defined by EEA (2007, 2011). However, the situation of the endemic plants is itself rather a good indicator for assessing the biodiversity and ecology of the ecosystems: Do we know enough about the threats, about the ecology and distribution patterns of endemic species? And, do we know enough about possible ways to protect rare, endemic and endangered species?

Clearly, this is not the case. As discussed, for example, in Bruchmann (2011) the geographical range dimension of European (endemic) species has been quite extensively examined, whereas the ecological range dimension, which is the knowledge of species' habitat affinities, was only meagrely represented in the available data sets. Bruchmann found that about one quarter of European's endemic plants were data-deficient in the ecological dimension as these endemics could not be assigned to coarsely predefined habitat categories (see also Hobohm and Bruchmann 2009; Bruchmann and Hobohm 2011).

The knowledge on demographic range is even worse. Even in Europe, where the efforts of monitoring population trends have been quite strong for years, there is little authentic monitoring data available on the population trends of species or habitat types. The works on the latest version of the European Red List of vascular plants (Bilz et al. 2011), for example, showed that only 8 % of Europe's plant species have been assessed sufficiently and suggest to set 'a priority for fieldwork and research' in monitoring data-deficient species.

Many activities worldwide have been initiated to protect the last individuals or populations of different vascular plant species in the wild, or, at least, in botanical gardens. In contrast to animals, the reintroduction to the wild is much less problematic if the conditions are favourable. Ex-situ protection might be the last step to be used if other activities in the wild do not turn out well or if the relating habitat is completely destroyed.

References

Alroy J (2008) Dynamics of origination and extinction in the marine fossil records. Proc Natl Acad Sci USA 105(Suppl 1):11536–11542

Arens NC, West ID (2008) Press-pulse: a general theory of mass extinction? Paleobiology 34(4):456

Barrett S, Shearer B, Crane C, Cochrane A (2007) The risk of extinction resulting from disease caused by *Phytophthora cinnamomi* to threatened flora endemic to the Stirling Range National Park, Western Australia. In: 11th international Mediterranean ecosystems (MEDECOS) conference 2007, Perth, 2–5 Sept 2007

Bilz M, Kell SP, Maxted N, Lansdown RV (2011) European red list of vascular plants. Publications Office of the European Union, Luxembourg

Broennimann O, Thuiller W, Hughes G, Midgley GF, Alkemande JMR, Guisan A (2006) Do geographic distribution, niche property and life form explain plants' vulnerability to global change? Glob Chang Biol 12:1079–1093

Bruchmann I (2011) Plant endemism in Europe: spatial distribution and habitat affinities of endemic vascular plants. Dissertation, University of Flensburg, Flensburg. URL: www.zhb-flensburg.de/dissert/bruchmann

Bruchmann I, Hobohm C (2010) Halting the loss of biodiversity: endemic vascular plants in grassland of Europe. Grassland Sci 15:776–778
Bruchmann I, Hobohm C (2011) Über Grenzen hinweg: Schutz endemischer Gefäßpflanzen in Europa? Treffpunkt Biologische Vielfalt 10:119–125
Burney DA, Flannery TF (2005) Fifty millennia of catastrophic extinctions after human contact. Trends Ecol Evol 20(7):395–401
Butterfield NJ (2007) Macroevolution and macroecology through deep time. Palaeontology 50(1):41–55
Caldecott JO, Jenkins MD, Johnson T, Groombridge B (1994) Priorities for conserving global species richness and endemism. World Conservation Press, Cambridge
Cassey P (2005) Insights into ecology. In: Sax DF, Stachowicz JJ, Gaines SD (eds) Species invasions: insights into ecology, evolution and biogeography. Sinauer Associates, Sunderland, pp 9–12
Ceballos G, Garcia A, Ehrlich PR (2010) The sixth extinction crisis. J Cosmol 8:1821–1831
Commission of the European Communities (2009) Report from the commission of the council and the European parliament: composite report on the conservation status of habitat types and species as required under Article 17 of the habitats directive. Commission of the European Communities, Brussels, pp 17
Cox NA, Temple HJ (2009) European red list of reptiles. Office for Official Publications of the European Communities, Luxembourg
Cuttelod A, Seddon M, Neubert E (2011) European red list of non-marine molluscs. Publications Office of the European Union, Luxembourg
EEA (ed) (2007) Europe's environment: an assessment of assessments. Publications Office of the European Union, Copenhagen
EEA (ed) (2011) Europe's environment: an assessment of assessments. Publications Office of the European Union, Copenhagen
Ellenberg H (1996) Vegetation Mitteleuropas mit den Alpen, 5th edn. Ulmer, Stuttgart
Fontaine B, Bouchet P, Van Achterberg K, Alonso-Zarazaga MA, Araujo R, Asche M, Aspock U, Audisio P, Aukema B, Bailly N, Balsamo M, Bank RA, Barnard P, Belfiore C, Bogdanowicz W, Bongers T, Boxshall G, Burckhardt D, Camicas JL, Chylarecki P, Crucitti P, Davarveng L, Dubois A, Enghoff H, Faubel A, Fochetti R, Gargominy O, Gibson D, Gibson R, Gomez Lopez MS, Goujet D, Harvey MS, Heller K-G, Van Helsdingen P, Hoch H, De Jong H, De Jong Y, Karsholt O, Los W, Lundqvist L, Magowski W, Manconi R, Martens J, Massard JA, Massard-Geimer G, Mcinnes SJ, Mendes LF, Mey E, Michelsen V, Minelli A, Nielsen C, Nieto Nafria JM, Van Nieukerken EJ, Noyes J, Papa T, Ohl H, De Prins W, Ramos M, Ricci C, Roselaar C, Rota E, Schmidt-Rhaesa A, Segers H, Zur Strassen R, Szeptycki A, Thibaud J-M, Thomas A, Timm T, Van Tol J, Vervoort W, Willmann R (2007) The European Union's 2010 target: putting rare species in focus. Biol Conserv 139:167–185
Frankham R (2005) Genetics and extinction. Biol Conserv 126:131–140
Garfì G (2006) *Zelkova sicula*. In: IUCN (ed.) 2012. IUCN Red List of Threatened Species. Version 2012.1. www.iucnredlist.org. Downloaded on 24 Sept 2012
Gaston KJ, Kunin WE (1997) Rare-common differences: an overview. In: Kunin WE, Gaston KJ (eds) The biology of rarity: causes and consequences of rare-common differences. Chapmann and Hall, London, pp 11–29
Gentry AH (1986) Endemism in tropical versus temperate plant communities. In: Soulé ME (ed) Conservation biology: the science of scarcity and diversity. Sinauer Associates, Inc.-Publisher, Sunderland, pp 153–182
Gray A, Pelembe T, Stroud S (2005) The conservation of the endemic vascular flora of Ascension Island and threats from alien species. Oryx 39:449–453
Hobohm C, Bruchmann I (2009) Endemische Gefäßpflanzen und ihre Habitate in Europa – Plädoyer für den Schutz der Grasland-Ökosysteme. RTG-Berichte 21:142–161
Imbert E, Youssef S, Carbonell D, Baumel A (2011) Do endemic species always have a low competitive ability? A test for two Mediterranean plant species under controlled conditions. J Plant Ecol 4:1–8

Işik K (2011) Rare and endemic species: why are they prone to extinction? Turk J Bot 35:411–417
Jamieson IG (2007) Has the debate over genetics and extinction of island endemics truly been resolved? Anim Conserv 10:139–144
Jansson R (2003) Global patterns in endemism explained by past climatic change. Proc R Soc Lond 270:583–590
Kalkman VJ, Boudot J-P, Bernard R, Conze K-J, De Knijf G, Dyatlova E, Ferreira S, Jovic M, Ott J, Riservato E, Sahlén E (2010) European red list of dragonflies. Publications Office of the European Union, Luxembourg
Keith DA, Akcakaya HR, Thuiller W, Midgley GF, Pearson RG, Phillips SJ, Regan HM, Araujo MB, Rebelo TG (2008) Predicting extinction risks under climate change: coupling stochastic population models with dynamic bioclimatic habitat models. Biol Lett 4:560–563
Kowarik I (2003) Biologische Invasionen – Neophyten und Neozoen in Mitteleuropa. Ulmer, Stuttgart
Lambdon PW, Stroud S, Gray A, Niissalo M, Renshaw O, Sarasan V (2010) Anogramma ascensionis. In: IUCN 2012. IUCN red list of threatened species. Version 2012.2. www.iucnredlist.org. Downloaded on 20 June 2012
Lansdown RV, Medina Domingo L (2011) Marsilea batardae. In: IUCN 2012. IUCN red list of threatened species. Version 2012.2. www.iucnredlist.org. Downloaded on 20 June 2012
Lavergne S, Thompson JD, Garnier E, Debussche M (2004) The biology and ecology of narrow endemic and widespread plants: a comparative study of trait variation in 20 congeneric pairs. Oikos 107:505–518
Leakey RE, Lewin R (2008) The sixth extinction: patterns of life and the future of humankind. Paw Prints, New York
Louys J, Curnoe D, Tong H (2007) Characteristics of *Pleistocene megafauna* extinctions in Southeast Asia. Palaeogr Palaeocl Palaeoecol 243:152–173
Marshall C, Grace J (1992) Fruit and seed production: aspects of development, environmental physiology and ecology. Cambridge University Press, New York
Martin PS, Klein RG (1989) Quaternary extinctions: a prehistoric revolution. University Arizona Press, Tucson
Mendelson JR, Lips KR, Gagliardo RW, Rabb GB, Collins JP, Diffendorfer JE, Daszak P, Ibáñez DR, Zippel KC, Lawson DP, Wright KM, Stuart SN, Gascon C, da Silva HR, Burrowes PA, Joglar RL, La Marca E, Lötters S, du Preez LH, Weldon C, Hyatt A, Rodriguez-Mahecha JV, Hunt S, Robertson H, Lock B, Raxworthy CJ, Frost DR, Lacy RC, Alford RA, Campbell JA, Parra-Olea G, Bolaños F, Domingo JJ, Halliday T, Murphy JB, Wake MH, Coloma LA, Kuzmin SL, Price MS, Howell KM, Lau M, Pethiyagoda R, Boone M, Lannoo MJ, Blaustein AR, Dobson A, Griffiths RA, Crump ML, Wake DB, Brodie ED (2006) Confronting amphibian declines and extinctions. Science 313:48
Millennium Ecosystem Assessment Board (ed) (2005) Ecosystems and human well-being: synthesis. Island Press, Washington, DC
Mucina L, Wardell-Johnson GW (2011) Landscape age and soil fertility, climate stability, and fire regime predictability: beyond the OCBIL framework. Plant Soil. doi:10.1007/s11104-011-0734-x
Murray BR, Thrall PH, Gill AM, Nicotra AB (2002) How plant life-history and ecological traits relate to species rarity and commonness at varying spatial scales. Austral Ecol 27:291–310
Myers N (1988) Threatened biotas: hotspots in tropical forests. Environmentalist 8:1–20
Myers N (1990) The biodiversity challenge: expended hotspots analysis. Environmentalist 10:243–256
Nieto A, Alexander KNA (2010) European red list of saproxylic beetles. Publications Office of the European Union, Luxembourg
Ohlemüller R, Anderson BJ, Araujo MB, Butchart SHM, Kudrna O, Ridgely RS, Thomas CD (2008) The coincidence of climatic and species rarity: high risk to small-range species from climate change. Biol Lett 4:568–572
Olesen JM, Rønsted N, Tolderlund U, Cornett C, Mølgaard P, Madsen J, Jones CG, Olsen CE (1998) Mauritian red nectar remains a mystery. Nature 393:529

Olesen JM, Eskildsen LI, Venkatasamy S (2002) Invasion of pollination networks on oceanic islands: importance of invader complexes and endemic super generalists. Divers Distrib 8:181–192

Ozinga WA, Schaminée JHJ (eds) (2005) Target species – species of European concern. Alterra-report 1119. Alterra, Wageningen, 193pp

Payne JL, Finnegan S (2007) The effect of geographic range on extinction risk during background and mass extinction. Proc Nat Acad Sci USA 104(25):10506–10511

Pompe S, Hanspach J, Badeck F, Klotz S, Thuiller W, Kühn I (2008) Climate and land use change impacts on plant distributions in Germany. Biol Lett 4:564–567

Sax DF, Stachowicz JJ, Gaines SD (eds) (2005) Species invasions: Insights into ecology, evolution, and biogeography. Sinauer Associates, Inc. Publication, Sunderland

Secretary of the Convention on Biological Diversity (2010) Global biodiversity outlook 3. Progress Press Ltd., Montréal

Strahm W (1993) The conservation and restoration of the flora of Mauritius and Rodrigues. Reading University, Reading

Temple HJ, Terry A (2007) The status and distribution of European mammals. Office for Official Publications of the European Communities, Luxembourg

Temple HJ, Cox NA (2009) European red list of amphibians. Office for Official Publications of the European Communities, Luxembourg

Thuiller W, Lavorel S, Araújo MB, Sykes MT, Prentice IC (2005) Climate change threats to plant diversity in Europe. Proc Natl Acad Sci USA 102

Thuiller W, Midgley GF, Hughes G, Bomhard B, Drew G, Rutherford MC, Woodward FI (2006) Endemic species and ecosystem sensitivity to climate change in Namibia. Glob Chang Biol 12:759–776

Traveset A, Richardson DM (2006) Biological invasions as disruptors of plant reproductive mutualisms. Trends Ecol Evol 21:208–216

Van Swaay C, Cuttelod A, Collins S, Maes D, López Munguira M, Šašic M, Settele J, Verovnik R, Verstrael T, Warren M, Wiemers M, Wynhof I (2010) European red list of butterflies. Publications Office of the European Union, Luxembourg

Wignal PB (2001) Large igneous provinces and mass extinctions. Earth Sci Rev 53(1–2):1–33

Part III
Endemic Vascular Plants in Space

The following case studies present examples from the Tropics, Mediterranean regions, temperate and boreal-arctic regions, mainland regions, and continental and oceanic islands. We still know little about vascular plant endemism in relation to the habitat in many ecoregions, however, and especially in tropical and subtropical mainland regions.

Chapter 5
Biogeography of Endemic Vascular Plants – Overview

Carsten Hobohm, Monika Janišová, Jan Jansen, Ines Bruchmann, and Uwe Deppe

5.1 Latitudinal, Longitudinal, Radial and Altitudinal Gradients in Endemism

The overall taxonomic diversity of vascular plants is low towards high latitudes and high towards the tropics. This latitudinal gradient is, in general, also true for endemic vascular plant taxa. A large number of processes which might be responsible for latitudinal gradients of species diversity have been discussed and tested. These include processes such as productivity, competition or mutualism, as well as dispersal, genetical isolation, geographical separation, speciation, survival and extinction processes (Jansson 2003; Gaston 1996; Pianka 1966). Thus, it is clear that environmental heterogeneity and contemporary and historical climate parameters are important factors explaining regional to continental patterns of diversity, independent of the fact that the underlying processes are not yet fully understood.

In contrast to latitudinal or altitudinal gradients, longitudinal and radial gradients of species diversity (cf. Huston 1994) are often not recognised or non-existent. However, hardly any biodiversity hotspots and endemic-rich regions are located far from the sea. Exceptions to this rule are found in Asia (Caucasus, Himalaya and SW China; cf. Huang et al. 2011a, b; Mittermeier et al. 2005; Davis et al. 1994, 1995, 1997).

C. Hobohm (✉) • I. Bruchmann • U. Deppe
Ecology and Environmental Education Working Group, Interdisciplinary Institute of Environmental, Social and Human Studies, University of Flensburg, Flensburg, Germany
e-mail: hobohm@uni-flensburg.de

M. Janišová
Institute of Botany, Slovak Academy of Sciences, Banská Bystrica, Slovakia

J. Jansen
Institute for Water and Wetland Research, Radboud University, Nijmegen, The Netherlands
e-mail: jan.jansen@science.ru.nl

C. Hobohm (ed.), *Endemism in Vascular Plants*, Plant and Vegetation 9,
DOI 10.1007/978-94-007-6913-7_5, © Springer Science+Business Media Dordrecht 2014

The inverse richness gradients of endemism from the coasts to the central parts of the continents often reflect a climatic gradient of decreasing humidity and increasing amplitude between summer and winter temperature.

Latitudinal and altitudinal patterns and gradients have often been compared with regard to ecology and biodiversity. The richness of all vascular plant taxa declines with latitude and altitude as well, and biogeographers have long recognised that the altitudinal gradient of biodiversity and climate mirrors the latitudinal gradient (Körner 2000). Stevens (1992) has shown that total maximum species diversity along an altitudinal gradient is often found towards the lower end of the elevation. This finding is compatible with the view that tropical climates harbour the highest biodiversity.

However, the mirror does not work perfectly (see Körner 2000 also for differences of latitudinal and altitudinal richness patterns). The alpine zone in mountain areas of the tropics is climatically different from arctic regions in the North or South because, for example, the amplitudes of temperature, light, vegetation period and also precipitation can differ remarkably if one compares average winter and summer or night and day values in these regions. One of the biologically most important differences between the climate of tropical alpine zones and arctic regions is that the tropics have warm days, cold nights and long vegetation periods whereas the winter of arctic regions is cold and long and the vegetation period is short.

In many temperate, subtropical and tropical regions endemism peaks at mid-altitudes; this does not reflect the altitudinal distribution of vascular plants in total which, in general, shows highest diversity at the lower end of the altitudinal gradient.

The interval of maximum endemic species richness – in absolute numbers – in Nepal is between 3,800 and 4,200 m (Vetaas and Grytnes 2002), in SW China 1,000–2,000 m (Huang et al. 2011a, b), in Taiwan 0–1,500 m (Hsieh 2002), in Peru 2,500–3,000 m (Van der Werff and Consiglio 2004), and in Ecuador 1,300–2,500 m (assessment of the list given in Valencia et al. 2000). Endemics of Corsica, in the Mediterranean, are also concentrated at intermediate levels of elevation between 1,000 and 1,800 m (see figure 48 in Hobohm 2000a). In Europe, endemics of rocks and screes are concentrated between 500 and 1,800 m, grassland endemics peak between 550 and 2,100 m, endemics of scrub and heath formations between 200 and 1,550 m, forest endemics between 300 and 1,500 m, endemics of coastal and saline habitats, arable land, ruderal habitats and settlements at altitudes between 0 and 1,200 m, endemics connected to freshwater habitats, mires, swamps, fens and bogs between 100 and 1,750 m (Hobohm and Bruchmann 2009).

We assume that the higher altitude of maximum endemism in contrast to total taxonomic diversity is related to both the stricter separation/isolation of the high mountain zones and the higher total species diversity at lower altitudes which is the basis for evolution and dispersal (cf. Hobohm 2008b; Vetaas and Grytnes 2002). Interestingly, in continental SW China and in Taiwan (Huang et al. 2011a, b; Hsieh 2002) the number of endemics and the total number of vascular plant taxa seem to peak at the same altitudinal zone in the lowlands, whereas in most other regions endemism peaks at higher elevations than overall diversity.

The number of endemics found in an altitudinal zone is also a function of the area and climate of each zone and depends on maximum elevation. These factors might

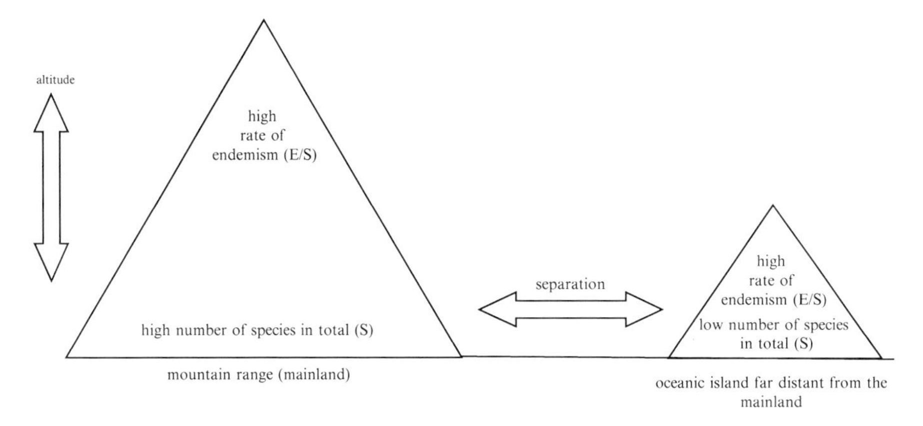

Fig. 5.1 High rates of endemism and total species richness. In the case of oceanic islands far distant from the mainland, a high ratio (E/S) is often combined with a relatively low number of species in total (S)

explain the differences in the maxima in the two neighbouring countries Ecuador and Peru. The coastal zone of Ecuador is relatively large relative to the whole size of the country. This zone is part of the Tumbes-Chocó-Magdalena Hotspot (Mittermeier et al. 2005) and very rich in endemics. The altitude of maximum endemism in Peru is more elevated because the coastal zone is small, extremely dry and relatively poor in endemics (Van der Werff and Consiglio 2004).

The rate of endemism (as a percentage of all taxa) in a mountain range normally increases continuously with altitude. This finding is not in conflict with the patterns discussed above showing that absolute numbers of endemics peak at intermediate or lower altitudes in most cases. The increasing rates in high mountain areas are comparable with values for oceanic islands which are highest far distant from the mainland (Fig. 5.1). However, the altitudinal gradient of the proportion of endemics (in %) is normally not mirrored by a latitudinal gradient (Hsieh 2002; Dhar 2002; Vetaas and Grytnes 2002; Hobohm 2000a; Talbot et al. 1999).

Stebbins and Major (1965: 16) analysed the endemic flora of California and concluded that

> *"…a high degree of endemism is found in those regions having a great variety of plant habitats."*

Hendrych (1982) examined the diversity of endemic plant species in Europe. He concluded that endemics are concentrated in high mountain areas such as the Alps, Sierra Nevada and Greater Caucasus, which are characterised by high habitat diversity.

At regional to continental scales, the diversities of higher plant species and endemic species increase with increasing habitat diversity. Areas with high habitat diversity generally also have high species diversity, although this is not always the case (cf. Cowling and Lombard 2002; Trinder-Smith et al. 1996; Huston 1994).

The main reasons are positive effects such as the number of speciation processes caused by a greater species pool and the avoidance of negative influences, including anthropogenic impacts, which are more effective in habitat-poor than habitat-rich environments. It is more likely that suitable habitats will remain in habitat-rich as compared to habitat-poor regions.

Habitat diversity is not an obligatory precondition for the coexistence of (endemic) species under stable ecological conditions. This finding does not question the positive effects of spatial heterogeneity for endemism. Spatial heterogeneity might be primarily important for the survival of species under changing climate conditions.

On the other hand, it is very easy to find high-elevation and habitat-rich ecoregions with relatively poor endemism. For example, mountain areas in northern North America, South Chile and Argentina, Norway, Central Australia, and the Cape Verde Islands are such regions. All these areas have been influenced by strong climate change, the northern and southern territories by Pleistocene glaciation cycles, Central Australia and the Cape Verde Islands by several dramatic cycles of wet and dry periods during the Quaternary (Crisp et al. 2001; Brochmann et al. 1997; Hendrych 1982).

Size of area, which is a factor often strongly correlated with the diversity of (endemic) species (see e.g. discussion in Körner 2000; Rosenzweig 1995a, b), has similar effects. According to Rosenzweig's area hypothesis the size of geographic area is responsible for the latitudinal gradient of species diversity. This hypothesis is based on the fact that the tropics, as the largest zone and having a continuous cover stretching into the northern and the southern hemisphere, can host species with larger range sizes, resulting in higher speciation rates and lower extinction rates (cf. Rohde 1992; Rapoport 1982). But, as the discussion of Chown, Gaston and Storch (Chown and Gaston 2000; Storch 2000) has shown, too little is currently known about range sizes of species in a given latitudinal gradient or in a humidity gradient. And actually, such gradients may divide the tropics into a larger number of different wet to dry ecozones than higher latitude areas (e.g. Bailey 1998). We are not aware of any studies that have tried to examine range size differences of vascular plant taxa in relation to area size of different climate zones or ecoregions.

However, area as a driver of biological processes means very little (cf. Ricklefs and Lovette 1999; Buckley 1985; MacArthur and Wilson 1967; Abbott 1977). But, in the context of negative influences on biodiversity in combination with a variety of environmental conditions area becomes important for the probability of the survival of individuals and populations. Survival of the biota in large regions under changing conditions depends on having a sufficient number of patches so that changes in one patch will not affect other patches, and consequently there is only little or no net change over a large area. For example, almost all the mid- and high altitudes of the Alps were covered by Pleistocene glaciers. But, the Alps harbour many endemic vascular plants which survived in the valleys or in the foothills of the Alps close to the Mediterranean Basin. Stability at this scale is a stochastic phenomenon where local and regional processes such as catastrophes, gap dynamics and successions are involved, and is different from the deterministic mathematical equilibrium of the competitive exclusion principle.

5.2 Oceanic Islands, Continental Islands and Mainland Regions

Oceanic islands are defined by their origin and development within the marine environment. When an oceanic island rises above sea level, colonisation from other islands or mainland regions can begin. All plants which are endemic to an oceanic island must have been developed and evolved in three steps: (i) dispersal of ancestors, which in the case of a great distance is an improbable event, (ii) establishment of a founder-population which normally involves finding adequate ecological conditions and overcoming a genetic bottleneck, and (iii) speciation processes on the basis of a reduced genepool and genetic drift in an environment which is more or less different from the original habitat. Because of reduced dispersal across the sea, oceanic islands are often relatively poor in total species numbers. On the other hand, isolation favours speciation. Thus, the percentage of endemics can be high.

In contrast, continental islands were by definition part of the mainland before they became separated. When they first became islands they were normally already covered by vegetation. This is the reason why continental islands do not normally reach high rates of endemism. Because of the larger species pool the conditions for increasing endemism are favourable compared to oceanic islands, and continental islands, in general, are not poorer in endemic vascular plant taxa than oceanic islands.

Unlike islands, mainland regions do not have clear natural borders, such as a coastline, everywhere. Biogeographical information about countries such as Austria or Switzerland is often much better than information related to mountain ranges which belong to two or several countries, such as the Alps. This is the reason why terrestrial ecoregions are often analysed with a strong focus on countries.

5.3 Habitats

Endemic vascular plants are found in diverse habitat types in almost all tropical, subtropical, temperate, boreal and arctic climate zones of the world. Tables 5.1, 5.2, 5.3, 5.4, and 5.5 present examples of regions with the respective numbers of endemics and information about the main physiognomic habitat types to be found there. Many of the examples show a relatively high total number of endemic species. This might be a result of the major scientific effort, especially in species-rich areas, and overestimations due to taxonomic double, or even triple, identity or to the revised taxonomic status of species (Kier et al. 2005; Yena 2007). We assume that the scientific effort into biogeography in general is greater where biodiversity is high or unique.

Table 5.1 Examples of biodiversity hotspots, ecoregions and landscapes with numbers of endemic vascular plants and the main habitats in various parts of the Americas (partly nested or overlapping), from North to South or West to East, respectively

Region (country or archipelago, area in km^2)	Total numbers of endemic vascular plant taxa (note that the term species is often used in the meaning of species plus subspecies)	Description of biomes, main formations, or vegetation units	Reference
California Floristic Province (Oregon, California, Mexico; 293,804)	2,124–2,133 species	Many different forest types, shrubland, coastal dunes, coastal salt marshes, maritime chapparal, savanna, coastal prairie scrub, vernal pools, freshwater marshes; 189 endemics are strictly confined to serpentine areas.	Kruckeberg (1992), Raven and Axelrod (1978), Davis et al. (1997), Mittermeier et al. (2005)
Klamath-Siskiyou Region (Oregon, California; 55,000)	280 taxa	Mixed evergreen, montane and subalpine forests, also serpentine vegetation.	Davis et al. (1997)
Vernal Pools (California, Mexico; 20,000)	c. 140 species	Ephemeral pools and vernal lakes with concentric rings of standing water-connected vegetation types (many annuals).	Lazar (2004), Keeler-Wolf et al. (1998), Zedler (1990), Holland and Jain (1988)
Sonoran Desert (USA, Mexico; 310,000)	650 species	Semi-desert vegetation dominated by succulent plants, shrubs, hemicryptophytes and therophytes.	Major (1988), Davis et al. (1997), Breckle (2000)
Mojave Desert (USA; 140,000)	138 taxa	Semi-desert vegetation dominated by shrubs, perennials and therophytes. Creosote bush in which the most common dominants are *Larrea tridentata* and *Ambrosia dumosa*, covering 70 % of the desert.	Stebbins and Major (1965), Davis et al. (1997), McGinley (2008i)
Central Highlands of Florida (USA; 10,000)	8 species and subspecies, plus 35 species and subspecies endemic or subendemic to Florida and whose centre of distribution is the Central Highlands	Xerophytic scrub vegetation with evergreen oaks and sand pine.	Davis et al. (1997), McGinley (2007c)

California Floristic Province of Baja California (California, Mexico; 14,000)	172 taxa, 67 near-endemics	Succulent scrub and chaparral, at higher elevations also rocky habitats, only few endemics in riparian habitats or permanent oases (such as *Brahea armata*).	O'Brien in Sula Vanderplank (Chap. 7, this book)
Baja California (Mexico; 143,396)	740 species	Tropical deciduous dry forest, conifer forest, winter-rain Mediterranean chaparral, coastal sage scrub, microphyllous scrub, coastal vegetation, desert scrub communities.	Davis et al. (1997), Riemann and Ezcurra (2005, 2007)
Madrean Pine-Oak Woodland (Mexico, small parts of the USA; 461,265)	<3,975 species	*Pinus*, oak and *Abies* forest, mixed pine-oak forest, cloud forest.	Mittermeier et al. (2005)
Montane Forests of Sierra Madre del Sur (South Pacific region of Mexico; 9,000)	16 species (to Guerrero) and 161 species (to Mexico)	Oak forest (1,900–2,500 m); cloud forest (2,300 m); pine-oak forest (2,400–2,500 m).	McGinley (2007g)
Coahuila and adjacent areas (Mexico; c. 17,500)	350 (subendemic) species	Desert shrubland is the dominant vegetation type (highest endemism) pine and oak forest are also rich in endemic species.	Villarreal-Quintanilla and Encina-Dominguez (2005)
Mesoamerica (from central Mexico to Panama; 1,130,019)	>2,941 species	Tropical and subtropical rain forest, moist, seasonal, dry forest formations, montane forest, grassland, mangrove, semi-desert and thornscrub.	Mittermeier et al. (2005)
Uxpanapa-Chimalapa Region (Mexico; 7,700)	>47 species	Evergreen, semi-evergreen and semi-deciduous tropical rain forests, montane rain forest, pine and pine-oak forests, xeric vegetation.	Davis et al. (1997)
Tehuacan-Cuicatlan Region (Mexico; 9,000)	810 taxa	Several dryland scrub formations with many species succulent, spiny or thorny, or forming rosettes; deciduous forest.	Davis et al. (1997)

(continued)

Table 5.1 (continued)

Region (country or archipelago, area in km²)	Total numbers of endemic vascular plant taxa (note that the term species is often used in the meaning of species plus subspecies)	Description of biomes, main formations, or vegetation units	Reference
Cuatro Cienegas Region (Mexico; 2,000)	23 species	Grassland with aquatic, semi-aquatic and gypsum-dune habitats in valley, desert scrub and chaparral, oak-pine woodlands, and montane forests of pine, fir and Douglas fir.	Davis et al. (1997)
La Amistad Biosphere Reserve (Costa Rica, Panama; >10,000)	3,000 species	Ten life zones in altitudinal gradient from tropical humid forest to subalpine rain paramo.	Davis et al. (1997)
Cerro Azul-Cerro Jefe Region (Panama; 53)	45 species	Tropical premontane wet forest, tropical wet forest, tropical premontane rain forest.	Davis et al. (1997)
Sierra Nevada de Santa Marta (Colombia; 12,232)	125 species	Three endemics are located in equatorial forest, 32 species in sub-andean-forest; 29 endemic species in Andean forest, 61 endemic species in paramo.	Davis et al. (1997), Carbonao and Lozano-Contreras (1997)
Coastal Cordillera (Venezuela; 45,000)	c. 500 taxa	Mangroves, coastal thorn scrub, hill savanna, deciduous forest, semi-deciduous lower montane forest, evergreen lower montane forest, evergreen montane cloud forest (highest endemism), upper montane elfin forest, upper montane scrub.	Davis et al. (1997)
Tropical Andes (from western Venezuela to northern Chile and Argentina; 1,542,644)	c. 15,000 species	Tropical wet, cloud, moist and dry forests, woodlands, scrub ecosystems, high altitude grasslands, many other habitats.	Mittermeier et al. (2005)
Tepuis ecoregion (Venezuela, Guyana, Suriname, Brazil; 3,260–7,000)	766 species	Forest, savanna, montane shrubland, grassland, rivers, waterfalls, cliffs, rock communities.	Steyermark (1986), Huber (1995), Davis et al. (1997), McGinley (2008g)

Tumbes-Choco-Magdalena (Peru, Ecuador, Colombia, Panama; 274,597)	2,750 species	Dry, moist and wet forest, rain forest, scrub, mangrove areas, beaches, rocky shoreline, desert vegetation.	Mittermeier et al. (2005) (see also Davis et al. (1997))
Continental Ecuador (275,680)	3,834 species	Mangroves, different coastal forest and scrub types, lowland rainforest, other wet to dry forest types, wet to dry and scrub paramo. Most endemics are associated in forest.	Valencia et al. (2000), Davis et al. (1997)
Caatinga of north-eastern Brazil (Brazil; 70,000–1,000,000)	c. 360 species	Low shrubby caatinga (up to 1 m tall) to tall caatinga forest (up to 30 m tall).	Davis et al. (1997), McGinley (2007b)
Cerrado (Brazil; 2,031,990)	4,400 species	Dry forest, woodland savanna, other savanna types, scrub grassland, open grassland; the herbaceous species are almost totally endemic.	Mittermeier et al. (2005), Pennington et al. (2006)
Atlantic Forest (Mata Atlantica: Brazil, Paraguay, Argentina, Uruguay; 1,233,875)	8,000 species	Lowland forest, woodland of the coastal plain, slope forest, high- altitude woodland, grassland or ‘campo rupestre’ (rocky grassland).	Mittermeier et al. (2005)
Chilean Winter Rainfall Valdivian Forests (Chile, Western Argentina; 397,142)	1,957 species	Dry desert communities (30 % of the area), sclerohyllous matorral and forest (30 %), decidous forest (15 %), Andean-Patagonean forest (4 %), broad-leaved rainforest (1 %), high elevation alpine vegetation (20 %).	Mittermeier et al. (2005)
Anconquija Region (Argentina; 6,000)	80–400 species	Different forest types from winter-dry rain forest to temperate cloud forest, Andean paramo-grassland, high Andean vegetation, spiny shrubland with tree cacti, semi-arid shrubland.	Davis et al. (1997)
Peninsula Valdes (Argentina; 3,600)	38 species	Predominant vegetation is Patagonian desert steppe, including e.g. tussock grasslands, xerophytic cushion grasses.	Clough (2008)

Table 5.2 Examples of biodiversity hotspots, ecoregions and landscapes with numbers of endemic vascular plants and the main habitats in various parts of the Atlantic Ocean and Caribbean (partly nested or overlapping), from North to South or West to East, respectively

Region (country or archipelago, area in km^2)	Total numbers of endemic vascular plant taxa (note that the term species is often used in the meaning of species plus subspecies)	Description of biomes, main formations, or vegetation units	Reference
Caribbean Islands (228,000–240,000)	6,550–7,920 species	Dry evergreen bushland and thicket, savanna, woodland, seasonal, montane, moist tropical forest, lowland rain forest, brackish and freshwater swamp, mangrove forest, many other habitats.	Davis et al. (1997), Mittermeier et al. (2005), Gillespie and Clague (2009), Anádon-Irizarry et al. (2012)
Cuba (108,722–110,860)	3,226–3,421 species	Different wet to dry forest types, woodland, swamps and gallery forests, thorny thickets, savanna, wetlands, coastal habitats, cultivated land. 23 % of the area is covered by forest, 24 % by shrublands and savanna, 44 % by cropland and crop/natural vegetation mosaic, 0.4 % by urban and built-up habitats, and 9 % by wetlands and water bodies including sea, brackish and freshwater habitats.	Davis et al. (1997), WRI (2003), Republica de Cuba (2009), Foster et al. (2012)
Cajalbana Tableland and Preluda Mountain region (Cuba; 100)	40 species	Pine forests, thorny xerophytic thickets, some gallery forests.	Davis et al. (1997)
Coast from Juragoa to Casilda Peninsula; Trinidad Mountains; Serra del Escambray (Cuba; 2,700)	40 species	Succulent and evergreen scrub thickets, including cacti, evergreen and semi-deciduous forests, seasonal and montane forests at higher elevations.	Davis et al. (1997)
Jamaica (10,990–11,425)	800–923 species	Different wet to dry forest types, sandy savanna (small areas), swamps dominated by palms, tall grasses and sedges, cultivated ground, coastal habitats with mangroves, dunes and rocks.	Davis et al. (1997)

		Fifty-eight percent of the area is covered by forest, 23 % by shrubland, savanna and grassland, 11.4 % by cropland and urban habitats, and 7 % by wetlands and water bodies including sea, brackish and freshwater habitats.	WRI (2003), Pryce et al. (2008)
Blue and John Crow Mountains (Jamaica; 782)	87 species	High altitude shrubby "elfin" woodland with bryophytes and lichens (2,000 m and higher), lower and upper montane rain forest, montane scrub, tall-grass montane savanna, cliff and landslide vegetation, cultivated ground.	Davis et al. (1997)
Cockpit Country (Jamaica; 430)	101 species	Evergreen seasonal forest, mesic limestone forest and degraded limestone forest, limestone cliff and landslide vegetation, pastures and agricultural crops in valleys.	Davis et al. (1997)
Hispaniola (76,190–76,261)	c. 1,445–2,016 species	Different wet to dry forest types (33 % of the area), thickets with cacti, humid savanna and primary dry grassland (together 25 % of the island), rocky and sandy coastal habitats. 27 % of the area is cropland and crop/natural vegetation mosaic, and urban habitats, and 13 % is wetlands and water bodies.	Bolay (1997), Davis et al. (1997), WRI (2003)
Bermuda (UK; 53)	15 species	Mangrove (without endemics) and woodland. Most endemics are connected to woodland, 1 to caves.	Sterrer (1998), McGinley (2007a), WRI (2003)
Greenland (2,180,000)	15 taxa	Different tundra vegetation types dominated by cryptogams, hemicryptophytes or chamaephytes, coastal vegetation including salt marsh and patches of mires, small stands of birch trees up to 7 m in the South of the island.	Groombridge and Jenkins (2002), Daniels et al. (2005), Lepping and Daniels (2006)

(continued)

Table 5.2 (continued)

Region (country or archipelago, area in km^2)	Total numbers of endemic vascular plant taxa (note that the term species is often used in the meaning of species plus subspecies)	Description of biomes, main formations, or vegetation units	Reference
Atlantic Ocean Islands (Ascension, Azores, Canary Islands, Cape Verde, Iceland, Madeira Islands, St. Helena, Tristan da Cunha; 117,804)	881 species	Laurisilva forest, dry conifer forest, other forest types, succulent scrub, other scrub communities, heath, swamps, rocky habitats, cliffs, coastal habitats.	Davis et al. (1994), Pott et al. (2003), Schäfer (2005a, b)
Azores (Portugal; 2,250)	80 taxa (72 species, subspecies and varieties, 8 hybrids)	Laurisilva (very reduced), montane cloud forest, shrubland, grassland, several communities on volcanics, wetlands, littoral communities. Fifty-seven or more endemics inhabit rocky habitats, cliffs, lava flows, ravines, steep slopes or volcanic craters, 34 are connected to juniper, laurel or *Pittosporum* forest and scrub, 18 endemics occur in littoral habitats, 3 in grasslands, 6 in freshwater habitats, 2 in bogs, and 4 in anthropogenic habitats (many of them in more than one habitat group).	Davis et al. (1994), Schäfer (2005a, b), Gillespie and Clague (2009)
Madeira Archipelago incl. Salvage Islands (Portugal; 815)	141 taxa (125 species, 26 subspecies)	Fourteen or more taxa in freshwater habitats and swamps, 36 in coastal habitats, 12 in cropland, ruderal and urban habitats, 4 in grasslands, 88 in rocky habitats and cliff communities, 23 in scrub and heath, 43 in forests (many of them in more than one habitat group).	Davis et al. (1994), Borges et al. (2008), Hobohm, data base in progress

Canary Islands (Spain; 7,542)	545–650 species and subspecies	Coastal vegetation, dry lowland with succulent scrub, thermophilous lowland, laurel and montane pine forest, montane scrub, rocky habitats.	Davis et al. (1994), Pott et al. (2003), Hobohm, data base in progress, counts based on Izquierdo et al. (2004)
		One endemic species occurs in freshwater habitats, 44 taxa in coastal areas, 22 in cropland, ruderal and urban habitats, 388 in rocky habitats and cliff communities, 239 in scrub, heath and succulent dominated communties, and 96 in forests (many of them occur in more than one habitat type, other endemics are not confined to habitat types, yet).	
Tenerife (Canary Islands, Spain; 2,043)	134 species and subspecies	Rocks and screes (largest proportion of endemics), scrub, heath, succulents (second large proportion), forest (laurel forest, *Pinus canariensis* forest and other types), coastal and saline habitats.	Hobohm and Moreira-Muñoz (this book) based on counts in Izquierdo et al. (2004)
Cape Verde Islands (4,033)	82 (sub) species	Dominant vegetation types are open grassland and semi-desert, coastal habitats, shrubland and other vegetation types subdominant or in small patchesOf the endemic taxa, 31 are hygrophytic, 34 mesophytic and 17 xerophytic.	Davis et al. (1994), Brochmann et al. (1997), Duarte et al. (2008)
Sao Tomé and Principe (1,001)	148 taxa	Mangroves, low altitude forest (<800 m asl.), secondary forest, dry forest, shade forest for cocoa and coffee, shrublands and grasslands, mountain forest (800–1,400 m asl.), mist forest (1,400–2,024 m asl.).	Juste and Fa (1994), Davis et al. (1994), Nau (2003), Ministry for Natural Resources and the Environment (2007)
St. Helena (British Dependent Territory; 122)	50 species	Tree fern thickets, forestry plantations, rocks, semi-desert, various introduced shrub communities, cattle pasture and abandoned New Zealand flax (*Phormium tenax*) plantations.	Davis et al. (1994), Gillespie and Clague (2009)

(continued)

Table 5.2 (continued)

Region (country or archipelago, area in km^2)	Total numbers of endemic vascular plant taxa (note that the term species is often used in the meaning of species plus subspecies)	Description of biomes, main formations, or vegetation units	Reference
Tristan da Cunha Islands (UK; 159)	40 species	Tussock grassland in coastal areas, fern bush vegetation interspersed with single trees (300–500 m), wet heath with ferns, sedges, grasses, angiosperms, and mosses (up to 800 m), peat bogs with *Sphagnum* in depressions (above 600 m), feldmark and montane rock communities (above 600 m).	Davis et al. (1994), Ashworth et al. (2000)
Gough Island Wildlife Reserve (South Atlantic Ocean, including Inaccessible Island, UK; 75)	4 species	Tussock grassland (100–300 m), vegetation of fernbush with shrub (up to 500 m), wet heath (800 m), peat bogs (above 600 m).	Cooper and Ryan (1994), McGinley (2008c)
Falkland Islands (UK; 12,173)	15 species	Coastal margins with rocks, dunes and tussac grassland, acid grassland, dwarf shrub heath with ferns, scrub, feldmark and inland rocks, bogs and fens, standing and running waters, non-native habitats (arable and horticultural land, coniferous woodland).	Ashworth et al. (2004)

Table 5.3 Examples of biodiversity hotspots, ecoregions and landscapes with numbers of endemic vascular plants and habitats in various parts of Europe and Africa (partly nested or overlapping), from North to South or West to East, respectively

Region (country or archipelago, area in km^2)	Total numbers of endemic vascular plant taxa (note that the term species is often used in the meaning of species plus subspecies)	Description of biomes, formations, or vegetation units	Reference
Europe (from Canary Islands to Ural Mountains, from Iceland and Svalbard to Cyprus, Caucasus and Asian Turkey excluded; c. 10,500,000)	6,250–6,500 taxa (c. 165 groups of microspecies, 5,250–5,500 species, 850 subspecies)	At least 2,772 endemics occur (also) in rocky habitats, 1,320 in grassland, 1,125 in scrub/heath, 773 in forest, 450 in coastal habitats, 413 in urban and agricultural lands, 266 in freshwater habitats (including connected habitats such as embankments), 104 in bogs/mires/fens (wide overlap, not all endemics assigned to habitat types yet). Large areas in Europe are covered by agricultural land, urban habitats and forests (together ≫50 % of the area) whereas grasslands, coastal and saline habitats, freshwater habitats, bogs, mires and fens, or rocks and screes only represent sizes of <10 %.	Hobohm (2004, 2008a, b), Hobohm and Bruchmann (2009)
		Many more taxa – 4–7 times – are basiphilous than acidophilous, but many are indifferent.	
		Mean number (median value) of regions per endemic is 1 (of 42) for rocky habitats, scrub/heath and coastal habitats, 2 for forests, agricultural lands and urban habitats, and freshwater habitats, and 3 for grasslands and bogs/mires/fens.	Hobohm, EvaplantE in progress
Nordic regions in Europe (from Denmark to Svalbard and Iceland to NW Russia; c. 1,750,000)	180 species, subspecies, varieties and hybrids	More or less open ground in the alpine belts (Scandes: 47 endemics) and on the alvars (Gotland and Öland: 20 endemics), heathland or rocky habitats (Iceland: 10 endemics), coastal habitats (Arctic coasts: 8 endemics, Baltic shores: 30 endemics), dunes and sea-facing hillsides (Jylland: 4 endemics). Another 61 endemics occur in Norden.	Jonsell and Karlsson (2004)

(continued)

Table 5.3 (continued)

Region (country or archipelago, area in km^2)	Total numbers of endemic vascular plant taxa (note that the term species is often used in the meaning of species plus subspecies)	Description of biomes, formations, or vegetation units	Reference
Pyrenees (France, Spain and Andorra; 30,000)	120–200 species	Mediterranean evergreen forests, semi-deciduous submediterranean forests, deciduous Atlantic forests, montane and subalpine coniferous forests, Mediterranean montane scrub, meadows and pastures, rock and scree habitats, freshwater habitats, bogs and fens.	Davis et al. (1994), Burga et al. (2004)
Alps (from France to Slowenia; 200,000)	350–400 species, 50 subendemic species in whole area	Broadleaved and conifer forests, dwarf shrub communities and moorland, alpine grasslands, moraine, rock and scree communities, arable land, meadows and pastures, running and standing waters. 35–60 % of endemic and subendemic species live in rocky places, stony slopes and grassland of mid to high altitudes, few in freshwater and amphibic habitats or woods (59 % of the Austrian taxa are basiphytes, 28 % are acidophytes.).	Pawlowski (1969), Davis et al. (1994), Dullinger et al. (2000), Casazza et al. (2005, 2008), Rabitsch and Essl (2009)
Carpathians (Slovakia, Hungary, Poland, Ukraine, Romania; 190,000)	100–120 endemic species, 30 subendemics	Forest, grassland, rock communities. Most of the (sub)endemics are inhabited in grassland, 10–15 % of endemic and subendemic species live in rocky places/stoney slopes, very few species belong to forest.	Pawlowski (1969), Davis et al. (1994)
South Crimean Mountains and Novorossia (Ukraine, Russia; 80,500)	Some 220 species	Maquis and shiblyak, oak, pine, pine-beech and juniper woodland, dry grassland.	Davis et al. (1994)

Mediterranean Basin (including the Azores, Madeira, Canary, Cape Verde Islands and parts of North-Africa and West Asia; 2,085,292)	11,700–13,000 species	Evergreen sclerophyllous, lauriphyllous, coniferous and deciduous forest types, hard-leafed or sclerophyllous shrublands, softleaved and drought phrygana, different types of coastal vegetation (saltmarsh, saline shrubland, dunes, rocky shores), many fire adapted vegetation types Endemic vascular plant taxa in the French Mediterranean region occur in habitats on steeper slopes, with higher rock cover and in lower and more open vegetation than their widespread congeners, but they are not more stress-tolerant.	Mittermeier et al. (2005), Gillespie and Clague (2009)
Southeast France (France, including Corse; 38,750)	215 (sub) species	Most of the (sub)endemics belong to calcareous rocky grasslands, cliffs and rocks at mid and high altitudes.	Médail and Verlaque (1997)
Corse (France; 8,750)	126–130 species c. 162 subendemic species	Most of the (sub)endemics belong to siliceous rocky grassland and ridges, to maritime rock crevices and to riparian grassland, moist banks, to torrents at low and mid altitudes, many endemic taxa occur in mid and high altitudes above 700 m.	Davis et al. (1994), Gamisans and Marzocchi (1996), Medail and Verlaque (1997), Hobohm (2000b)
Crete (Greece; 8,700)	248 taxa (species groups, species, subspecies)	Evergreen scrub, fragments of evergreen oak, pine and cypress forest, montane rock and cliff vegetation; island extensively cleared or modified by agricultural management. 9 endemics occur in freshwater habitats, bogs, mires or fens, 21 in coastal, brackish and saline habitats, 36 in cropland, ruderal and urban habitats, 171 in rocky habitats and screes, 74 in scrub and heath, 28 in forests (many of them in more than one habitat type).	Davis et al. (1994), Hobohm, data base in progress

(continued)

Table 5.3 (continued)

Region (country or archipelago, area in km^2)	Total numbers of endemic vascular plant taxa (note that the term species is often used in the meaning of species plus subspecies)	Description of biomes, formations, or vegetation units	Reference
Cypres (9,250)	108 species and subspecies	1 % of the area is covered by forest (26 endemic taxa), 13 % by shrublands including grassy habitats (46 endemics), 82 % by cropland and urban area (28 endemics), and 3 % by wetlands and water bodies (12 freshwater and 6 coastal endemics). Few taxa occur in more than one group of habitats.	Hobohm, data base in progress, WRI (2003)
Troodos Mountains (Cypres; 1,800)	36 species	Evergreen scrub, evergreen oak, pine, Cyprus cedar forest, rock and cliff communities; extensively cleared or modified by agriculture in lowlands.	Davis et al. (1994)
Mediterranean conifer and mixed forests of the Middle Atlas Range (Morocco, Algeria, Tunesia)	>450 species: 237 species in Middle Atlas Range, 190 s in the Rif Massif, 91 species in the Tellien Atlas	Mediterranean conifer and mixed forest in relatively humid, medium to high elevations.	McGinley (2007e)
High Atlas (Morocco; 7,000)	160 species	Forest (28 % of the area), meadows (pozzines) and various scrub communities (30 %), eroded rock, alpine pseudo-steppe, dwarf spiny scrub and scree (over 30 %).	Davis et al. (1994)
Nile Delta (Egypt; c. 22,750)	8 species	Dominated by swamp (of *Cyperus papyrus* or of *Phragmites australis* with *Typha capensis*, *Juncus maritimus*, with some small sedges), freshwater wetlands and salt-tolerant marsh vegetation.	McGinley (2008f)
Sinai (Egypt including Saint Catherine mountains; 60,000)	28–30 species	Irano-Turanian steppe vegetation, 16 endemics to gorge habitats of the rugged mountainous districts which contains Saint Catherine mountains.	Davis et al. (1994), Moustafa and Zaghloul (1996)

Southern Sinai (Saint Catherine Protectorate; 4,350)	19 species	Arid, mountain ecosystem that forms an island of central Asian steppe vegetation along with Irano-Turanian biota.	Grainger (2003)
Saharan Desert (North Africa; 9,000,000)	162–375 species	Desert and semi-desert vegetation with therophytes, hemicryptophytes and shrubs including rock and scree habitats, scrub, dunes and saline habitats. Only few tree species as relicts of wet palaeoclimate.	Major (1988), Breckle (2000)
Sudanian Regional Centre of Endemism (from Senegal to the foothills of the Ethiopian Highlands; 3,731,000)	c. 960 species	Woodland, wooded grassland, other grasslands, rocky inselberg vegetation, riparian forest, upland dry or semi-evergreen forest, extensive wetlands, different secondary vegetation types (scrub), farmland.	Davis et al. (1994)
Eastern Afromontane (from Saudi Arabia and Yemen to Zimbabwe; 1,017,806)	2,356 species	Montane forest, other forest types, woodland and savanna, alpine moorland and bog, zones with heather, bamboo vegetation, *Papyrus* and *Carex* wetland, hot springs, sclerophytic vegetation on old lava flows.	Mittermeier et al. (2005)
Horn of Africa (from NE Kenya to SE Oman and SW Saudi Arabia; 1,650,000)	2,750 species	Bushland, succulent shrubland, dry evergreen forest and woodland, semi-desert grassland, low-growing dune and rock vegetation, mangrove and riverian vegetation.	Mittermeier et al. (2005)
Ethiopian Highlands (Ethiopia, Erithrea, Sudan, Egypt; 519,278)	555 species	Moist and dry forests, woodland vegetation at foothills and lower elevations, montane bamboo, 'dega' and 'weyna dega' (vegetation dominated by conifers), heathland scrub, Afroalpine ecosystems at high altitudes. The majority of endemics is associated with open grassland, dry woodland, and heathland.	Mittermeier et al. (2005)

(continued)

Table 5.3 (continued)

Region (country or archipelago, area in km^2)	Total numbers of endemic vascular plant taxa (note that the term species is often used in the meaning of species plus subspecies)	Description of biomes, formations, or vegetation units	Reference
Guinean Forests of West Africa (from Guinea to Cameroon, including Sao Tomé and Príncipe; 620,314)	1,800 species	Moist forest, freshwater swamp forest, semi-deciduous forest.	Mittermeier et al. (2005)
Upper Guinea (from Guinea to Togo; 340,000)	650 species	Tropical rain forest.	Holmgren and Poorter (2007)
Tai National Park (Cote d'Ivoire; 3,500)	150 species	Evergreen rain forest of the Guinea zone.	Davis et al. (1994)
Mont Nimba (Guinea, Cote d'Ivoire, Liberia; 480)	c. 13 species	Lowland and transitional rain forest, grasslands.	Davis et al. (1994)
Cross River National Park (Nigeria; 4,227)	130 species	Lowland rain forest, freshwater swamp forest, montane forest, grassland.	Davis et al. (1994)
Mount Cameroon (Cameroon; 1,100)	49 species (with another 50 subendemics)	Submontane and montane forest, subalpine grassland.	Davis et al. (1994), Cable and Cheek (1998), Blom (2001)
Cristal Mountains (Gabon; 9,000)	>100 species	Wet evergreen coastal Guineo-Congolian rain forest.	Davis et al. (1994)
Salonga National Park (Zaire; 36,560)	150–200 species	Tropical evergreen Guineo-Congolian rain forest, swamp and riverine forest, secondary vegetation, grassland.	Davis et al. (1994)
Haut Shaba (Zaire; 496,871)	>300 species	Steppic savannas dominated by grasses, with many shrubs and bulbous plants, woodland, localized swamps, ravine and riparian gallery forests.	Davis et al. (1994)

Garamba National Park (Zaire; 56,727)	c. 50 species	Dominant vegetation types: shrub savanna, savanna woodland, and long grass savanna.	Davis et al. (1994) McGinley (2008b)
Coastal Forests of Eastern Africa (from Mozambique to Somalia, including islands lying immediately offshore; 291,250)	1,360–1,750 species	Moist forest, drier forest, with coastal thicket, fire-climax savanna woodland, seasonal and permanent swamp, littoral habitats that include mangrove vegetation; 70 % of all endemics have been recorded from forest habitats; 550–554 in coastal forest.	Burgess et al. (1998), Burgess et al. (2003), Mittermeier et al. (2005)
Rondo Plateau (Tanzania; 250)	>200 strict or near-species	Dry semi-deciduous lowland forest, woodland, thicket, secondary communities, large parts converted to farmland and plantations.	Davis et al. (1994)
Maputaland-Pondoland-Albany (from South Africa to southern Mozambique; 274,316)	1,900 species	Warm temperate forest, different types of thickets, bushveld, grassland, many stem succulents.	Davis et al. (1994), Mittermeier et al. (2005)
Albertine Rift (including parts of Uganda, Rwanda, Democratic Republic of Congo, Burundi, Tanzania; 313,051)	>551 species (taxa) with 92 species restricted to Virunga Vocanoes	Glaciers and rocks at the top of the Rwenzori Mountains (5,100 m), alpine moorland (3,400–4,500 m), giant *Senecio* and *Lobelia* vegetation (3,100–3,600 m), giant heather (3,000–3,500 m), raised bogs (3,000–4,000 m), bamboo forest (2,500–3,000 m), montane forest (1,500–2,500 m), lowland forest (600–1,500 m), savanna woodland (600–2,500 m) and savanna grassland (600–2,500 m). *Papyrus* and *Carex* wetland, together with lakes and streams, have their own unique habitat types varying from the rocky and sandy edges to the benthic zones of the lakes.	Owiunji et al. (2005), Mittermeier et al. (2005), Plumptre et al. (2007)
Eastern Arc Mountains and Southern Rift (from southern Kenya to North Mozambique; 16,123)	1,200 species	Lowland, submontane, montane and upper montane forest, grassland, heathland, bog.	Davis et al. (1994), Mittermeier et al. (2005)

(continued)

Table 5.3 (continued)

Region (country or archipelago, area in km^2)	Total numbers of endemic vascular plant taxa (note that the term species is often used in the meaning of species plus subspecies)	Description of biomes, formations, or vegetation units	Reference
Namib Desert (Namibia; 134,400)	492 species	Desert and xeric shrublands.	Cowling and Hilton-Taylor (1999)
Nama-Karoo (Namibia; 198,500)	377 species	Desert and xeric shrublands.	Cowling and Hilton-Taylor (1999)
Succulent Karoo (Western Cape Domain; South Africa, Namibia; 102,691)	1,750–2,539 species	Low to dwarf succulent shrubland dominated almost entirely by leaf succulents, stem succulents and fine-leaved evergreen shrubs, clustered in broken, rocky habitats rather than sandy or loamy flats.	Davis et al. (1994), Cowling and Hilton-Taylor (1999), Mittermeier et al. (2005)
Kaokoveld (Damaraland-Kaokoland Domain; Namibia, Angola; 70,000)	116 endemic or near-endemic taxa	Desert vegetation, xeric shrubland, montane savanna, escarpment vegetation.	Davis et al. (1994), Cowling and Hilton-Taylor (1999)
Cape Floristic Region (South Africa; 78,555)	6,210 species	Fynbos (shrubland which covers c. 60 % of the whole area characterized by evergreen rush or reed-like plants, restioids, small leafed shrubs, ericoids, relatively tall shrubs, proteoids, and bulbs or geophytes) and several non-fynbos vegetation types, endemism largely confined to fynbos. Most endemics grow on nutrient-poor substrates with low pH.	Cowling (1992), Davis et al. (1994), Mittermeier et al. (2005)

Cape Peninsula (South Africa; 471)	88 species (+ 2 subspecies)	Forest and thicket harbour 11 endemics, dune asteraceous fynbos 29, wet restioid fynbos 32, ericaceous fynbos and upland restioid fynbos 38, sandplain proteoid fynbos 40, coastal scree asteraceous fynbos and mesic oligotrophic proteoid fynbos 63, mesic mesotrophic proteoid fynbos 49, undifferentiated cliff community 17, wetland 22, wet and oligotrophic proteoid fynbos 23, wet mesotrophic proteoid fynbos 24, Renosterveld/grassland 8 endemic taxa (many endemics occur in more than one vegetation type).	Trinder-Smith et al. (1996), Mittermeier et al. (2005)
Sneeuberg (South Africa; 20,000)	33 endemics, 13 subendemic species (5 of them previously reported as Drakensberg Alpine Regions)	Karoo Escarpment Grassland above 1,600 m, arid Fynbos above ca. 2,100 m on rocky peaks and stony plateaux, southern Mistbelt Forest on southern slopes, elements of temperate forest, moist and other thickets, dominant on foothills at altitudes of ca. 1,000–1,400 m, elements of the Nama-Karoo biome.	Clark et al. (2008)
Drakensberg Alpine Region (Lesotho, South Africa; 40,000)	334–394 species	Chiefly subalpine and alpine grassland, shrubland (heathland), with scrub and savanna at lower altitudes, extensive wetlands in the alpine belt.	Davis et al. (1994), Van Wyk and Smith (2001), Mucina and Rutherford (2006), Carbutt and Edwards (2006)

Table 5.4 Examples of biodiversity hotspots, ecoregions and landscapes with numbers of endemic vascular plants and the main habitats in various parts of the Arctic, Asia and Indian Ocean Islands (partly nested or overlapping), from North to South or West to East, respectively

Region (country or archipelago, area in km²)	Total numbers of endemic vascular plant taxa (note that the term species is often used in the meaning of species plus subspecies)	Description of biomes, formations, or vegetation units	Reference
Arctic (circumpolar; c. 4,500,000)	96 species	Different tundra vegetation types (e.g. dominated by rocks, screes, arctic-alpine dwarf shrubs, herbs, grasses, bryophytes and/or lichens).	Talbot et al. (1999)
Central and Northern Asia (from Ural Mountains eastwards to the Pacific Ocean, and southwards to the Caucasus and the Middle Asian Mountains; 16,794,400)	c. 2,500 species	Different forest types (33 % of the region), tundra, *Sphagnum* bogs, forest steppe and steppe, semi-desert and desert vegetation, montane and alpine vegetation.	Davis et al. (1995)
Caucasus (Armenia, Azerbaijan, Georgia, Russia; 532,658)	1,200- > 1,600 species	Lowland broadleaved forests, montane broadleaved and coniferous forests, swamp forest, montane steppe and woodlands, subalpine meadows, semi-desert to desert vegetation, subalpine and alpine meadows.	Davis et al. (1995), Burga et al. (2004), Mittermeier et al. (2005)
Mountains of Central Asia (including Pamir and Tien Shan; 863,362)	1,500 species	Desert, semi-desert, and steppe vegetation types, subalpine and alpine meadows, tundra-like vegetation at high altitudes, riverine woodland, shrub communities, spruce and other forest types.	Davis et al. (1995), Mittermeier et al. (2005), Nowak and Nobis (2010)
Altai Mountains (Russia, Mongolia, Kazakhstan, China; 261,700)	200–390 (sub) species	Mixed coniferous forests, steppe and forest steppe, subalpine and alpine vegetation, tundra.	Davis et al. (1995), Burga et al. (2004), Pyak et al. (2008)
Irano-Anatolian area (from Central Turkey to Turkmenistan; 899,773)	2,500–3,000 species	Mountainous forest steppe, primary steppe, salt steppe, halophytic marsh, subalpine vegetation including thorn-cushion (tragacantic) formations, alpine vegetation, riparian forest.	Mittermeier et al. (2005), Şekercioğlu et al. (2011)

Mountains of South-East Turkey, North-West Iran and Northern Iraq (Turkey, Iran, Iraq; 147,332)	c. 500 species	Oak forest at lower altitudes, mountain steppe between 2,200 and 2,700 m, thorn-cushion zone of spiny plants between 2,700 and 3,300 m, alpine zone including grassland, scree and rock vegetation above 3,300 m.	Davis et al. (1994)
Isaurian, Lycaonian and Cilician Taurus (Turkey; 45,120)	235 species	Littoral dune communities, halophytes and remnants of alluvial forest, wetland vegetation, maquis and phrygana at low altitudes, oak and pine forest, some stands of *Cedrus libani* at mid altitudes, mountain steppe and scree communities at higher altitudes.	Davis et al. (1994), Vogiatsakis (2012)
Gypsum areas in Sivas (Turkey, Central Anatolia, Eastern part of Cappadocia; 4,000)	122 species (to Turkey)	Gypsum habitats only with vegetation of low density: many hemicrytophytes, annuals, chamaephytes and shrubs.	Akpulat and Celik (2005)
Badkhiz-Karabil Semi-desert (Turkmenistan; 133,600)	75 species	Xeric savanna-like vegetation with therophytes, perennial grasses, sedges and forbs, shrubs and arid pistachio scrub.	Atamuradov et al. (1999), McGinley (2008a)
The Levantine Uplands (Turkey, Syria, Lebanon, Israel, Jordan; 96,675)	c. 635 species	Oak forest and scrub, other forest types, *Juniperus* scrub, subalpine tragacanthic communities and alpine vegetation, degraded vegetation with maquis, garigue and batha.	Davis et al. (1994)
Alpine Regions of Iran (Iran; c. 10,000)	357 (sub) species	Communities dominated by large herbs and thorn-cushion plants, alpine meadow communities, high alpine xerophytic communities covered by graminoids, rocky and scree habitats.	Noroozi et al. (2008), Chap.7 by Noroozi, this book
Negev Desert (Israel; 11,000)	80 species	Desert and semi-desert vegetation.	Breckle (2000)
Highlands of South-Western Arabia (Saudi Arabia, Yemen; 70,000)	c. 170 species	Deciduous bushland and thicket from 200 m to c. 2,000 m altitude, evergreen bushland and thicket and *Juniperus* scrub, above c. 2,000 m.	Davis et al. (1994)

(continued)

Table 5.4 (continued)

Region (country or archipelago, area in km^2)	Total numbers of endemic vascular plant taxa (note that the term species is often used in the meaning of species plus subspecies)	Description of biomes, formations, or vegetation units	Reference
Central Desert of Oman (Oman; 150,000)	12 species	Vegetation can be broadly classified as part of the Acacia-*Zygophyllum-Heliotropium* savanna (in the central plains) and *Prosopis-Calligonum* semi-desert (in the dune desert).	Ghazanfar (2004)
Dhofar Fog Oasis (Oman, Yemen; 30,000)	c. 60 species	Dry deciduous shrubland on the seaward-facing escarpment, semi-deciduous thicket and grassland at higher altitude.	Davis et al. (1994)
Himalaya (from North Pakistan to North Myanma; 741,706)	3,160 species	Open woodland savanna, mixed evergreen forest at low altitudes, temperate humid forest, subtropical evergreen broadleaf and pine forest, alluvial and alpine grassland, shrubby rhododendron stands, alpine scrub communities, rocky vegetation.	Mittermeier et al. (2005)
Indian Subcontinent (India, Pakistan, Nepal, Bhutan, Bagladesh, Myanma, Sri Lanka, northeastern Indian Ocean Islands; 4,850,000)	>11,330 species	Tropical evergreen rain forest to high-altitude alpine vegetation, mixed broadleaved and coniferous forests, woodland, savanna, thorn scrub, dwarf shrub, alpine grassland, scree communities.	Davis et al. (1995)
Thar-Desert (India; 200,000–300,000)	36 taxa (24 species, 12 subspecies)	Semi-desert and desert vegetation.	Bhandari (1979), Khan (1997)
Western Ghats (India and Sri Lanka; 182,500)	1,700 species	Scrub forests, deciduous, moist and tropical rain forest, montane forest and rolling grassland in higher altitudes. Tropical rainforest covers 11 % of the area, dry, moist deciduous, and scrub forests cover another 11 %.	Mittermeier et al. (2005)
Agastyamalai Hills (India; 3,500)	100 species	Tropical thorn forest, tropical dry decidous forest, tropical moist deciduous forest, tropical evergreen rain forest and subtropical montane forest, lowland and montane grasslands.	Davis et al. (1995)

Nilgiri Hills (India; 5,520)	100 species	Tropical evergreen rain forest, tropical semi-evergreen forest, tropical moist deciduous, tropical dry thorn forest, tropical montane evergreen shoal forest and grassland.	Davis et al. (1995)
Mountains of Southwest China (China and small parts of Myanma; 262,466)	3,500 species	Broadleaved and coniferous forest, bamboo grove, scrub communities, savanna, meadow, prairie, freshwater wetland, and alpine scrub and scree communities.	Mittermeier et al. (2005)
Mount Emei (Sichuan Province, China; 154)	>100 (subendemic?) species	Subtropical evergreen broad-leaved forest (below 1,500 m), evergreen and deciduous broad-leaved mixed forest, coniferous and broad-leaved mixed forest, subalpine coniferous forest, subalpine shrubs (above 2,800 m).	McGinley (2008e)
Xishuangbanna Region (Yunnan Province, China; 19,220–19,690)	120–153 species	Tropical lowland seasonal rain forest, tropical evergreen dipterocarp forest, seasonal rain forest over limestone, monsoon forest, montane seasonal rain forest, montane evergreen broadleaved forest.	Davis et al. (1995), Shou-qing (1988)
Phong Nha-Ke Bang National Park (Central Vietnam; 2,746)	13 species (to Vietnam) 1 species (to the National Park)	Mostly tropical dense moist evergreen forest (>800 m) and low tropical montane evergreen forest (<800 m).	Clough (2008)
Indo-Burma (from South China and eastern Bangladesh to North Peninsular Malaysia; 1,938,745–2,373,285)	7,000 species	Wet to dry broadleaf evergreen forest, deciduous and montane forest, coniferous forest, savanna forest, coastal heath forest, shrubland and woodland, xerophytic formations at high altitudes, lowland floodplain swamp, mangrove, seasonally inundated grassland.	Mittermeier et al. (2005), Pawar et al. (2007), Tordorff et al. (2012)
North Myanmar (Myanmar; 115,712)	c. 1,250 species	Lowland tropical forest, various types of monsoon forest, mixed and conferous temperate forest, Rhododendron forest and alpine meadows.	Davis et al. (1995)
Bago (Pegu) Yomas (Myanmar; 40,000)	c. 100 species	Wet evergreen dipterocarp forest, pyinkado forest, moist teak forest, dry teak forest, dry dipteropcarp forest, pyinma forest, bamboo where canopy has been opened.	Davis et al. (1995)

(continued)

Table 5.4 (continued)

Region (country or archipelago, area in km²)	Total numbers of endemic vascular plant taxa (note that the term species is often used in the meaning of species plus subspecies)	Description of biomes, formations, or vegetation units	Reference
Taninthayi (Tenasserim) (Myanmar; 73,845)	c. 500 species	Wet evergreen dipteropcarp forest, montane rain forest, mangrove forests, fresh and brackish swamp vegetation and pinle-kanazo tidal forest, bamboo where forest has been cleared and on poor soils.	Davis et al. (1995)
Dong Phayayan Khao-Yai Forest Complex (Thailand; 6,155)	16 species	Evergreen forest (73.8 % of the area), mixed dipterocarp/deciduous forest (3 %), deforested scrub, grassland and secondary growth (18 %).	McGinley (2009)
Korean Peninsula (South Korea, North Korea; 220,759)	242–340 species (s. str.)	Evergreen broadleaved forest, deciduous and mixed forests, alpine forest (taiga), many other habitat types.	Davis et al. (1995), Kim et al. (2009)
Madagascar and the Indian Ocean Islands (600,461)	9,000–>10,800 species	Tropical rain forest, humid to dry deciduous forest, other dry and moist forest formations, several high elevated mountain ecosystems which are characterized by forests with mosses and lichens (21 % of the area is covered by forest), shrubland, savanna, grassland and spiny desert (63 % of the area), littoral vegetation including mangroves. 2 % is wetlands and water bodies.	Davis et al. (1994), Mittermeier et al. (2005), WRI (2003)
Socotra (Yemen; 3,625)	c. 230–296 species	Semi-desert and dry-deciduous shrubland on coastal plains and lower slopes of mountains, semi-deciduous thicket and grassland at higher altitudes.	Davis et al. (1994), Miller and Morris (2004)
Andaman and Nicobar Islands (India; 8,249)	c. 227 species	Tropical evergreen rain forest, semi-evergreen rain forest, moist deciduous forest, beach forest, swamp forest, bamboo scrub.	Davis et al. (1995)

Seychelles (450)	72–120 species	Mangrove and coastal forest, lowland rain forest, palms, pandan and hardwood characterize the natural forest of the granitic islands (below 610 m), cloud forest with tree ferns and mosses (above 610 m); Most endemics are found at intermediate and high altitudes whereas only two species are confined to the coastal zone and no endemics occur in mangrove vegetation.	Robertson (1989), Fleischmann (1997), Fleischmann et al. (2003), McGinley (2008d), Gillespie and Clague (2009)
Madagascar (585,000–587,000)	c. (6,400-)7,570–10,000 (-11,400) taxa, 8,884 species s. str.	Several evergreen and deciduous forest types (23 % of the area) including primary rain forest, xerophyllous forest, gallery forest, mangroves, secondary forest and other forest types, scrub (42 % of the area) including spiny and succulent thicket, degraded scrub (more than 25 % of the scrub area) and rupicolous shrubland, wetlands (2 % of the area) including swamps, rice cropland, other natural or semi-natural freshwater and wet brackish habitats, rock outcrops, grasslands including secondary savanna or pseudosteppe, coastal and urban habitats. Endemism is high in primary forest (89 %), rocky habitats (82 %), swamps (56 %), and lower in coastal habitats (21 %), grassland (little) and urban habitats (0 %).	Davis et al. (1994), Goodman and Benstead (2003), Burga and Zanola (2007), Moat and Smith (2007), Gillespie and Clague (2009), Cribb and Hermans (2009), Callmander et al. (2011)
Mascarene Islands (Mauritius with Rodrigues, Reunion, France; 4,481)	700–749 species	Tropical to subtropical vegetation, ranging from mangroves, lowland coastal and dry forests to moist montane and upland mist forests, and high-altitude heath.	Davis et al. (1994), Gillespie and Clague (2009)

Table 5.5 Examples of biodiversity hotspots, ecoregions and landscapes with numbers of endemic vascular plants and the main habitats in various parts of Australasia and Pacific Ocean Islands (partly nested or overlapping), from North to South or West to East, respectively

Region (country or archipelago, area in km^2)	Total numbers of vascular plant taxa (note that the term species is often used in the meaning of species plus subspecies)	Description of biomes, formations, or vegetation units	Reference
Japan (373,490)	(>222-)1,950–2,000 species	Boreal mixed or conifer forest to subtropical broadleaf evergreen forest with many different forest types (57 % of the area), 17 % of the area is covered by shrubland and grassland, 21.3 % cropland and urban habitats, 5 % wetland and water bodies.	Davis et al. (1995), Mittermeier et al. (2005), Conservation International and Duffy (2008), Gillespie and Clague (2009), Natori et al. (2012)
Yakushima (Japan; 503)	45 species and 27 subspecies/varieties – 94 species plus subspecies	Different forest types, with zones of evergreen forest, mixed *Cryptomeria* forest, *Rhododendron-Juniperus* scrub and subalpine vegetation.	Davis et al. (1995), Tokumara (2003), United Nations Environment Programme-Wo and Clough (2008)
Bonin-(Ogasawara) Islands (Japan; 73)	150–215 species	Subtropical evergreen broadleaved forests, scrub.	Davis et al. (1995), Mueller-Dombois and Fosberg (1998), Ito (1998)
Taiwan (including further 72 small islands; 36,210)	1,041–1,075 species	Tropical rain forest, evergreen broadleaved forest, mixed forest, coniferous forest, alpine grassland.	Davis et al. (1995), Mittermeier et al. (2005), Hsieh (2002), Huang (1993–2003)
Hainan Island (China; 34,000)	397 (−505) species	Tropical lowland seasonal rain forest, monsoon forest, savanna, montane seasonal rain forest, mangrove forest.	Fuwu et al. (1995), Francisco-Ortega et al. (2010)
Philippines (297,179)	3,470–5,832 ->6,091 species	Rain forest, montane forest, pine dominated cloud forest, seasonal forest (9 % of the area covered by forest), cropland and urban habitats represent 86 % of the area, wetlands and water bodies 5 %.	Davis et al. (1995), Myers et al. (2000), Mittermeier et al. (2005), WRI (2003), Gillespie and Clague (2009)

Palawan (Philippines; 14,896)	200–400 species	Lowland evergreen dipterocarp rain forest, lowland semi-deciduous forest, montane forest, forest over limestone, forest over ultramafic rocks, mangroves, beach forest.	Davis et al. (1995)
Sibuyan Island (Philippines; 445)	54 species	Lowland primary rain forest, montane forests, summit grassland, heath forest, mangroves, beach vegetation.	Davis et al. (1995)
Sundaland (from South Thailand to Java, Sumatra and Borneo; 1,500,000)	15,000 species	Lowland rain forest, beach forest, mangrove, peat swamp forest, alluvial bench forest and freshwater swamp, montane forest, scrubby, subalpine forest, bare exposed peaks of high mountain areas.	Mittermeier et al. (2005)
Sumatra (Indonesia; 472,610)	>1,200	Lowland evergreen rain forest, peat swamp forest, mangroves, montane rain forest, limestone vegetation, heath forest, montane graslands, subalpine vegetation.	Davis et al. (1995), Kalla (2003)
Borneo (Brunei Darussalam, Malaysia, Indonesia; 10,000,000)	6,000–7,500 species	Lowland tropical evergreen rain forest, montane forest, heath forest (kerangas), limestone forest, ultramafic vegetation, alluvial and peat swamp forests, mangroves, beach forest, subalpine and alpine vegetation including scrub and rocky habitats.	Davis et al. (1995), Barthlott et al. (2005), Gillespie and Clague (2009)
Java (Indonesia; 138,204)	c. 230–250 species	Submontane and montane forests, lowland forests.	Davis et al. (1995), Caujape-Castells et al. (2010)
Wallacea (Indonesia between Java, New Guinea and Australia; 338,494)	1,500 species	Tropical rain forest, savanna woodland, *Eucalyptus* forest, lowland forest on ultrabasic soils dominated by the myrtle family.	Mittermeier et al. (2005)
New Guinea (Irian Jaya, Papua New Guinea; 790,000)	9,900 (−16,000) species	Mangrove, freshwater, swamp and peatswamp forest, lowland alluvial and hill tropical rain forest, mountain rain forest, sedgeland, heath communities, subalpine forest and scrub, alpine grasslands and rocky habitats. Broadleaf forest covers about two-thirds of the island.	Myers (1988), Davis et al. (1995), Nau (2003), Barthlott et al. (2005), Gillespie and Clague (2009)

(continued)

Table 5.5 (continued)

Region (country or archipelago, area in km^2)	Total numbers of vascular plant taxa (note that the term species is often used in the meaning of species plus subspecies)	Description of biomes, formations, or vegetation units	Reference
East Melanesian Islands (from Bismarck Islands to Vanuatu; 99,384)	3,000–3,500 (–4,000) species	Coastal vegetation, mangrove forest, freshwater swamp forest, lowland rain forests, seasonally dry forests and grasslands, montane rain forest.	Mueller-Dombois and Fosberg (1998), Mittermeier et al. (2005), WWF Australia (2009)
Polynesia-Micronesia (47,000)	3,074 species	Strand vegetation, mesic forest, tropical rain forest, cloud forest, savanna, open woodland of grassland and scrub, wetland including mangrove forest.	Davis et al. (1995), Mueller-Dombois and Fosberg (1998), Allison and Eldridge (2004), Mittermeier et al. (2005)
Hawaiian Islands (USA; 16,641–16,760)	927–1,198 taxa (937 species, 124 subspecies, 137 varieties)	Tropical dry and moist forest, tropical low and high shrubland, herbland, grassland, coastal vegetation, wetland and bog.	Davis et al. (1995), Mueller-Dombois and Fosberg (1998), Wagner et al. (2005), Barthlott et al. (2005), Caujape-Castells et al. (2010)
Fiji Islands (18,275)	812–814 species	Rain forest, montane forest, dry forest and open woodland, shrub and grassland, mangroves.	Davis et al. (1995), Mueller-Dombois and Fosberg (1998)
Islands of Samoa (14 volcanic islands; Samoa, USA; 3,039)	150–165 species	Tropical rain forest most extensive is lowland forest, followed by montane forest, and cloud forest (>650 m). Some other minor habitat types including montane scrub, *Pandanus* scrub, littoral scrub, montane swamp forests, and summit scrub.	Davis et al. (1995), Mueller-Dombois and Fosberg (1998), Nau (2003), Honeycutt and McGinley (2008), Whistler (2011)
Kingdom of Tonga (scattered distribution of 170 islands; 750)	11–13 species	Dominated by tropical moist forest, <500 m: lowland broadleaf rain forest, >500 m: subtropical rain forest.	Mueller-Dombois and Fosberg (1998), World Wildlife Fund (2011)

Marquesas Islands (France; 1,275)	163–166 taxa (150 species, 2 subspecies, 14 varieties)	Coastal vegetation, para-littoral and lowland forest, dry forest at low or mid elevation (100–1,000 m), grassland (>600 m), moist and wet forest (up to 800 m), high elevation cloud forest(>1,000 m), wet shrubland or heathland (>1,200 m).	Davis et al. (1995), Florence and Lorence (1997), Mueller-Dombois and Fosberg (1998), Wagner and Lorence (1997), Lorence and Wagner (2011)
Australia (7,682,428)	14,260–19,870 species	Closed forest (0.4 %), open forest (4.6 %), woodlands (13.9 %), open woodlands (25.8 %), tall shrublands (31.1 %), low shrublands (6.2 %), hummock grassland (= spinifex grass steppe; 0.6 %), tufted grasses/graminoids (9.2 %), other herbaceous plants (6.4 %), littoral complex (0.3 %), other (1.5 %).	Green (1985), Davis et al. (1995), WRI (2003), Australian National Herbarium 2012: www.anbg.gov.au/aust-veg/australian-flora-statistics.html (assessed 20/9/2012)
Northern Territory (Australia; 1,420,970)	567 species	Rain forest (with 41 endemic sp.), non-rain forest habitats (with 526 endemic sp.).	Woinarski et al. (2006)
Kakadu-Alligator Rivers Region (Northern Territory, Australia; 30,000)	40 species	Tropical sclerophyll forest, woodland, rain forest, swamp forest, mangroves, saltmarsh, sedgeland, grassland.	Davis et al. (1995)
Central Australian Mountain Ranges (Australia; 168,000)	120 species	Grasslands, shrub steppe in saline areas, shrublands, woodlands, riparian vegetation in gorges and gullies, rock and cliff vegetation.	Davis et al. (1995)
Queensland Wet Tropics (Eastern Australia; 18,487)	576–654 species	Rain forest, sclerophyll shrubland, sclerophyll woodland, tall sclerophyll forest, woodlands, swamp forest, eucalypt forest, sclerophyll shrubland, dune vegetation, sedge swamp communities.	Davis et al. (1995), Mittermeier et al. (2005), UNEP (2011)
Northern Province of Western Australia (Australia; c. 320,000)	241 species	Savanna with Poaceae and scattered *Eucalyptus* trees.	Beard et al. (2000)
North Kimberley Region (Northern Province of Western Australia; 99,100)	102 species	High grass savanna, other savannas with short grasses, eucalypt woodland, mangroves, rain forest patches.	Davis et al. (1995), Government WA (2001)

(continued)

Table 5.5 (continued)

Region (country or archipelago, area in km^2)	Total numbers of vascular plant taxa (note that the term species is often used in the meaning of species plus subspecies)	Description of biomes, formations, or vegetation units	Reference
Eremaean Province (Western Australia; c. 1,600,000)	432 species	Vegetation cover varies from dry *Spinifex* grassland in the north to low *Acacia*-woodlands in the South.	Beard et al. (2000)
Southwest Australia (Australia, including Southwest Province; 356,717–489,944)	2,472–3,620 species	Woodland, *Eucalyptus* forest, Mediterranean shrubland (Kwongan); all vegetation types are dominated by almost entirely woody species (there is no grassland).	Davis et al. (1995), Beard et al. (2000), Hopper and Gioia (2004), Mittermeier et al. (2005), WWF Australia (2006)
Sydney Sandstone Region (New South Wales, Australia; 24,000)	c. 50 species	Evergreen sclerophyll forests, woodlands, shrublands, grasslands, coastal dunes and swamps, mangroves, small areas of rain forest.	Davis et al. (1995)
Kangaroo Island (Australia; 4,405)	36 species	*Eucalyptus* woodland, sclerophyll woodland, open and grassy woodland, and grassland.	Cleveland (2007)
New Caledonia (France; 18,972)	2,432–2,582 species	>1,750 endemics of the archepelago occur in wet evergreen forest (which recently covers c. 18–22 % of the area), 987 or more endemic species in natural and semi-natural maquis vegetation (which covers c. 30 % of the area), 223 endemic species in sclerophyllous forest (pristine stands cover less than 0.5 % of the area), 6 endemic species in savanna, 1 endemic species in mangrove (1 % of the area), 23 endemic species in littoral vegetation. Secondary vegetation including cropland (20 % of the area), savanna and secondary forest covers c. 50 % of the area. Endemism is much higher on ultrabasic substrates than elsewhere. Many endemics occur in more than one habitat.	Schneckenburger (1991), Morat (1993), Davis et al. (1995), Lowry (1998), Mueller-Dombois and Fosberg (1998), Mittermeier et al. (2005), WRI (2003), Gillespie and Clague (2009)

Norfolk and Lord Howe Islands (Australia; 54)	154 species and subspecies	Pine forest, mixed hardwood forest, palm/hardwood forest, palm/tree fern forest, evergreen rain forest, scrub vegetation, grassland, mangroves.	Davis et al. (1995), Mueller-Dombois and Fosberg (1998)
New Zealand (270,197)	1,865–1,944 species	Cover (area in %): high-productive exotic grassland 33–34, low-productive grassland 6, native forest 24, exotic forest 7, wetlands and water bodies 6, cropland and urban habitats 8–9. Forest and woody scrub harbour 506 endemics. Alpine habitats house 530 endemics (alpine and other grasslands some 180), mud banks and wetland margins 87, bogs 135, swamp, lagoons and lake margins 32, aquatic habitats 10.	Davis et al. (1995), McGlone et al. (2001), Connor (2002), Mittermeier et al. (2005), WRI (2003), De Lange et al. (2006, 2009)
Northland (New Zealand; 14,000)	60 taxa	Evergreen warm-temperate to subtropical forest, scrub, wetland communities, including mangrove forest, coastal communities.	Davis et al. (1995)
New Zealand areas with ultramafic soils (New Zealand)	34 species	North Cape: 15 endemics in low-growing shrubland (widespread) and forest; Nelson-Marlborough: 18 endemics in low forest, small leaved shrub, low scrub and heathland, tussock grassland; Otago-Southland: 1 endemic in stunted and patchy forest, low woodland and shrubs, subalpine and alpine tussock grassland.	Lee (1992), Mueller-Dombois and Fosberg (1998)
Chatham Islands (New Zealand; 963)	40 taxa	Evergreen cool temperate forest, mixed broadleaved forest, scrub, rush/scrubland, grasslands, peatlands, wetland and coastal communities.	Davis et al. (1995)
Subantarctic Islands (Auckland Islands, Campbell Islands, Antipodes Islands, Macquarie Island, Australia and New Zealand; 949)	35 taxa	Grasslands, fellfields, herbaceous communities, wetlands, forests, coastal vegetation.	Davis et al. (1995)

(continued)

Table 5.5 (continued)

Region (country or archipelago, area in km^2)	Total numbers of vascular plant taxa (note that the term species is often used in the meaning of species plus subspecies)	Description of biomes, formations, or vegetation units	Reference
Macquarie Island (Australia; 150)	3 species	Tundra vegetation dominated by herbaceous angiosperms and bryophytes (endemic species are not serpentinite species although serpentinite outcrops are common).	Adamson et al. (1993), Parks and Wildlife Service (2006)
Rapa Nui (Easter Island; 166)	3 species	Almost completely covered by herbaceous grassland, except for a few isolated stands of ornamental trees and shrubs.	Mueller-Dombois and Fosberg (1998), Honeycutt and McGinley (2007)
Galapagos Islands (Ecuador; 7,880)	180 species or 229 taxa (224 species, 5 subspecific taxa)	Mangroves and other coastal communities in the littoral zone, drought-tolerant scrub with cacti in the arid zone (up to 80–400 m), humid zones above 300 m with scrub and open forest, and fern-sedge vegetation and bogs with *Sphagnum* spp. above 900 m.	Lawesson et al. (1990), Davis et al. (1995), Mueller-Dombois and Fosberg (1998), McGinley (2007d), Gillespie and Clague (2009)
Islas Desventuradas (Chile; 3.9)	21 species	Dwarf forest (<4 m), scrub and patches of annual forbs and grasses.	Mueller-Dombois and Fosberg (1998)
San Ambrosio (Islas Desventuradas, Chile; 2.2)	11 species	Dwarf forest (up to 4 m) in the moister parts of the island, matorral, low scrubland, patches of annuals and herbaceous plants.	Mueller-Dombois and Fosberg (1998), McGinley (2007f)
San Felix (Islas Desventuradas, Chile; 1.4)	2 species	Dwarf shrub (matorral), patches of annuals.	Mueller-Dombois and Fosberg (1998), McGinley (2007f)
Juan Fernandez Islands (Chile; 100)	121–130 species	Subtropical montane rain forest, tall dry forests, tree fern forests, brushwood and scrub, subalpine heath-fern vegetation, primary and secondary grasslands, ridge, cliff and alpine communities, seashore vegetation.	Davis et al. (1995), Mueller-Dombois and Fosberg (1998), Gillespie and Clague (2009), Moreira-Muñoz (this book)

Most regions and landscapes with endemic vascular plants represent more than a single group of physiognomic habitats. Furthermore, many endemics occur in more than a single vegetation unit. Thus, it is very often difficult to determine how many endemics are related to a particular habitat type. Even those scientists who have expert knowledge of a certain region, such as a biodiversity hotspot or an ecoregion, might find it difficult to estimate where endemics are concentrated or which habitat type harbours most endemics.

We were only able to obtain information for a few regions on both the relationship between endemism and habitat type and the importance (size) of different habitat types within the region. These regions are e.g. Europe, Cape Floristic Region, Madagascar, New Caledonia, and New Zealand (cf. Tables 5.1, 5.2, 5.3, 5.4, and 5.5).

In Europe, most endemics are found in habitats with rocks and screes or in grasslands and scrub which together cover much smaller areas than cropland, forest or urban habitats. Most endemics in Europe are basiphytes or indifferent to soil-pH, but acidic substrates are dominant throughout the continent; acidic substrates harbour fewer endemics (Ewald 2003; Hobohm 2008a).

The predominant habitat in the Cape Floristic Region is fynbos. This habitat type harbours the largest number of endemics, almost all of which grow on nutrient-poor and acidic soils.

In New Caledonia, the majority of endemics inhabit wet evergreen forest which nowadays has been reduced to a fifth of its former range. A further large proportion of the endemics is found in the maquis vegetation which covers a third of the archipelago's area. Endemism on this archipelago is much higher on ultrabasic substrates than elsewhere.

On Madagascar, most endemics are woody plants that inhabit forest and woodland. The majority inhabit tropical and subtropical humid forest. Many others inhabit more or less dry woodlands and thickets. Today, secondary grassland and wooded grassland cover most of the island. These open vegetation types are extremely poor in endemics.

In New Zealand, the majority of the endemics are found in alpine habitats although low and mid-altitude habitats are dominant in the landscape. Forest, woody scrub, wet and aquatic habitats also house many endemics in New Zealand.

In the following we group habitats according to Davies et al. (2004) or Song and Xu (2003), respectively.

5.3.1 Coastal and Saline Habitats

Many coastal and saline habitats that extend in narrow strips along coasts or shorelines represent very old arrays of environmental conditions. The abiotic factors of certain regions in this transition zone might be as old as the ocean. However, marine currents, winds and migrating birds support dispersal processes which lead to a low likelihood of genetic isolation.

At a regional scale, the database for endemism in these habitats is largely fragmentary.

Analyses of the European endemic flora show that at least 450 endemic species and subspecies are restricted to coastal habitats (Hobohm 2008a; Hobohm and Bruchmann 2009, Hobohm, database EvaplantE in progress).

Van der Maarel and Van der Maarel-Versluys (1996) found that many of the coastal endemics are not necessarily restricted to coastal habitats but that the latter represent their optimal habitat. They found that about 30 % of the endemic species are dune and beach species and another 30 % are species of maritime rocks. It should be noted that this distribution pattern varies greatly, depending on the different coastal regions of Europe, as the coastlines extend from Arctic and Subarctic to Mediterranean regions (further analyses in Van der Maarel and Van der Maarel-Versluys 1996).

The Eastern mangroves of the Indian, Indo-Pacific and Pacific Ocean on the coasts of Indonesia, Madagascar, Southeast Asia and North Australia are more species-rich than western mangroves on the Atlantic Ocean coasts of Africa and America. This might also be the case for endemic species (Schroeder 1998; Wikramanayake et al. 2002). However, at present we can only speculate on this, and further research and data are indispensable.

5.3.2 Inland Waters, Mires, Bogs and Fens

5.3.2.1 Overview

Only few comprehensive analyses on endemic vascular plant taxa deal with lakes, rivers, bogs, mires, swamps, or other aquatic habitats. These are, for example, associated with the vernal pools of California, wetlands, mires, bogs, fens and swamps of Europe, aquatic and semiaquatic vegetation on Madagascar and New Zealand, and swamps and other freshwater habitats of the Nile Delta in Egypt (Bruchmann 2011; Hobohm and Bruchmann 2009; McGinley 2008h; Hobohm 2008b; Lazar 2004; Gautier and Goodman 2003; Ranarijaona in Goodman and Benstead 2003; McGlone et al. 2001; Ferry et al. 1999; Davis et al. 1994, 1995, 1997). Information about single endemic plants and their (aquatic or wet) habitats can be found on a number of websites (e.g. IUCN Red List data, www.iucnredlist.org) and many regional floras.

Only a few regions show moderate or high endemism associated with water bodies or wetlands. The vernal pools of California harbour c. 140 endemics in ephemeral freshwater communities and related vegetation types (Lazar 2004; Keeler-Wolf et al. 1998; Zedler 1990; Holland and Jain 1988).

In Baja California only few taxa are associated with water habitats or wetlands, e.g. the palm *Brachea armata* which is found in permanent oases (Vanderplank in lit.).

In the swamp habitats and other freshwater wetlands of the Nile Delta in Egypt McGinley (2008f) identified 8 endemic species (of a total of 553 plant species).

According to Ranarijaona (in Goodman and Benstead 2003: 251) c. 128 aquatic and semiaquatic vascular plant taxa are endemic to Madagascar. This number is relatively low compared to the total number of endemics on the island (8,000–10,000; see Gautier and Goodman 2003). Ferry et al. (1999) suggest that the low number of endemic aquatic plants on Madagascar may be a result of Quaternary climate fluctuations; during dry periods, freshwater and semi-aquatic habitats were essentially dry, eliminating local floras which depended on wet conditions. The small total area of water bodies and wetlands and the fact that lakes and running waters are normally relatively young and discrete units are not the best conditions for promoting endemism. Compared to other regions in the world the percentage of wetland inhabitants that are endemic to Madagascar is high. Of the 338 Malagasy vascular plants that belong to the flora of aquatic and semiaquatic vegetation, 128 (or 38 %) are endemics. The relatively high ratio can be explained by the fact that Madagascar is most likely the oldest island in the world and separated from the mainland of Africa about 165 million years ago (Goodman and Benstead 2003).

For New Zealand a figure of 264 endemic species is given for wetlands (McGlone et al. 2001). The relatively high number for the much smaller areas of New Zealand (269,000 km^2) compared to Europe (275 endemic taxa, 10,500,000 km^2) or Madagascar (128 endemic taxa, 587,000 km^2) might be explained by less marked effects of climate fluctuations, higher precipitation rates, and the higher proportion of wetland areas in New Zealand in general (De Lange et al. 2006, 2009; Connor 2002; Davis et al. 1994, 1995).

We assume that the global proportion of endemics in wetlands and water bodies is indeed very low (Photos 5.1, 5.2, and 5.3).

In the following we give an overview of the endemics which are associated with the wetlands of Europe.

5.3.2.2 Wet Habitats of Europe

We here present an analysis based on an earlier publication (Hobohm and Bruchmann 2011). The improvement of the list of endemics and discussions with colleagues resulted in a few changes. However, the main result of the earlier publication remain valid.

Analysis of Endemism Associated with Wet Habitats in Europe

The area covered here is Europe as defined in Fontaine et al. (2007). We divided Europe into 42 geographical units representing islands or groups of islands, nations, groups of small nations or, in the case of the former Soviet Union, parts of a nation (cf. Bruchmann 2011; Hobohm and Bruchmann 2009; Tutin et al. 1968–1993, see Fig. 5.2). The details are given in Bruchmann (2011).

Photo 5.1 *Colchicum figlalii*, extremely rare and restricted to temporarily inundated depressions in serpentinite areas of Turkey (Photographed by Gerhard Pils)

The data base EvaplantE (cf. Bruchmann 2011; Bruchmann and Hobohm 2010; Hobohm and Bruchmann 2009; Hobohm 2008a) contains information about most endemic vascular plant species or subspecies of Europe. The version EvaplantE 11/2012 shows 6,244 vascular plant taxa as endemics of Europe.

The taxonomic status and our species concept is primarily based on Flora Europaea (Tutin et al. 1968–1993) plus floras and lists of the Canary Islands, the Madeira archipelago and Cyprus (Borges et al. 2008; Izquierdo et al. 2004; Press and Short 1994; Meikle 1977, 1985). Many other regional and national floras were also used to gather information about habitats, ecological conditions, altitudes, and so on (cf. Bruchmann 2011; Hobohm 2008a, b; Kliment 1999).

In general, biogeographical analyses based on field data of plant compositions show some degree of bias. An imbalance in the biogeographical data arises as a result of geographically different perceptions and activities in the field and from the use of different taxonomies (different floras). We tried to minimise this problem by using a broad species concept and international floras and checklists, such as Flora Europaea (Tutin et al. 1968–1993, reprints 1996) and Euro + Med plantbase (cf. e.g. Greuter and Raab-Straube 2012) as primary sources.

We filtered and analysed the recent version of the data base (EvaplantE; version 11/2012) with a focus on European endemic vascular plants occurring under wet conditions or in the succession stages following inundation. We excluded taxa which primarily inhabit coastal habitats such as saltmarshes or rocky habitats

Photo 5.2 Nutrient-poor mire with carnivorous *Nepenthes madagascariensis* (S Madagascar near Fort Dauphin; photographed by Carsten Hobohm)

Photo 5.3 Freshwater lagoon with *Pandanus peyrierasii*, *Pandanus rollotii*, *Pandanus longistylus*, *Typhonodorum lindleyanum* (large leaves), *Barringtonia racemosa* (to the *left*). All but one Malagasy *Pandanus* spp. are endemic to the island (Mandena, S Madagascar; photographed by Carsten Hobohm)

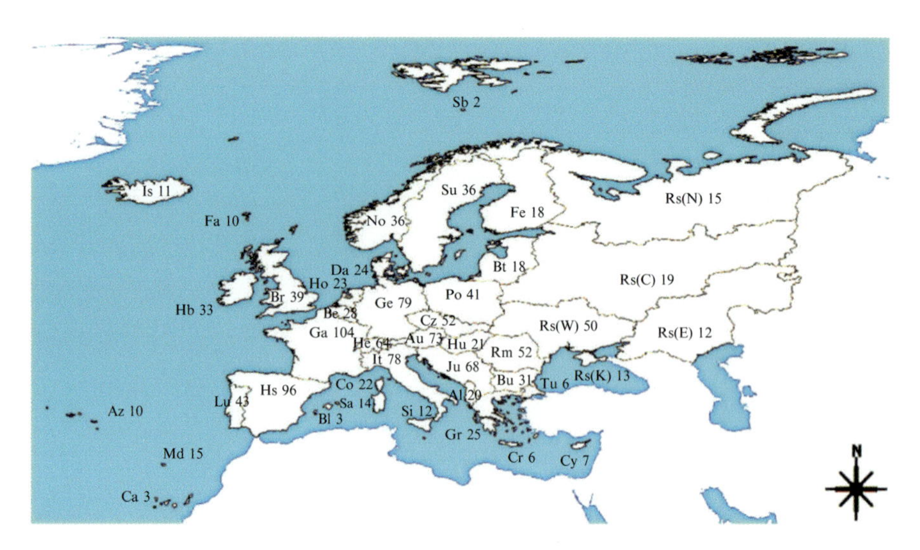

Fig. 5.2 Distribution of 275 endemic vascular plant taxa associated with wet or regularly inundated habitats in Europe

near the sea, taxa that only exceptionally occur in wet or inundated habitats, and apomictic microspecies (*Alchemilla*, *Taraxacum*, *Ranunculus auricomus*). Recently, we excluded *Marsilea azorica* which is conspecific with *M. hirsuta* from Australia. Thus, this species was reclassified from *endemic* to *neophytic* (Schaefer et al. 2011).

Additionally, we verified or corrected the nomenclature and geographical distribution for most taxa on our list using the Euro+Med plantbase on the internet (Euro+Med 2006–2011). In a few cases, this procedure had the effect that taxa changed their designation from *endemic* to *subendemic* because they have since been found in North African countries, such as Morocco, Algeria, Tunesia, or in parts of western Asia. These taxa were also eliminated from the list.

The analysis is founded on absolute numbers of taxa per region and descriptive statistics. As a first step, we summed up numbers of species or subspecies for the whole of Europe, and numbers of species and subspecies per region. The regions are of different sizes. At the moment it is impossible to reliably determine numbers of European endemic vascular plant taxa in relation to habitat for an artificially defined grid cell. This is the reason why certain statistical methods cannot be applied (see also the discussions of species-area and endemics-area relationships, e.g. in He and Hubbell 2011; Dengler 2009; Werner and Buszko 2005; Green and Ostling 2003). Preliminary investigations, for example, showed a negative correlation between area and endemism of vascular plants in Europe (log-log-space, all endemics) because many small regions in the South of Europe show high endemism, whereas very large regions in Scandinavia and Russia are poor in endemics. Further problems and restrictions related to the statistics are discussed in Bruchmann (2011).

If we compare numbers of regions of the same area then we can use these values as direct measurements for the density of endemics (Bruchmann 2011). Also in other cases we can compare density values directly (see Chap. 2). In our division of Europe this allows a direct comparison of density values of many pairs of regions.

Endemic Vascular Plants in Wetlands of Europe

The data base EvaplantE comprises at least 275 endemic plant taxa – species and subspecies – which occur more or less regularly in wetland communities (Table 5.6). All of these are restricted to the boundaries of Europe as defined in Fontaine et al. (2007). This number is a minimum value because c. 20 % of the listed taxa in EvaplantE have not yet, or have only inadequately, been characterized in relation to ecological conditions or habitat.

However, compared with endemics of other habitat types the number of 275 is low (Bruchmann 2011).

Only few of the taxa listed are hydrophytes living in standing or running waters. Many more taxa are not strongly associated with wet habitats and occur in both wet and other habitats (Hobohm and Bruchmann 2011, Table 5.6).

Figure 5.2 shows distribution patterns of endemic vascular plants related to wet habitats. The numbers in the South-west of Europe are higher than in the North-east. The numbers for a triangle-shaped region between Spain, former Yugoslavia and Germany are much higher than for the rest of Europe. The highest absolute numbers of taxa were found in France and Spain. Austria, Italy and Germany also have high numbers.

We assume that this fact reflects a combination of processes and conditions which favour endemism in general and endemism of wet habitats in particular (e.g. oceanic climate, high precipitation rate, humidity). Southern France is located within and between two high-mountain ranges with high environmental heterogeneity, the Pyrenees and the Alps. The country is also located between two marine environments which influence and stabilise the climate. Therefore, three major climate regimes occur in France: Mediterranean, temperate Atlantic and high-mountain climate. Mainland France is connected with other species-rich regions – e.g. the Iberian Peninsula, the Alps, Italy. Some of the Pleistocene refugia are located in the country or not far away. Thus, migration distances after glacial cycles might have been relatively short (Medail and Diadema 2009; Krebs et al. 2004).

The East of Europe shares many taxa with the West of Asia. Many landscapes, habitat types and ecological conditions are quite similar West and East of the border between Europe and Asia. The southern and south-eastern part of the continental border, in particular, is artificially defined (along the narrow Bosporus and along the Ural River). Thus, the marginal regions bordering Asia necessarily have fewer endemics than they would have if these regions were located in another part of Europe. This applies particularly to the European part of Turkey, which is rather small.

Table 5.6 Comparison of the density (E/A) of wetland endemics in different parts of Europe (Left column: increasing area; abbreviations refer to the geographic areas mapped in Fig. 5.2)

Higher density of endemics	Lower density of endemics (area increasing to the right side)
Madeira archipelago (Md)	Fa, Az, Bl, Ca, Cr, Cy, Tu, Sa, Si, Rs (K), Sb, Is, Rs (E), Rs (N)
Faroe Islands (Fa)	Az, Bl, Ca, Cr, Cy, Tu, Sb
Azores (Az)	Bl, Ca, Cr, Cy, Tu, Sb
Balearic Islands (Bl)	Ca, Sb
Canary Islands (Ca)	Sb
Crete (Cr)	Tu, Sb
Corsica (Co)	Cy, Tu, Sa, Si, Rs (K), Al, Sb, Hu, Is, Bt, Fe, Rs (E), Rs (N), Rs (C)
Cyprus (Cy)	Tu, Sb
Turkey (Tu)	Sb
Sardinia and Malta (Sa)	Si, Rs (K), Sb, Is, Rs (E)
Sicily (Si)	Sb, Is, Rs (E)
Crimean Peninsula Rs (K)	Sb, Is, Rs (E)
Albania (Al)	Sb, Is, Bt, Fe, Rs (E), Rs (N), Rs(C)
Belgium and Luxembourg (Be)	Ho, Da, Sb, Hu, Is, Gr, Bt, Fe, Rs (E), Rs (N), Rs (C)
Netherlands (Ho)	Sb, Hu, Is, Bt, Fe, Rs (E), Rs (N), Rs (C)
Switzerland (He)	Da, Sb, Hb, Lu, Hu, Is, Bu, Gr, Cz, Bt, Br, Rm, Po, No, Fe, Su, Rs (W), Rs (E), Rs (N), Rs (C)
Denmark (Da)	Sb, Hu, Is, Bt, Fe, Rs (E), Rs (N), Rs (C)
Ireland (Hb)	Hu, Is, Bu, Gr, Bt, Fe, Rs (E), Rs (N), Rs (C)
Austria (Au)	Lu, Hu, Is, Bu, Gr, Cz, Bt, Br, Rm, Ju, Po, No, Fe, Su, Rs (W), Rs (E), Rs (N), Rs (C)
Portugal mainland (Lu)	Hu, Is, Bu, Gr, Bt, Br, Po, No, Fe, Su, Rs (E), Rs (N), Rs (C)
Hungary (Hu)	Is, Bt, Fe, Rs (E), Rs (N), Rs (C)
Bulgaria (Bu)	Gr, Bt, Fe, Rs (E), Rs (N), Rs (C)
Greece without Crete (Gr)	Bt, Fe, Rs (E), Rs (N), Rs (C)
Czech Republik and Slovakia (Cz)	Bt, Br, Rm, Po, No, Fe, Su, Rs (W), Rs (E), Rs (N), Rs (C)
Estonia, Latvia, Lithuania and Oblast Kalingrad (Bt)	Fe, Rs (E), Rs (C), Rs (N)
Great Britain (Br)	No, Fe, Su, Rs (W), Rs (E), Rs (N), Rs (C)
Romania (Rm)	Po, No, Fe, Su, Rs (W), Rs (E), Rs (N), Rs (C)
Italy (It)	Ju, Po, No, Fe, Su, Rs (W), Rs (E), Rs (N), Rs (C)
former Yugoslavia (Ju)	Po, No, Fe, Su, Rs (W), Rs (E), Rs (N), Rs (C)
Poland (Po)	No, Fe, Su, Rs (E), Rs (N), Rs (C)
Norway (No)	Fe, Su, Rs (E), Rs (N), Rs (C)
Finland (Fe)	Rs (E), Rs (N)
Germany (Ge)	Su, Rs (W), Rs (E), Rs (N), Rs (C)
Sweden (Su)	Rs (E), Rs (N), Rs (C)
Spain (Hs)	Rs (W), Rs (E), Rs (N), Rs (C)
France (Ga)	Rs (W), Rs (E), Rs (N), Rs (C)
Former Soviet Union, southwestern division (Rs (W))	Rs (E), Rs (N), Rs (C)

These factors together might explain the relatively high numbers of European endemic vascular plant taxa related to wet habitats in France and neighbouring countries (cf. Rull 2004; Rosenzweig 1995a, b) and relatively low numbers to the East. However, this cannot easily been verified at present.

Table 5.6 shows the result of all (250) pairwise comparisons of the density of endemism (E/A) that can be calculated directly. Some regions, such as the Madeira archipelago, Corsica, Switzerland, Austria, Italy, Germany, mainland Spain and France are only to be found in the left column. These countries show relatively high density values. Others occur only in the right column: Regions such as Svalbard and the eastern, northern and central divisions of the former Soviet Union have relatively low density values. All other regions occur in both columns.

The comparison shows that the density of endemics increases roughly from the continental regions in eastern Europe to the more oceanic regions towards the west. However, Switzerland has a higher density value than Ireland. Austria, the Czech Republic and Slovakia have higher values than Great Britain.

The highest density values on a N-S gradient are in general not represented by the Mediterranean regions, e.g. Austria has a higher density than Portugal, Greece or former Yugoslavia, and even the Faroe Islands towards the North-west have a higher density than the Canary Islands, Crete, or Cyprus, to the South. In general, temperate regions do not have lower values than Mediterranean regions. However, the boreal-arctic regions towards the north, such as Svalbard, Iceland, Finland and northern Russia, have extremely low values.

We obtained altitudinal range data for 156 taxa on the list. Wetland endemics occur at all altitudes between sea level and alpine zones. The average of the minima (median) is 300 m above sea level, the average of the maxima 1,800 m. This means that most endemics occur in the montane and subalpine zones. *Gentiana bavarica* occurs in damp places, e.g. moors with spring water, wet alpine meadows, and snow patches. This species represents the absolute maximum of 3,600 m above sea level in our list.

Forty-one endemics are basiphytes, 31 acidophytes. Many species are indifferent to pH or have not yet been characterized (Table 5.7).

5.3.3 Forest and Woodland

Many publications cited in Tables 5.1, 5.2, 5.3, 5.4, and 5.5 name forest or woodland as the only or as one of the main habitat types – often in the warm-temperate and tropical zones of the world. Data for the boreal-arctic and cold temperate zones is scarce and seldom provides quantified information on patterns of endemism.

In the coastal forest of the Eastern Africa Biodiversity Hotspot many endemics have been recorded from forest habitats. However, it is difficult to give quantified data because numbers for the areas studied diverge considerably between sources. While Burgess et al. (1998, 2003, 2004) state that of the 1,750 strictly endemic species 554 species are confined to the coastal forest habitat and 812 species inhabit

Table 5.7 List of European endemics that are connected with wet habitats (Excerpt from EvaplantE, version 11/2012)

European endemic vascular plants which occur in wet habitats or after inundation (according to EvaplantE, last updated version 11/2012)	Habitats, ecology (plant communities)
Achillea asplenifolia	Wet lowland meadows (Molinion) and swampy grasslands
Achillea ptarmica	Damp grassy places, wet meadows, marshy fields, banks of running waters, tall forb communities and swamps (Molinietalia)
Achillea oxyloba	Mountain rocks, pastures, stoney river banks and screes, calcicole
Aconitum corsicum	Tall herb communities, edges of moors and swamps
Aconitum variegatum ssp. variegatum	Scrub, river banks, riparian forests, subalpine tall herb communties, moist or wet, nutrient- and base-rich substrates
Agropyron tanaiticum	Sandy river-banks, upper stream terraces, pioneer grass that profits from sodding activities
Allium schmitzii	River banks and moist rock-crevices
Allium suaveolens	Damp meadows and moors
Alyssum wulfenianum	Calcareous rocks and screes, river banks and gravel, secondary localities metalliferous substrates
Angelica heterocarpa	Muddy banks of tidal rivers
Angelica razulii	Banks of streams (tall herb communities), also in forest clearings and wet pastures
Arabis kennedyae	Shaded rocks near streams
Arabis soyeri	Alpine and subalpine spring vegetation and banks with permanent cold water
Arenaria gothica	Dry limestone pavement and lake shores
Armeria maritima ssp. purpurea	Marshy meadows and inundated lake-shores (Deschampsietum rhenanae, Cratoneurion, Primulo-Schoenetum)
Armoracia macrocarpa	Reeds, wet meadows, wet pastures and salt steppe (Scirpo-Phragmitetum, Phalaridetum arundinaceae, Agrostio-Alopecuretum pratensis), preferably on alkaline and nutrient-rich soils
Asparagus pseudoscaber	Temporarily inundated meadows
Baldellia alpestris	Ponds, glacial lakes, streamlets and peat bogs, acidic substrates or waters
Brassica glabrescens	Calcareous river banks, on gravel (pioneer communities)
Calamagrostis scotica	Bogs, marshes and fens
Callitriche brutia	Still, often shallow water
Callitriche hamulata	Base-poor, cool, flowing water and lakes (Batrachion fluviatilis, Nanocyperion flavescentis)
Callitriche platycarpa	Fresh water (rarely brackish water), flowing or still, often base-rich water
Callitriche truncata	Always submerged, including brackish waters

(continued)

Table 5.7 (continued)

European endemic vascular plants which occur in wet habitats or after inundation (according to EvaplantE, last updated version 11/2012)	Habitats, ecology (plant communities)
Calycocorsus stipitatus	Montane and subalpine zones, moorland, spring vegetation and river banks (Caricion fuscae, Cardamino-Montion)
Campanula pulla	Rocks and screes, stony slopes and mountain pastures, spring vegetation, secondary localities on river banks and river gravels, somewhat calcicole
Cardamine amara ssp. austrica	Streamsides, marshes and flushes
Cardamine amara ssp. balcanica	Streamsides, marshes and flushes
Cardamine amara ssp. opicii	Streamsides, marshes and flushes
Cardamine amara ssp. pyrenaea	Spring vegetation, streamsides, wet pastures
Cardamine asarifolia	Streamsides and other damp places, calcifuge
Cardamine crassifolia	Peatbogs and other wetlands
Cardamine raphanifolia	River banks and spring vegetation, also wet grassland, herb communties and humid forests
Carduus crispus ssp. multiflorus	Roadsides, waste places and streamsides
Carduus personata	Streamsides, meadows, scrubland and woods, tall herb communties (Arunco-Petasition, Rumicion alpini, Adenostylion, Alnetum incanae)
Carex bergrothii	Fens and wet woods
Carex camposii	Flushes and streamsides
Carex cretica	Damp places, banks, riparian *Platanus orientalis* woodland, steep wet banks (Adiantion capilli-veneris), spring vegetation and seepage meadows (Brachypodio-Holoschoenion)
Carex durieui	Damp grassland, pastures and moors
Carex frigida	Streamsides, trickling waters and wet mountain grassland
Carex fuliginosa ssp. fuliginosa	Alpine zone, rock-ledges, stream-sides and wet, stony places (Festucion variae)
Carex jemtlandica	Fens
Carex lowei	Among rocks and along ditches and small rivers in damp wooded valleys
Carex nevadensis	Wet places, river banks, wet slopes and mires
Carex pulicaris	Damp and marshy meadows and moors (Caricion nigrae, Caricion davallianae, Molinietalia)
Carex randalpina	Banks and wet slopes, preferably associated with *Carex acutiformis* or *Phalaris arundinacea* (Caricetum oenensis, Magnocaricion)
Carex trinervis	Damp maritime sands, wet dune slacks, wet heaths and fens (Caricetum trinervi-nigrae and Empetro-Ericetum)
Caropsis verticillatoinundata	Ponds and temporary ponds/lakes, peat bogs and other locations in sandy places and temporarily soaked near the seaside

(continued)

Table 5.7 (continued)

European endemic vascular plants which occur in wet habitats or after inundation (according to EvaplantE, last updated version 11/2012)	Habitats, ecology (plant communities)
Centaurea appendiculata	River sands
Centaurea arenaria ssp. sophiae	Sandy river-banks
Centaurea donetzica	River sands; pine forests
Centaurea konkae	River sands
Centaurea macroptilon	Molinietalia caeruleae
Centaurea paczoskii	Sandy steppes, open river sands
Centaurea protogerberi	River sands
Centaurea savranica	River sands and sandy steppe slopes
Centaurium microcalyx	Wet grasslands,temporarily inundated places, moors and other wet places
Cephalaria litvinovii	Ravines, among scrub, along river valleys, gullies
Cephalorhynchus cyprius (Cicerbita cypria)	In moist, shaded positions, by streams and moist hillsides among pine and riverine forest on igneous formations
Cerastium azoricum	Humid rocks in rivines and cliffs and near waterfalls
Cerastium brachypetalum ssp. doerfleri	Wet roadsides, screes, stony mountain slopes, river banks in the mountains
Ceterach lolegnamense	On roadside walls in hills, and on stream banks
Chaerophyllum elegans	Wet meadows, river banks, spring vegetation, tall herb communities, nutrient-rich, wet substrates
Chondrilla chondrilloides	River gravels, calcicole
Chrysosplenium alpinum	Mountain spring vegetation and flushes
Chrysosplenium oppositifolium	Spring vegetation in woods, wet ground near small rivers (Cardaminion amarae, Alnenion glutinoso-incanae)
Cirsium bourgaeanum	Marshes
Cirsium brachycephalum	Fens and wet meadows, saltmarshes (Cirsion brychycephali-Bolboschoenion, Loto-Trifolenion, Magnocaricion elatae)
Cirsium creticum ssp. triumfetti	Wet meadows and marshes
Cirsium dissectum	Wet places, usually on peaty soils; fens, bogs, wet fields on peaty soil (Juncion acutiflori)
Cirsium glabrum	Damp screes and streamsides
Cirsium rivulare	Damp places, calcifuge; wet meadows, swamps, forest glades, banks of streams and rivers, forest gullies
Cirsium tymphaeum	Spring vegetation and other damp places
Cochlearia pyrenaica	Near spring vegetation and permanently flowing, cold rivers, also in moors and wet forest
Cochlearia tatrae	Rocks, screes, spring vegetation, along streems, on gravel banks, restricted to granite (Oxyrio digynae-Saxifragetum carpaticae)
Coronilla globosa	Cliffs and river gravels
Coronopus navasii	Banks of ponds and ephemeral waters which occur during winter times

(continued)

Table 5.7 (continued)

European endemic vascular plants which occur in wet habitats or after inundation (according to EvaplantE, last updated version 11/2012)	Habitats, ecology (plant communities)
Cymbalaria hepaticifolia	Shady places and on rocks by streams
Cyperus cyprius	By streams and ditches with flowing water on igneous formations
Dactylorhiza alpestris	Damp meadows and fens
Dactylorhiza cordigera	Mountain grassland, also marshes
Dactylorhiza incarnata ssp. coccinea	Marshes, fens and bogs, dune-slacks and other damp alkaline sandy areas near sea, damp inland lake-shores
Dactylorhiza incarnata ssp. pulchella	Bogs and other neutral to acidic wet peaty places
Dactylorhiza islandica	Moorland, grassland and damp woods, mainly on acid soils
Dactylorhiza lapponica	Fens and marshes, wet meadows and banks, sometimes in open woods
Dactylorhiza maculata ssp. schurii	Moorland, grassland and damp woods, mainly on acid soils
Dactylorhiza majalis ssp. occidentalis	Damp meadows and fens
Dactylorhiza praetermissa	Damp meadows and fens
Dactylorhiza pseudocordigera	Fens and calcareous grasslands
Dactylorhiza purpurella	Damp meadows and fens
Dactylorhiza sphagnicola	Peat bogs, wet meadows, dune slacks and nutrient-poor reeds
Deschampsia littoralis	Periodically inundated lake shores, also on river banks
Deschampsia wibeliana	Estuaric marshes; sandy banks influenced by North sea tides and freshwater
Diphasiastrum madeirense	Moorlands, juniper forests and damp heathland
Doronicum cataractarum	Stream-sides and other shady places, preferably damp and wet tall herb communities in contact with cold freshwater of small rivers
Dryopteris aitoniana	Moist woods and along shady levadas in laurisilva regions
Dryopteris maderensis	Damp forest, beside levadas and streams
Eleocharis carniolica	River-banks and seasonally flooded grassy places
Elymus alaskanus ssp. subalpinus	River-banks, rocks and stony slopes
Elymus scandicus	River-banks, rocks and stony slopes
Epilobium alsinifolium	Humid and marshy places, near brooks, lakes and snow patches, and spring vegetation ((hygrophilous herb communities, Cardamino-Montion, Adenostylion, Cardaminion amarae)
Epilobium fleischeri	Pioneer-communities on subalpine river banks and moraines
Epilobium nutans	Subalpine wet spring vegetation and moors (Cardamino-Montion, Adenostylion, Cardaminion amarae, Drepanocladion exannulati)

(continued)

Table 5.7 (continued)

European endemic vascular plants which occur in wet habitats or after inundation (according to EvaplantE, last updated version 11/2012)	Habitats, ecology (plant communities)
Erica tetralix	Bogs, wet heaths and pinewoods; in the more humid and oceanic regions also in dryer habitats such as damp, shady walls
Erucastrum palustre	Marshes and rice fields
Eryngium viviparum	Places liable to winter flooding
Erysimum creticum	Gravelly riverbeds, roadsides and garigues
Euphorbia uliginosa	Temporary ponds, wet heath and other temporarily flooded or wet locations
Euphrasia calida	Apparently associated with warm spring vegetation
Euphrasia scottica	Wet moorland, fens and flushes
Festuca nitida	Calcicole; Petasition paradoxi
Festuca rubra ssp. thessalica	Mountain springs
Fraxinus pallisiae	River-banks and flood-plains
Galanthus nivalis	Open places wettet in spring, deciduous woodland and shady streamsides, forest edges, scrublands
Galeopsis pyrenaica	Sandy or gravelly ground by mountain streams, calcifuge
Genista berberidea	Damp meadows and bogs, wet banks with shrub communities
Gentiana bavarica	Damp places, spring vegetation and moors, snow patches and wet or damp alpine grassland
Gentiana clusii	Alpine grassland, screes, rocky habitats, moors
Geranium palmatum	Rocky cliffs and along levadas, usually in moist, shady places
Geranium yeoi	Woodland, on banks, and along levadas
Geum rhodopeum	Wet and marshy places, wet meadows
Goniolimon graminifolium	Sandy ground by rivers
Hemerocallis lilioasphodelus	Rocky mountain woods and riversides; naturalised by rivers and in wet meadows
Herniaria ciliolata	Coastal dunes and rocks, secondary localities at the edge of ephemeral ponds
Hierochloe hirta ssp. hirta	Nutrient-poor wet meadows, fens and lake-shores, also river banks, on sand or gravel
Hierochloe odorata ssp. baltica	Wet meadows, fens, riversides and lake-margins
Holcus gayanus	Ephemeral sources of rocky habitats in high mountain regions (Holco-Bryetum, Isoeto-Nanojuncetea)
Huperzia dentata	Damp, shady, rocky places in wooded slopes, gullies and banks above levadas, steep, often sandy slopes
Hypericum elodes	Damp or shallow water, standing waters in peat bogs
Inula helvetica	Woods and streamsides
Isoetes azorica	Pools and small lakes
Isoetes boryana	Shallow lakes
Isoetes heldreichii	Schistose soil of lake margin

(continued)

Table 5.7 (continued)

European endemic vascular plants which occur in wet habitats or after inundation (according to EvaplantE, last updated version 11/2012)	Habitats, ecology (plant communities)
Isoetes longissima	Rapidly flowing water
Isoetes malinverniana	Rapidly flowing water of irrigation channels
Isoetes setacea	Ponds and small lakes
Isoetes velata ssp. asturicense	Shallow lake-margins and seasonal pools
Isoetes velata ssp. tenuissima	Shallow lake-margins and seasonal pools
Jasonia tuberosa	Rock-crevices and river gravels
Juncus jacquinii	Alpine streamsides and damp grasslands, acidic substrates
Juncus requienii	Fens and streamsides
Juncus thomasii	Mountain forests, alpine meadows, also in bogs
Knautia godetii	Damp meadows and bogs
Lathyrus neurolobus	Woodland streams, steep wet banks (Adiantion capilli-veneris), spring vegetation and seepage meadows (Brachypodio-Holoschoenion)
Lathyrus palustris ssp. nudicaulis	River margins, ponds, streams, marshes, other damp or inundated places on all kinds of soils
Leontodon berinii	River gravels and banks (Epilobion fleischeri)
Leucojum vernum	In meadows, foothills, and upper limit of beech forests, in forest edges, river banks, and riparian meadows and herb communties
Limosella tenella	Muddy lake-shores
Lonicera nigra	Mountain spruce and beech forests, on forest edges, in scrublands, along riverbanks
Luzula sylvatica ssp. henriquesii	Damp woods, moorland and damp rocky places
Malcolmia graeca	Gravelly or rocky places, river banks, open phrygana, open *Abies cephalonica* forest, subalpine meadows, rock crevices
Malus praecox	Decidous woodland along rivers
Mentha longifolia ssp. cypria	Streams and spring vegetation
Myosotis lamottiana	Wet mountain meadows
Myosotis rehsteineri	Summer-inundated lake-margins, dry during spring and autumn (Deschampsion littoralis)
Najas mircrocarpa	Shallow water (associated with *Chara fragilis*)
Narcissus cyclamineus	Damp mountain pastures and banks of permanent watercourses, preferably with tree crown cover of alders and others
Narcissus jonquilla	Riparian zones, in wet meadows on the banks of rivers and in stony areas on streambed and river margins
Narcissus longispathus	In or near mountain streams
Narthecium ossifragum	Bogs and wet acidic heaths (Ericion tetralicis)
Narthecium reverchonii	Damp places by mountain streams
Noccaea cypria	Moist rocky slopes and igneous rocks near streams
Odontites kaliformis	Dry places, coastal pioneer communities and salt marshes, secondary localities inland at the edge of ephemeral waters

(continued)

Table 5.7 (continued)

European endemic vascular plants which occur in wet habitats or after inundation (according to EvaplantE, last updated version 11/2012)	Habitats, ecology (plant communities)
Oenanthe conioides	Estuaric pioneer communities; muddy banks influenced by North sea tides and freshwater
Oenanthe divaricata	By streams and levadas and among wet rocks
Oenanthe fluviatilis	In still or slowly flowing water
Oenanthe lisae	Marshes
Oenanthe tenuifolia	Marshes
Oenothera ammophila	Open, sandy and sometimes ruderal habitats, especially on seashores, dunes, sandy river banks
Oxytropis triflora	Alpine grassland and rocky habitats, secondary localities on river banks (pioneer communities)
Papaver laestadianum	Screes, barren rock outcrops, open gravel in the middle alpine belt, secondary localities along river banks in the lowland
Papaver sendtneri	Moving screes and river gravels, sometimes in rock crevices
Paradisea lusitanica	Woods, damp meadows and marshes
Pastinaca kochii ssp. latifolia	Riverbanks and rocky places
Pedicularis foliosa	Meadows, stream-sides and scrub (Caricion ferrugineae; Caricion davallianae, Erico-Pinion), calcicole
Pedicularis limnogena	Streamsides, spring vegetation and wet grassland
Pedicularis pyrenaica	Alpine humid pastures and screes; preferably on acidic substrates
Pedicularis recutita	Mountain meadows, river banks (Alnion viridis, willow scrub) and damp or shady places
Pedicularis sylvatica ssp. hibernica	Wet heaths and bogs, often also in dryer habitats
Petagnia gussonei	Alongside woodland streams
Petasites kablikianus	Wet gravel, stream-banks and wooded gorges (Petasition officinalis)
Petasites paradoxus	Stream-banks and wet stony ground, calcicole
Peucedanum gallicum	Nutrient-poor grassland, herb and scrub communities, river banks and fringe communties, preferably on acidic substrates
Peucedanum lancifolium	Wet meadows, marshes and bogs
Pilularia globulifera	Shallow water, marshy ground, pools and lake margins
Pilularia minuta	Seasonally wet hollows and at margins of ditches
Pinguicula grandiflora	Bogs and damp moorland, also wet rocky habitats, planted and persistent in scattered places of SW England
Pinguicula leptoceras	Wet places in the mountains, spring vegetation and moors
Pinguicula nevadensis	Bogs and wet places in the mountains
Plagius flosculosus	Damp meadows, pastures, swamps and river banks (Molinio-Juncetea, Juncetea maritimi)
Polygala amara	Subalpine and alpine rocks and stony slopes, moor-meadows, bogs and spring vegetation

(continued)

Table 5.7 (continued)

European endemic vascular plants which occur in wet habitats or after inundation (according to EvaplantE, last updated version 11/2012)	Habitats, ecology (plant communities)
Polygala amarella	Dry grassland, wet meadows, moorland, near spring vegetation
Primula clusiana	Rocks and stony slopes, nutrient-poor mountain grassland, moors, snow-patches, calcicole
Primula deorum	Wet places, peat bogs, banks of lakes, wet grassland (hygrophilous herb vegetation and pastures in the subalpine and alpine belts)
Primula frondosa	Marshes and damp meadows, wet, grassy places along brooks and peat bogs, shady cliffs, rock crevices, snow patches
Pseudorchis albida ssp. albida	Damp and marshy meadows, marshy tundra, peat bogs, pastures and grassy heaths, on mountain slopes
Ranunculus aconitifolius	Spring vegetation, streamlets and wet grasslands in montaneous regions, herb communities and open forests (Calthion, Adenostylion, Salicion herbaceae)
Ranunculus barceloi (*R. chaerophyllus var. balearicus*)	Banks and brooksides
Ranunculus cordiger	Moist moorland, grassland, and river banks
Ranunculus flammula ssp. minimus	All kinds of wet places
Ranunculus flammula ssp. scoticus	All kinds of wet places
Ranunculus fluitans	Running waters, non-freezing ditches, winter-green hydrophyte communities
Ranunculus hederaceus	Water plant communities, slow flowing gullies and small rivers
Ranunculus kykkoensis	Moist slopes with open pine, or golden oak stands and road banks
Ranunculus longipes	Shallow ponds and seasonal inundated places
Ranunculus montanus	Subalpine and alpine pastures, moor-meadows, open pine or deciduous forests on mountain slopes, eutrophic and often calcareous subtrates
Ranunculus ololeucos	Oligotrophic ditches, ponds and glacier lakes
Ranunculus platanifolius	Damp wooded ravines, tall herb communities (Alnetum viridis, Sorbo-Calamagrostietum, Betulo-Adenostyletea, Aceri-Fagetum, Tilio-Acerion, Berberidion)
Ranunculus revelieri	Small temporary waters (Isoetion)
Ranunculus wilanderi	Damp moss tundra below a bird cliff
Rheum rhaponticum	Wet mountain rocks; wet, rocky and stony places, along brooks and river beds, on silcate places
Rhododendron ponticum ssp. baeticum	In woods or by streams in the mountains; calcifuge
Romulea revelierei	Damp or seasonally inundated ground near the sea
Rumex balcanicus	Beside mountain streams
Sagina pilifera	Wet and damp pastures and moors, also other wet places (Saginetea piliferae)

(continued)

Table 5.7 (continued)

European endemic vascular plants which occur in wet habitats or after inundation (according to EvaplantE, last updated version 11/2012)	Habitats, ecology (plant communities)
Salix appennina	Damp woods and marshes
Salix bicolor	Subalpine spring vegetation and banks of small rivers
Salix cantabrica	Along mountain rivers
Salix daphnoides	Near rivers, ponds, lakes, and other wet places
Salix glabra	Subalpine and alpine forb and scrub communities, often close to small rivers or spring vegetation
Salix mielichhoferi	Along small rivers in the mountains (Alnion viridis), secondary localities in rocks and screes of high mountan regions, also in swamps
Salix repens ssp. repens	Coastal dunes, moorland and heaths, also beside sandy routes; ssp. *repens* in wet places such as dune slacks, ssp. *arenaria* also in wet places such as coastal reeds dominated by *Phragmites communis*.
Salix salviifolia	River margins with occasionally high water levels
Salix silesiaca	Riversides and damp clearings
Sanguisorba dodecandra	Subalpine meadows and streams
Sanguisorba lateriflora	Wet pastures, scrub, river banks, also near rocks and in screes
Saponaria cypria	Rocky mountainsides with forest openings, road banks, by streams and streamlets
Saponaria ocymoides	Subalpine, dry calcareous srees and pine forest, river gravel
Saussurea alpina ssp. macrophylla	Meadows, forest glades and edges, rocks and stony slopes, banks of rivers and streams
Saussurea esthonica	Meadows, forest glades and edges, rocks and stony slopes, banks of rivers and streams
Saussurea porcii	Subalpine meadows, swamps in alpine zone
Saxifraga aquatica	Margins of fast-flowing streams
Saxifraga clusii	Damp, shady rocks, springs and by mountain streams
Saxifraga hostii	Rock crevices and wet calcareous sinter, calcicole
Saxifraga hypnoides	Moist grassland, screes, streamsides and mountain-ledges, damp rock ledges, boulders and dunes and by mountain streams
Saxifraga mutata	Moist and wet rocks and screes, on river gravels
Saxifraga oppositifolia ssp. amphibia	During summer inundated lake-margins, on gravel (Deschampsietum rhenanae; most likely extinct)
Saxifraga spathularis	Damp rocks in the mountains, humid forests, small river banks, somewhat calcifuge
Saxifraga umbrosa	Shady banks, streamsides and mountain grassland, also in wet and shady forests
Schedonorus uechtritzianus	Damp grassland, river-banks and sea-shores
Scorzonera fistulosa	Wet meadows and other seasonally wet places

(continued)

Table 5.7 (continued)

European endemic vascular plants which occur in wet habitats or after inundation (according to EvaplantE, last updated version 11/2012)	Habitats, ecology (plant communities)
Scrophularia alpestris	Damp woods and by streams
Scrophularia hirta	Cliffs, banks, rocks, walls, along levadas and roadsides, usually in damp or wet places
Scrophularia racemosa	Wet rocks, steep slopes, along levadas and in other damp or wet places
Scrophularia trifoliata	Streamsides and other damp, shady places
Scutellaria minor	Damp places (Molinietalia caeruleae)
Securinega tinctoria	Sandy riverbanks and streambeds (shrub communities)
Sedum aetnense	Ephemeral ponds
Senecio doria ssp. legionensis (*S. fontanicola*)	Meadows and marshes, spring vegetation and fens
Senecio subalpinus	Damp places, wet meadows, river banks and forest glades in middle and upper mountain zones (Alnion viridis, Calthion, Caricion fuscae, Alnion incanae)
Sibthorpia peregrina	Woodland, along levadas, and on banks, in damp, shady places
Silene asterias	Marshy places
Silene laconica	Rocky limestone slopes, streambeds, roadside gravel, olive groves
Silene pusilla	Moist mountain rocks, screes and stream sides, subalpine and alpine spring vegetation, also on shady rocks in lower altitudes (preferably associated with bryophytes)
Silene saxifraga	Mountains, rocky habitats and sunny slopes, river gravels, prefering alkaline substrates
Sisymbrella aspera ssp. praeterita	Alluvial gravels and damp soils
Sisymbrium supinum	Muddy and gravelly lake shores (Agropyro-Rumicion, Chenopodion rubri)
Solanum patens	Near river beds and ravines, wet slopes of laurisilva forest
Soldanella pindicola	Close to mountain spring vegetation
Symphytum officinale ssp. uliginosum	River-banks and damp grasslands, tall herb communities on wet ground
Syringa josikaea	Along rivers, on wet, shady slopes, in ravines
Thalictrum morisonii	Marshes and wet grassland
Thalictrum speciosisimum	Humid and damp river banks, wet hay meadows
Tofieldia calyculata	Fen meadows and fens on alkaline ground (Caricion davallianae)
Tolpis azorica	Shady rocks; juniper forests and ravines, on grassy slopes and moorland
Tragopogon brevirostris ssp. bjelorussicus	River sands and pine woods

(continued)

Table 5.7 (continued)

European endemic vascular plants which occur in wet habitats or after inundation (according to EvaplantE, last updated version 11/2012)	Habitats, ecology (plant communities)
Tragopogon brevirostris ssp. longifolius	Damp and wet meadows
Tragopogon floccosus	River sands, maritime sands
Trichomanes speciosum	Near waterfalls and caves, and other damp or dark places such as levadas; this fern also is epiphytic
Trifolium saxatile	Screes, dry gravel in river beds and moraines
Trisetum fuscum	Moist mountain meadows and stream-sides
Vaccinium padifolium	On open slopes and moorland, also growing in light shade of humid wooded ravines
Veronica dabneyi	Steep slopes and rocks in the most humid regions, often close to waterfalls or crater lakes
Veronica repens	Damp places in mountains, nutrient-poor wet ground, moors and river banks
Viola cretica	Forests, stony pastures, river beds, spring vegetation

the surrounding non-forested habitats, Mittermeier et al. (2005) claim that some 1,225 (about 70 % of all endemic species in the hotspot region) have been recorded from forest.

Approximately one half of the 1,605–1,957 endemic species within the Chilean-Winter-Rainfall-Valdivian-Forest Hotspot belong to shrub and forest ecosystems, which constitute 50 % of the whole area (Mittermeier et al. 2005).

In some biodiversity hotspotswith high endemism, such as Wallacea or the Guinean forests of West Africa (Tables 5.3 or 5.5, respectively), tropical rain forest is also predominant (Mittermeier et al. 2005). In the wet tropics of Mesoamerica many endemic species inhabit rain forest, cloud forest and montane forest (e.g. Mittermeier et al. 2005) but it was not possible to find specific numbers at smaller scales for this region. However, according to Gentry (1986) in montane regions of Central America and especially the Andes, local endemism seems to result mostly from a veritable explosion of speciation in relatively few taxa, mostly of shrubs, herbs and epiphytes. These groups constitute almost half of the neotropical flora. Many of them are not associated with forest but can also be found there. And they account for most of the excess floristic diversity of the Neotropics compared with the Paleotropics. In lowland Amazonia endemism is prevalent to habitat islands. Most of the taxa involved are canopy trees and lianas, with derivative species in specialised habitats such as white sands or seasonally inundated swamp forests. In some cases, it is clear that local endemic habitat specialists are derived from wide-ranging ancestors of the terra firme forest. Habitat specialisation has obviously been the prevalent evolutionary pathway of giving rise to local endemics in Amazonia (Gentry 1986).

Photo 5.4 *Arbutus canariensis* (Los Tilos, La Palma) is endemic to the western Canary Islands (Photographed by Carsten Hobohm)

Photo 5.5 *Sonchus canariensis* (in a *Pinus canariensis* forest near Villaflor, Tenerife; photographed by Carsten Hobohm) is endemic to Tenerife and Gran Canaria, Canary Islands

Photo 5.6 The genus *Micronychia* (*Micronychia macrophylla* Ranomafana, Madagascar; photographed by Carsten Hobohm), including three species, is endemic to Madagascar (Schatz 2001)

Photo 5.7 *Ravenala madagascariensis* is a pioneer tree in humid forest regions of Madagascar. The endemic monotypic genus (Schatz 2001) is planted in many regions of the humid tropics today (Photographed by Carsten Hobohm; near Brickaville, Madagascar)

On New Caledonia (Table 5.5) endemism peaks at an extraordinarily high level, with the highest percentage numbers reached in humid evergreen and sclerophyllous forest (Lowry 1998; Morat 1993; Schneckenburger 1991).

Some authors claim that the wet tropics or the tropical rain forest harbour most (endemic) vascular plant species (e.g. Mittermeier et al. 2005; Parmentier et al. 2007). Very often, tropical rainforest is named as the centre of endemism, with extraordinarily high endemism; however, this assumption should be checked carefully, as results from several ecoregions show a contrary trend. For the Northern Territory of Australia, for example, Woinarski et al. (2006) showed that rain forest-associated species have less propensity for endemism (602 species with 41 endemics) than species associated with other habitats (3,611 species with 526 endemics). Knowledge of the occurrence, distribution patterns, and diversity of endemic vascular plants in forest ecosystems (all layers) is quite limited and sometimes fragmentary. The analyses of Kreft et al. (2004) indicate that knowledge of the structure and diversity of neotropical forests was for a long time based on incomplete floristic inventories which misjudged large groups of vascular plants.

5.3.4 Scrub, Heath and Succulent Shrubland

Shrub-dominated communities may either be a mature vegetation type (e.g. in subalpine zones or some types of drylands) or may occur as temporary communities (successional stages) that develop following intense disturbances or degradations of woodland or forest. If disturbances such as fire, clearing, or heavy grazing regimes are regular phenomena then the adapted shrub vegetation may remain stable over a long period of time, even if forest could grow there.

Several sources state that Mediterranean climate and semi-desert ecosystems in Southern Africa, Southwest Australia, California, the Mediterranean Basin, and Mexico harbour a high number of endemic vascular plants. Many endemic-rich areas of the world have been found to be areas of shrubland vegetation.

In the Cape Floristic Region the highest numbers of endemic taxa are mostly found in the dominant fynbos vegetation (Mittermeier et al. 2005; Davis et al. 1994; Cowling 1992).

On New Caledonia the second largest group of endemics is associated with maquis vegetation (Lowry 1998; Morat 1993; Schneckenburger 1991).

We assume that the ecological conditions in many shrub-dominated landscapes promote endemism, especially under soil conditions and precipitation rates that are relatively stable over long evolutionary time periods (Desmet 2007; Ojeda et al. 2001; Wisheu et al. 2000; Cowling and Hilton-Taylor 1999; Cowling et al. 1992, 1994, 1996). Strong influences of the major ocean surface currents in the neighbourhood might favour high climatic constancy in these regions; thus relatively low precipitation totals (including mist) in some shrub- and dwarf-shrub dominated semi-desert and Mediterranean-climate areas are highly predictable (Hobbs et al. 1995) (Photos 5.8 and 5.9).

Photo 5.8 Dry thicket (*above*) with *Adansonia rubrostipa* (with treads for gathering fruits), *Didierea madagascariensis* (*background*, comose) and *Euphorbia laro* (*right*) near Ifaty, Madagascar (Photographed by Carsten Hobohm)

Photo 5.9 *Euphorbia canariensis* (cactaceous habit), forming succulent scrub communities in semi-desert zones, is endemic to the Canary Islands (Barranco Seco, Tenerife, photographed by Carsten Hobohm)

5.3.5 Grassland and Habitats Dominated by Forbs, Mosses or Lichens

As Mucina and Rutherford stated, "*the term grassland is one of the most used, misused and abused terms of vegetation ecology*" ... (Mucina and Rutherford 2006: 350, see also Gibson 2009). In fact, the term grassland is often used as a generic term irrespective of the huge variety of environmental parameters, the different patterns of floristic or plant functional compositions, habitat extension and interface structures. Here we define grassland as a group of physiognomic habitats dominated by herbs and grasses which are not used as cropland. Sometimes shrubs or single trees occur but these cover less than 10 % of the area. This definition includes steppes, grassy savannas and paramo, and is consistent with other physiognomic classification systems (cf. EUNIS classification; Davies et al. 2004; White 1983).

Suttie et al. (2005) describe grassland ecosystems as one of the originally largest ecosystems in the world; estimates of the extent of this biome range from 31 to 40.5 % of the world's terrestrial area, depending on the definition of the term 'grassland'. Nowadays, large areas of grassland habitats are declining in quality and quantity, and some ecosystems are even dramatically threatened by extinction (Cremene et al. 2005; Mackay 2002; Dömpke and Succow 1998; Breymeyer 1990). Therefore, it is necessary that the great variety, and thus diversity, of grassland ecosystems should receive more attention than in the past (Photos 5.10, 5.11, 5.12, and 5.13).

The characteristic aspects of grassland-dominated ecosystems are widely influenced by their different developmental history. In the past, grassland was stabilised naturally either by specific climatic conditions such as low summer precipitation and winter drought, by periodical fire events, or by strong grazing regimes. Such habitats might be very old and thus characterised by long ecological continuity. On the other hand, there is the young semi-natural grassland that could potentially become forest if anthropogenic influences (cutting, burning, grazing, mowing) were less intensive (e.g. many grasslands at low- and mid altitudes in Europe).

The recent studies on occurrence of habitat specialists in grassland ecosystems show that, along with habitat persistence at the Holocene scale (Hájek et al. 2011), nutrient availability is also an important factor underlying specialist occurrence, while habitat-specialized grassland species tend to occur in nutrient-poor habitats (Fajmonová et al. 2012).

Mucina and Rutherford (2006) showed that some of the endemic-rich habitats of South Africa are linked to high altitude regions such as Drakensberg or Wolkensberg. The Drakensberg Alpine region, which is dominated by grassland, harbours about 334 endemic taxa (Mucina and Rutherford 2006) and for this reason it has been named a Centre of Plant Endemism (Van Wyk and Smith 2001). For many high altitude regions of the world, endemic-rich grassland such as alpine meadow, is mentioned in the literature, most often in connection with stony or rocky habitats (e.g. European Alps, Carpathians, Mountains of Central Asia and others). In the European flora, the group of grassland endemics is the second largest and

Photo 5.10 The composite *Cirsium dissectum* is endemic to W Europe and threatened by the decline in traditional management practices (Photographed by Carsten Hobohm; wet meadow in NW Germany)

Photo 5.11 *Orchis champagneuxii* is endemic to the Iberian Peninsula, S France and the Balearic Islands (Plus Morocco?; photographed by Carsten Hobohm in Serra da Estrela, Portugal)

Photo 5.12 *Phelypaea boissieri* is a parasite, endemic to Macedonia, Greece and Turkey and rare throughout almost all of its range (Photographed by Carsten Hobohm; subalpine grassland near Lake Prespa, Greece)

Photo 5.13 *Espeletia timotensis* is a character species of the paramos of the northern Andes (Photographed by Marinus Werger; Venezuela, Mucuchie, Piedras Negras, 4,100 m)

comprises many more taxa than e.g. forest ecosystems; this is not an area-effect because the area size of forest is much larger than that of grassland (Hobohm 2008a; Hobohm and Bruchmann 2009).

Beard et al. (2000) listed 241 endemic species for the savanna landscape of the Northern Province of Western Australia.

In summary, grasslands around the world are inhabited by a huge amount of endemics. Because this type of habitat is globally decreasing in area and quality many of the endemics are threatened with extinction (cf. Smolenice Grassland Declaration; EDGG 2010, and Hohhot Declaration 2008; on the internet).

5.3.6 Rock and Scree Habitats

5.3.6.1 Overview

Rock and scree habitats are found all over the world. Rocky habitats are dominant in many high mountain zones, in landscapes with steep slopes, and in coastal and desert regions. In these habitats differences in ecological conditions (light, water, wind speed, etc.) are much higher within small distances – a few meters – than in other habitat categories, e.g. grassland or forest. This also means that the impact of climate change might be small compared to habitats which are less heterogeneous (cf. Leuzinger et al. 2011).

The environmental conditions of rock and scree vegetation seem to have been relatively constant through time. Erosion and sedimentation patches are narrowly meshed. To find adequate ecological conditions under the pressure of climate change (glaciation periods and global warming) adapted vegetation types and endemic species would simply have to move a few hundred metres up or down the mountains (vertical displacement, Rull 2004). Rock and scree habitats as stepping stones that support the survival probability of the (endemic) species can be found at almost all altitudes of mountain ranges.

In the datasets analysed (Tables 5.1, 5.2, 5.3, 5.4, and 5.5) these habitats were not named very often as host areas for endemic species. On the other hand, in Europe, and possibly in many boreal-arctic and temperate regions worldwide, these are the habitats with very high or even the highest endemism. Studies on the endemic flora of Europe show that rocky habitats and screes represent the majority (>2,772 of c. 6,250–6,500 taxa) that are known to be endemic to Europe (Hobohm and Bruchmann 2009, data base EvaplantE in progress). In the Alps, 35–60 % of the 350–400 endemics species are found in habitats with rocks and screes, while in the Carpathians, which are not as high, only 10–15 % of 100–120 endemic species inhabit rocky habitats (Hobohm 2008a; Casazza et al. 2005, 2008; Dullinger et al. 2000; Pawlowski 1969). However, in Europe the highest endemism is confined to a habitat type which only represents a small percentage of the whole area (no exact data available yet). There is also data available on the endemic flora of the alpine regions of Iran (Noroozi et al. 2008) that reports a relationship between endemics

Photo 5.14 *Papaver sendtneri* is endemic to the Alps and shows preference for scree habitats (Photographed by Carsten Hobohm; Cogne Valley, Italian Alps)

and scree habitats. Talbot et al. (1999) list some endemic species of arctic regions that live in rock- and scree-dominated tundra vegetation.

Similarly high concentrations may exist in mountain areas of the tropics (e.g. the tropical Andes with about 15,000 endemic species) but no quantified data on endemics was found which could prove this assumption (Photo 5.14).

5.3.6.2 Comparison of High Endemism in Rocky Habitats with Low Endemism in Wetlands

In general, the number of endemics in wetlands is low compared to the numbers for rocky habitats and screes, grasslands, or scrub and heath landscapes (e.g. Bruchmann 2011; Bruchmann and Hobohm 2010 for Europe).

This fact should also be discussed in the context of zonal, extrazonal and azonal vegetation types. Walter (1954) defined zonobiomes as zonal vegetation types that are controlled by climate, orobiomes as different belts of mountain ranges, and pedobiomes as soil-dependent vegetation types that are azonal. For example, boreal spruce forest belongs to zonal vegetation, and aquatic vegetation, swamps, fens, bogs and riparian forest belong to azonal vegetation (cf. e.g. Bohn et al. 2000, 2003; Bailey 1998). However, the relationship between zonality and endemism is not clear. Some azonal vegetation types are very rich in endemics. Serpentine soils, for example, often give rise to sparse associations with many endemic plants (Chiarucci and Baker 2007; Alexander et al. 2007; Stevanović et al. 2003; Roberts and Proctor 1992). Rocky habitats which are strongly affected by both climate and substrate also harbour many endemic plants (Bruchmann 2011). In contrast, azonal vegetation types such as aquatic vegetation, reeds or bogs are normally poor in endemics (e.g. Parolly 2004; Meusel and Jäger 1992).

Wetlands cover a small part (a few percent) of the world's surface (Hobohm and Bruchmann 2009, 2011; Revenga et al. 2000). This is also true for rocks and screes which are inhabited by far more endemic taxa than is the case for wetlands. Thus, the differences in endemism cannot be explained by the size of area because both habitat types represent vegetation types which cover only a very small proportion of the earth.

In contrast to rocky habitats, most freshwater habitats, mires and swamps are very young, often much younger than 10,000 years. These habitats are characterised by low ecological continuity during the late Pleistocene and Holocene. Many lakes, for example, originated as dead ice holes during the late Pleistocene and silted up or developed into fens, bogs or peaty substrates covered by forest during the Holocene (Pott 2010; Hobohm and Bruchmann 2009). Furthermore, sudden changes in the physico-chemical conditions can impact whole water bodies. Perhaps as an adaptation to this, many aquatic and wetland plants, such as *Alisma plantago-aquatica*, *Eleocharis acicularis*, *Lemna minor*, *Phalaris arundinacea* or *Phragmites communis*, have long-distance dispersal abilities (wind, migratory birds) and very large ranges of distribution (Meusel and Jäger 1992; Meusel et al. 1978, 1965; Hultén 1971). In Europe, even the endemics of wetlands have, on average, larger ranges than endemics of rocks and screes (Hobohm and Bruchmann 2009). The relationship between the distribution patterns of specialists and refugial history of mires in the West Carpathians and Bulgaria, for example, has been shown by Hájek et al. (2011) and Horsák et al. (2007).

Rocky habitats, in general, are composed of different micro-habitats with very different environmental conditions with respect to light, water, organic material, soil, dynamics, etc. Under changing conditions, inhabitants of rocky habitats can normally find suitable survival conditions very close to the place where they are located. The higher rates of endemism here might be due to the relatively fixed structural complexity of their environment, compared with the dynamics of the lotic and lentic aquatic environment (see Scherrer and Körner 2011). Therefore, we assume that rocky habitats and screes are less affected by changing physico-chemical conditions than standing and running waters, banks, mires and swamps.

5.3.7 Arable, Horticultural and Artificial Habitats

In Europe a few regional endemics (e.g. *Anthemis lithuanica*, *Bromus secalinus ssp. multiflorus*, *Bromus interruptus*, *Carduus litigiosus*, *Centaurea polymorpha*, *Erucastrum gallicum*, and *Urtica atrovirens*; see Tutin et al. 1996a, b, c, d, e) might be largely restricted to anthropogenic habitats, such as arable land or ruderal

Photo 5.15 *Delonix regia*, endemic to W Madagascar, is widely cultivated in tropical regions (Private garden in S Madagascar near Fort Dauphin; photographed by Carsten Hobohm)

habitats, and so are not expected to occur in natural or seminatural habitats. Discussing the fact that certain plant species only occur in anthropogenic habitats, Gams (1938) and Pignatti (1978, 1979) explored the possibility that plant evolution is influenced by, or dependent on, human activities. Another explanation for the existence of endemic taxa in anthropogenic habitats could be that the original habitats of these taxa were destroyed, while associated endemics survived under similar environmental conditions. A combination of both explanations might also be correct.

413 European endemics occur in this group of relatively young but widespread habitats (Hobohm and Bruchmann 2009). However, we did not find much information about endemics in anthropogenic habitats outside of Europe.

Because of the duration of human influence we expect that this group of endemics might also exist in Africa and perhaps in Asia, but most probably neither in Australia nor in the Americas.

According to the IUCN Red List c. 63 vascular plant species on earth are extinct in the wild and still survive in horticulture. Rauer et al. (2000) have discussed the role of botanic gardens in protecting threatened vascular plant taxa (Photos 5.15 and 5.16).

Photo 5.16 *Melanoselinum decipiens*, species and genus endemic to Madeira, is widely cultivated for fodder on this island. Original habitats are shady rocks and banks in ravines of northern Madeira (Photographed by Carsten Hobohm near Fontes, Madeira)

References

Abbott I (1977) Species richness, turnover and equilibrium in insular floras near Perth, Western Australia. Aust J Bot 25:193–208

Adamson DA, Selkirk JM, Seppelt RD (1993) Serpentinite, harzburgite, and vegetation on subantarctic Macquarie Island. Arctic Alpine Res 25:216–219

Akpulat HA, Celik N (2005) Flora of gypsum areas in Sivas in the eastern part of Cappadocia in Central Anatolia, Turkey. J Arid Environ 61:27–46

Alexander EB, Coleman RG, Keeler-Wolf T, Harrison S (2007) Serpentine geoecology of western North America: geology, soils, and vegetation. Oxford University Press, New York

Allison A, Eldridge LG (2004) Polynesia-Micronesia. URL: http://multimedia.conservation.org/cabs/online_pubs/hotspots2. Downloaded 15 October 2012

Anádon-Irizarry V, Wege DC, Upgren A, Young R, Boom B, León YM, Arias Y, Koenig K, Morales AL, Burke W, Perez-Leroux A, Levy S, Koenig S, Gape L (2012) Sites for priority biodiversity conservation in the Caribbean Islands Biodiversity Hotspot. J Threatened Taxa 4(8):2806–2844

Ashworth AC, Vestal WD, Hokanson G, Joseph L, Martin M, McGlynn K, Newbrey MG, Schlecht N, Turnbull J, White A, Zimmerman T (2000) Tristan da Cunha Island Group and Gough Island. www.ndsu.edu/subantarctic. Last updated 2001

Ashworth AC, Vestal WD, Hokanson G, Joseph L, Martin M, McGlynn K, Newbrey MG, Schlecht N, Turnbull J, White A, Zimmerman T (2004) Biota Australis Terrestris. www.ndsu.edu/instruct/ashworth/subantarctic/index.htm. Last updated 01/2004.

Atamuradov H, Fet GN, Fet V, Valdez R, Feldman W (1999) Biodiversity, genetic diversity, and projected areas in Turkmenistan. J Sustain For 9:73–88

Bailey RG (1998) Ecoregions: the ecosystem geography of the oceans and continents. Springer, New York
Barthlott W, Mutke J, Rafiqpoor D, Kier G, Kreft H (2005) Global centers of vascular plant diversity. Nova Acta Leopoldina NF 92(342):61–83
Beard JS, Chapman AR, Gioia P (2000) Species richness and endemism in the Western Australian flora. J Biogeogr 27:1257–1268
Bhandari NN (1979) Phytogeography of the tropical flora of the Indian Desert. In: Larsen K, Holm-Nielsen LB (eds) Tropical botany. Academic, London, pp 143–152
Blom A (2001) Mount Cameroon and Bioko montane forests (AT0121). In: Word Wildlife Fund (ed) Wild world. URL: http://www.worldwildlife.org/wildworld/profiles/terrestrial/at/at0121_full.html
Bohn U, Neuhäusel R, Gollub G, Hettwer C, Neuhäuslová Z, Schlüter H, Weber H (2000/2003) Karte der natürlichen Vegetation Europas/Map of the natural vegetation of Europe. Maßstab/scale 1:2500000. Teil 1: Erläuterungstext mit CD-ROM. -Landwirtschaftsverlag, Bonn
Bolay E (1997) The Dominicain Republic. A country between rain forest and desert. Contributions to the ecology of a Caribbean Island. Markgraf, Weikersheim
Borges PAV, Abreu C, Aguiar AMF, Carvalho P, Jardim R, Melo I, Oliveira P, Sergio C, Serrano ARM, Vieira P (eds) (2008) A list of the terrestrial fungi, flora and fauna of Madeira and Selvagens archipelagos. Direccao Regional do Ambiente da Madeira and Universidade dos Acores, Funchal und Angro do Heroismo
Breckle S-W (2000) Biodiversität von Wüsten und Halbwüsten. Berichte der RTG 12:207–222
Breymeyer AI (ed) (1990) Managed grasslands (Ecosystems of the world 17A). Elsevier, Amsterdam
Brochmann C, Rustan OH, Lobin W, Kilian N (1997) The endemic vascular plants of the Cape Verde Islands, W. Africa. Sommerfeltia 24. Botanical Garden and Museum, University of Oslo, Oslo, 356pp
Bruchmann I (2011) Plant endemism in Europe: spatial distribution and habitat affinities of endemic vascular plants. Dissertation, University of Flensburg, Flensburg. www.zhb-flensburg.de/dissert/bruchmann/bruchmann_endemism.pdf
Bruchmann I, Hobohm C (2010) Halting the loss of biodiversity: endemic vascular plants in grassland of Europe. Grassland Sci Eur 15:776–778
Buckley RC (1985) Distinguishing the effects of area and habitat type on island species richness by separating floristic elements and substrate types and controlling for island isolation. J Biogeogr 12:527–535
Burga CA, Zanola S (2007) Madagaskar – Hot Spot der Biodiversität. Exkursionsbericht und Landeskunde. Schriftenreihe Physische Geographie Bodenkunde und Biogeographie 55:1–201
Burga CA, Klötzli F, Grabherr G (eds) (2004) Gebirge der Erde. Ulmer, Stuttgart, p 504 S
Burgess ND, Clarke GP, Rodgers WA (1998) Coastal forests of eastern Africa: status, endemism patterns and their potential causes. Biol J Linn Soc 64:337–367
Burgess N, Doggart N, Doody K, Negussie G, Sumbi P, Perkin A (2003) New information on the lowland coastal forests of Eastern Africa. Oryx 37:280–281
Burgess N, D'Amico Hales J, Underwood E, Dinerstein E, Olson D, Itoua I, Schipper J, Ricketts T, Newmann K (2004) Terrestrial ecoregions of Africa and Madagascar. Island Press, Washington, DC
Cable S, Cheek M (eds) (1998) Plants of mount Cameroon: a conservation checklist. Kew Publishing, Cumbria
Callmander MW, Phillipson PB, Schatz GE, Andriambololonera S, Rabarimanarivo M, Rakotonirina N, Raharimampionona J, Chatelain C, Gautier L, Lowry II, Porter P (2011) The endemic and non-endemic vascular flora of Madagascar updated. Plant Ecol Evol 144:121–125
Carbonao E, Lozano-Contreras G (1997) Endemismos y otras singularidades de la Sierra Nevada de Santa Marta, Colombia. Posibles causas de origen y necesidad de conservarlos. Revista de la Academia Colombiana de ciencias exactas, fisicas y naturales 21:409–419

Carbutt C, Edwards TJ (2006) The endemic and near-endemic angiosperms of the Drakensberg Alpine Centre. S Afr J Bot 72:105–132
Casazza C, Barberis G, Minuto L (2005) Ecological characteristics and rarity of endemic plants of the Italian Maritime Alps. Biol Conserv 123:361–371
Casazza G, Zappa E, Mariotti MG, Medail F, Minuto L (2008) Ecological and historical factors affecting distribution pattern and richness of endemic plant species: the case of the maritime and Ligurian Alps hotspot. Divers Distrib 14:47–58
Caujape-Castells J, Tye A, Crawford DJ, Santos-Guerra A, Sakai A, Beaver K, Lobin W, Vincent Florens FB, Moura M, Jardim R, Gomes I, Kueffer C (2010) Conservation of oceanic island floras: present and future global challenges. Perspect Plant Ecol Evol Syst 12:107–129
Chiarucci A, Baker AJM (2007) Advances in the ecology of serpentine soils. Plant Soil 293:1–2
Chown SL, Gaston KJ (2000) Reply from S.L. Chown and K.J. Gaston. In: Cleveland CJ (ed) Encyclopedia of earth. Trends Ecol Evol 15(12):514. URL: http://www.eoearth.org
Clark VR, Barker NP, Mucina L (2008) The Sneeuberg: a new centre of floristic endemism on the Great Escarpment, South Africa. S Afr J Bot 75:196–238
Clough LD (2008) Peninsula Valdés, Argentina. In: Cleveland CJ (ed) Encyclopedia of earth. URL: http://www.eoearth.org/article/Pennsula_Valds,_Argentina
Connor HE (2002) Regional endemism in New Zealand grasses. New Zeal J Bot 40(1):189–200
Conservation International, Duffy JE (eds) (2008) Biological diversity in Japan. In: Cleveland CJ (ed) Encyclopedia of earth. URL: http://www-eoaerth.org/article/Biological_diversity_in_Japan. Downloaded 20.10.2012
Cooper J, Ryan PG (1994) Management plan for the Gough Island Wildlife Reserve. Government of Tristan da Cunha, Edinburgh
Cowling RM (ed) (1992) The ecology of Fynbos. Nutrients, fire and diversity. Oxford University Press, Cape Town
Cowling RM, Hilton-Taylor CI (1999) Plant geography, endemism and diversity. In: Richard W, Dean J, Milton SJ (eds) The Karoo: ecological patterns and processes. Cambridge University Press, Cambridge, pp 42–56
Cowling RM, Lombard AT (2002) Heterogeneity, speciation/extinction history and climate: explaining regional plant diversity patterns in the Cape Floristic Region. Divers Distrib 8:163–179
Cowling RM, Holmes PM, Rebelo AM (1992) Plant diversity and endemism. In: Cowling RM (ed) The ecology of Fynbos. Nutrients, fire and diversity. Oxford University Press, Oxford, pp 62–112
Cowling RM, Witkowski ETF, Milewski AV, Newbey KR (1994) Taxonomic, edaphic and biological aspects of narrow plant endemism on matched sites in Mediterranean South-Africa and Australia. J Biogeogr 21:651–664
Cowling RM, Rundel PW, Lamont BB, Arroyo MK, Arianoutsou M (1996) Plant diversity in Mediterranean-climate regions. Trends Ecol Evol 11:362–366
Cremene C, Groza G, Rakosy L, Schileyko A, Baur A, Erhardt A, Baur B (2005) Alterations of steppe-like grasslands in Eastern Europe: a threat to regional biodiversity hotspots. Conserv Biol 19:1606–1618
Cribb P, Hermans J (2009) Field guide to the orchids of Madagascar. Kew Publishing, Kew
Crisp MD, Laffan S, Linder HP, Monro A (2001) Endemism in the Australian flora. J Biogeogr 28:183–198
Daniels FJA, Elvebakk A, Talbot SS, Walker DA (eds) (2005) Classification and mapping of arctic vegetation. Phytocoenologia 35(4):715–1079
Davies CE, Moss D, Hill MO (2004) EUNIS habitat classification, revised 2004. Report to European Environment Agency. European Topic Centre on Nature Protection and Biodiversity. Paris
Davis SD, Heywood VH, Hamilton AC (eds) (1994) Centres of plant diversity, vol 1, Europe, Africa, South West Asia and the Middle East. IUCN Publications, Unit, Cambridge
Davis SD, Heywood VH, Hamilton AC (eds) (1995) Centres of plant diversity, vol 2, Asia, Australasia and the Pacific. IUCN Publications, Unit, Cambridge

Davis SD, Heywood VH, Herrera-MacBryde O, Villa-Lobos J, Hamilton AC (eds) (1997) Centres of plant diversity, vol 3, The Americas. IUCN Publications, Unit, Cambridge
De Lange PJ, Sawyer JWD, Jensen CA (2006) New Zealand indigenous vascular plant checklist. New Zealand Plant Conservation Network, Wellington, 94pp
De Lange PJ, Norton DA, Courtney SP, Heenan PB, Barkla JW, Cameron EK, Hitchmough RA, Townsend AJ (2009) Threatened and uncommon plants of New Zealand. New Zeal J Bot 47(1):61–96
Dengler J (2009) Which function describes the species-area relationship best? A review and empirical evaluation. J Biogeogr 36:728–744
Desmet PG (2007) Namaqualand – a brief overview of the physical and floristic environment. J Arid Environ 70:570–587
Dhar U (2002) Conservation implications of plant endemism in high-altitude Himalaya. Curr Sci 82:141–148
Dömpke S, Succow M (1998) Cultural landscapes and nature conservation in Northern Eurasia. NABU, Bonn
Duarte MC, Rego F, Romeiras MM, Moreira I (2008) Plant species richness in the Cape Verde islands – eco-geographical determinants. Biodivers Conserv 17:453–466
Dullinger S, Dirnböck T, Grabherr G (2000) Reconsidering endemism in the North-eastern Limestone Alps. Acta Botanica Croatica 59:55–82
EDGG (ed) (2010) Smolenice grassland declaration. Bull Eur Dry Grassland Group 7:7
Euro+Med (2006–2011) Euro+Med PlantBase – the information resource for Euro-Mediterranean plant diversity. URL: www.bgbm.org/EuroPlusMed/. Downloaded 2/2012
Ewald J (2003) The calcareous riddle: why are there so many calciphilous species in the central European flora? Folia Geobot 38:357–366
Fajmonová Z, Zelený D, Syrovátka V, Vončina G, Hájek M (2012) Distribution of habitat specialists in semi-natural grasslands. J Veg Sci. doi:10.1111/jvs.12005
Ferry L, Robinson L, Ranarijaona H, Gasse F (1999) Les Lacs de Madagascar: présentation et typologie. Rapport Laboratoire Hydrologie, Montpellier
Fleischmann K (1997) Invasion of alien woody plants on the islands of Mahe and Silhouette, Seychelles. J Veg Sci 8:5–12
Fleischmann K, Héritier P, Meuwly C, Küffer C, Edwards PJ (2003) Virtual gallery of the vegetation and flora of the Seychelles. Bull Geobot Inst ETH 69:57–64
Florence J, Lorence DH (1997) Introduction to the flora and vegetation of the Marquesas Islands. Allertonia 7:226–237
Fontaine B, Bouchet P, Van Achterberg K, Alonso-Zarazaga MA, Araujo R, Asche M, Aspock U, Audisio P, Aukema B, Bailly N, Balsamo M, Bank RA, Barnard P, Belfiore C, Bogdanowicz W, Bongers T, Boxshall G, Burckhardt D, Camicas JL, Chylarecki P, Crucitti P, Davarveng L, Dubois A, Enghoff H, Faubel A, Fochetti R, Gargominy O, Gibson D, Gibson R, Gomez Lopez MS, Goujet D, Harvey MS, Heller K-G, Van Helsdingen P, Hoch H, De Jong H, De Jong Y, Karsholt O, Los W, Lundqvist L, Magowski W, Manconi R, Martens J, Massard JA, Massard-Geimer G, Mcinnes SJ, Mendes LF, Mey E, Michelsen V, Minelli A, Nielsen C, Nieto Nafria JM, Van Nieukerken EJ, Noyes J, Papa T, Ohl H, De Prins W, Ramos M, Ricci C, Roselaar C, Rota E, Schmidt-Rhaesa A, Segers H, Zur Strassen R, Szeptycki A, Thibaud J-M, Thomas A, Timm T, Van Tol J, Vervoort W, Willmann R (2007) The European Union's 2010 target: putting rare species in focus. Biol Conserv 139:167–185
Foster MN, Brooks TM, Cuttelod A, De Silva N, Fishpool LDC, Radford EA, Woodley S (2012) The identification of sites of biodiversity conservation significance: progress with the application of a global standard. J Threatened Taxa 4(8):2733–2744
Francisco-Ortega J, Wang Z-S, Wang F-G, Xing F-W, Liu H, Xu H, Xu W-X, Luo Y-B, Song X-Q, Gale S, Boufford DE, Maunder M, An S-Q (2010) Seed plant endemism on Hainan Island: a framework for conservation actions. Bot Rev 76(3):346–376
Fuwu X, Telin W, Zexian L, Huagu Y, Binghui C (1995) Endemic plants of Hainan Island. J Trop Subtrop Bot 3:1–12
Gamisans J, Marzocchi J-F (1996) La flore endémique de la Corse. Edisud, Aix-en-Provence

Gams H (1938) Die nacheiszeitliche Geschichte der Alpenflora. Jahrbuch der Vereinigung zum Schutze der Alpenpflanzen und -tiere 10:9–34

Gaston KJ (ed) (1996) Biodiversity: a biology of numbers and difference. Blackwell Science, Oxford

Gautier L, Goodman SM (2003) Introduction to the flora of Madagascar. In: Goodman SM, Benstedad JP (eds) The natural history of Madagascar. The University of Chicago Press, Chicago/London, pp 229–250

Gentry AH (1986) Endemism in tropical versus temperate plant communities. In: Soulé ME (ed) Conservation biology: the science of scarcity and diversity. Sinauer Associates, Inc.-Publisher, Sunderland, pp 153–182

Ghazanfar SA (2004) Biology of the central desert of Oman. Turk J Bot 28:65–71

Gibson DJ (2009) Grasses and grassland ecology. Oxford University Press, Oxford

Gillespie RG, Clague DA (eds) (2009) Encyclopedia of islands. University Press of California, Berkeley

Goodman SM, Benstead JP (eds) (2003) The natural history of Madagascar. The University of Chicago Press, Chicago/London

Grainger J (2003) 'People are living in the park'. Linking biodiversity conservation to community development in the middle east region: a case study from the Saint Katherine Protectorate, Southern Sinai. J Arid Environ 54:29–38

Green JW (1985) Census of vascular plants of western Australia, vol 2. W.A. Herbarium, Department of Agriculture, Perth

Green JL, Ostling A (2003) Endemics-area relationships: the influence of species dominance and spatial aggregation. Ecology 84(11):3090–3097

Greuter W, von Raab-Straube E (2012) Euro+Med Notulae, 6. Willdenowia 42:283–285

Groombridge B, Jenkins MD (2002) World atlas of biodiversity: earth's living resources in the 21st century. University of California Press, Berkeley

Hájek M, Horsák M, Tichý L, Hájková P, Dítě D, Jamrichová E (2011) Testing a relict distributional pattern of fen plant and terrestrial snail species at the Holocene scale: a null model approach. J Biogeogr 38:742–755

He F, Hubbell S (2011) Species-area relationships always overestimate extinction rates from habitat loss. Nature 473:368–371

Hendrych R (1982) Material and notes about the geography of the highly stenochoric to monotopic endemic species of the European flora. Acta Universitatis Carolinae-Biologica 335–372

Hobbs RJ, Richardson DM, Davis GW (1995) Mediterranean-type ecosystems: Opportunities and constraints for studying the function of biodiversity. In: Davis, GW, Richardson DM (eds) Mediterranean-type ecosystems. Ecol Stud 109:1–42. Springer, Berlin/Heidelberg/New York

Hobohm C (2000a) Biodiversität. Quelle & Meyer, Wiebelsheim

Hobohm C (2000b) Plant species diversity and endemism on islands and archipelagos, with special reference to the Macaronesian Islands. Flora 195:9–24

Hobohm C (2004) Ökologische Aspekte der Vielfalt endemischer Pflanzenarten in Europa unter besonderer Berücksichtigung großflächiger Beweidungsmaßnahmen als Instrumentarium für den Arten- und Biotopschutz. Schriftenreihe für Landschaftspflege und Naturschutz 78:281–292

Hobohm C (2008a) Ökologie und Verbreitung endemischer Gefäßpflanzen in Europa. Tuexenia 28:7–22

Hobohm C (2008b) Gibt es endemische Gefäßpflanzen in Mooren Europas? Mitt AG Geobot in Schl-Holst und Hambg 65:143–150

Hobohm C, Bruchmann I (2009) Endemische Gefäßpflanzen und ihre Habitate in Europa - Plädoyer für den Schutz der Grasland-Ökosysteme. Berichte der Reinh-Tüxen-Gesellschaft 21:142–161

Hobohm C, Bruchmann I (2011) Are there endemic vascular plants in wet habitats of Europe? Transylvanian Rev Syst Ecol Res 12:1–14

Holland RF, Jain SK (1988) Vernal pools. In: Barbour MJ, Major J (eds) Terrestrial vegetation of California, Special Publication 9. California Native Plant Society, Sacramento, pp 515–531

Holmgren M, Poorter L (2007) Does a ruderal strategy dominate the endemic flora of the West African forests? J Biogeogr 34:1100–1111
Honeycutt RL, McGinley M (2007) Rapa Nui and Sala-y-Gomez subtropical broadleaf forests. In: Cleveland CJ (ed) Encyclopedia of earth. URL: http://www.eoearth.org/article/Rapa_Nui_and_Sala-y-Gomez_subtropical_broadleaf_forests
Honeycutt RL, McGinley M (2008) Samoan tropical moist forests. In: Encyclopedia of earth. URL: http://www.eoearth.org/article/Samoan_tropical_moist_forests
Hopper SD, Gioia P (2004) The southwest Australian floristic region: evolution and conservation of a global hot spot of biodiversity. Annu Rev Ecol Evol Syst 35:623–650
Horsák M, Hájek M, Dítě D, Tichý L (2007) Modern distribution patterns of snails and plants in the Western Carpathian spring fens: is it a result of historical development? J Molluscan Stud 73:53–60
Hsieh CF (2002) Composition, endemisms and phytogeographical affinities of the Taiwan flora. Taiwania 47(4):298–310
Huang T-C (ed) (1993–2003) Flora of Taiwan, vols 1–6. National Taiwan University, Department of Botany, Taipai
Huang J-H, Chen B, Ying J-S, Ma K (2011a) Features and distribution patterns of Chinese endemic seed plant species. J Syst Evol 49(2):81–94
Huang J-H, Chen B, Liu C, Lai J, Zhang J, Ma K (2011b) Identifying hotspots of endemic woody seed plant diversity in China. Divers Distrib:1–16
Huber O (1995) Vegetation. In: Berry PE, Holst BK, Yatskievych K (eds) Flora of the Venzuelan Guayana. Missouri Botanical Garden and Timber Press, St. Louis, pp 97–160
Hultén E (1971) Atlas över växternas utbredning i Norden. Generalstabens Litografiska Anstalts Förlag, Stockholm
Huston MA (1994) Biological diversity. Cambridge University Press, Cambridge
Ito M (1998) Origin and evolution of endemic plants of the Bonin (Ogasawara) Islands. Res Popul Ecol 40:205–212
Izquierdo I, Martin JL, Zurita N & Arechavaleta M (eds) (2004) Lista de especies silvestres de Canarias (hongos, plantas y animales terrestres) 2004. Consejeria de Medio Ambiente y Ordenacion Territorial, Gobierno de Canarias
Jansson R (2003) Global patterns in endemism explained by past climatic change. Proc R Soc Lond 270:583–590
Jonsell B, Karlsson T (2004) Endemic vascular plants in Norden. In: Jonsell B (ed) Flora Nordica. General volume. Bergius Foundation, Stockholm, pp 139–159
Juste JB, Fa JE (1994) Biodiversity conservation in the Gulf of Guinea islands: taking stock and preparing action. Biodivers Conserv 3:759–771
Kalla J (2003) Submission for nomination of Tropical Rainforsést Heritage of Sumatra. URL: http://whc.unesco.org/uploads/nominations/1167.pdf
Keeler-Wolf T, Elam DR, Lewis K, Flint SA (1998) California vernal pool assessment: preliminary report, 161pp
Khan TI (1997) Biodiversity conservation in the Thar Desert, with emphasis on endemic and medicinal plants. Environmentalist 17:283–287
Kier G, Mutke J, Dinerstein E, Ricketts TH, Kuper W, Kreft H, Barthlott W (2005) Global patterns of plant diversity and floristic knowledge. J Biogeogr 32:1107–1116
Kim KOK, Hong SH, Lee YH, Na CS, Kang BH, Son Y (2009) Taxonomic status of endemic plants in Korea. J Ecol Field Biol 32(4):277–293
Kliment J (1999) Komentovaný prehlád vyšších rastlín flóry Slovenska, uvádzaných v literatúre ako endemické taxóny. Bull Slov Bot Spoločn Bratislava 21(suppl 4):1–434
Körner C (2000) Why are there global gradients in species richness? Mountains might hold the answer. Trends Ecol Evol 15(12):513
Krebs P, Conedera M, Pradella M, Torriani D, Felber M, Tinner W (2004) Quaternary refugia of the sweet chestnut (*Castanea sativa* Mill.): an extended palynological approach. Veg Hist Archaeobot 13:145–160

Kreft H, Köster N, Küper W, Nieder J, Barthlott W (2004) Diversity and biogeography of vascular epiphytes in Western Amazonia, Yasuni, Ecuador. J Biogeogr 31:1463–1476
Kruckeberg AR (1992) Plant life of western North America ultramafics. In: Roberts BA, Proctor J (eds) The ecology of areas with serpentinized rocks: a world view. Kluwer Academic Publishers, Dordrecht, pp 31–74
Lawesson JE, Hamann O, Rogers G, Reck G, Ochoa H (eds) (1990) Botanical research and management in Galapagos, vol 32, Monographs in systematic botany from the Missouri Botanical Garden. Missouri Botanical Garden, St Louis, 301pp
Lazar KA (2004) Characterization of rare plant species in the vernal pools of California. Master thesis, University of California, 344pp. www.vernalpools.org/documents/LazarThesisFinal.pdf
Lee WG (1992) New Zealand ultramafics. In: Roberts BA, Proctor J (eds) The ecology of areas with serpentinized rocks: a world view. Kluwer Academic Publishers, Dordrecht, pp 375–417
Lepping O, Daniels FJA (2006) Phytosociology of beach and saltmarsh vegetation in northern West Greenland. Polarforschung 76(3):95–108
Leuzinger S, Luo Y, Beier C, Dieleman W, Vicca S, Körner C (2011) Do global change experiments overestimate impacts on terrestrial ecosystems? TREE 26:236–241
Lowry II PP (1998) Diversity, endemism, and extinction in the flora of New Caledonia: a review. In: Peng CI, Lowry II PP (eds) Rare, threatened, and endangered floras of Asia and the Pacific Rim, Academia Sinica Monograph Series No. 16. pp 181–206
MacArthur RH, Wilson EO (1967) The theory of island biogeography. Princetown University Press, Princetown
Mackay R (2002) The atlas of endangered species: threatened plants and animals of the world. Earthscan, London
Major J (1988) Endemism: a botanical perspective. In: Myers AA, Giller PS (eds) Analytical biogeography. Chapman & Hall, London, pp 117–146
McGinley M (2007a) Bermuda subtropical conifer forests. In: Cleveland CJ (ed) Encyclopedia of earth. URL: http://www.eoearth.org/article/Bermuda_subtropical_conifer_forests
McGinley M (2007b) Caatinga. In: Cleveland CJ (ed) Encyclopedia of earth. URL: http://www.eoearth.org/article/Caatinga
McGinley M (2007c) Florida sand pine scrub. In: Cleveland CJ (ed) Encyclopedia of earth. URL: http://www.eoearth.org/article/Florida_sand_pine_scrub
McGinley M (2007d) Galapagos Islands xeric scrub. In: Cleveland CJ (ed) Encyclopedia of earth. URL: http://www.eoearth.org/article/Galpagos_Islands_xeric_scrub
McGinley M (2007e) Mediterranean conifer and mixed forests. In: Cleveland CJ (ed) Encyclopedia of earth. URL: http://www.eoearth.org/article/Mediterranean_conifer_and_mixed_forests
McGinley M (2007f) San Félix-San Ambrosio Islands temperate forests. In: Cleveland CJ (ed) Encyclopedia of earth. URL: http://www.eoearth.org/article/San_Flix-San_Ambrosio_Islands_temperate_forests
McGinley M (2007g) Sierra Madre del Sur pine-oak forests In: Cleveland CJ (ed) Encyclopedia of earth. URL: http://www.eoearth.org/article/Sierra_Madre_del_Sur_pine-oak_forests
McGinley M (2008a) Badkhiz-Karabil semi-desert. In: Cleveland CJ (ed) Encyclopedia of earth. URL: http://www.eoearth.org/article/Badkhiz-Karabil_semi-desert
McGinley M (2008b) Garamba National Park, Democratic Republic of Congo. In: Cleveland CJ (ed) Encyclopedia of earth. URL: http://www.eoearth.org/article/Garamba_National_Park,_Democratic_Republic_of_Congo
McGinley M (2008c) Gough Island Wildlife Reserve, United Kingdom. In: Cleveland CJ (ed) Encyclopedia of earth. URL: http://www.eoearth.org/article/Gough_Island_Wildlife_Reserve,_United_Kingdom
McGinley M (2008d) Granitic Seychelles forests. In: Cleveland CJ (ed) Encyclopedia of earth. URL: http://www.eoearth.org/article/Granitic_Seychelles_forests
McGinley M (2008e) Mount Emei and Leshan Giant Buddha, China. In: Cleveland CJ (ed) Encyclopedia of earth. URL: http://www.eoearth.org/article/Mount_Emei_and_Leshan_Giant_Buddha,_China

McGinley M (2008f) Nile Delta flooded savanna. In: Cleveland CJ (ed) Encyclopedia of earth. URL: http://www.eoearth.org/article/Nile_Delta_flooded_savanna
McGinley M (2008g) Tepuis. In: Cleveland CJ (ed) Encyclopedia of earth. URL: http://www.eoearth.org/article/tepuis
McGinley M (2008h) Venezuelan Andes montane forests. In: Cleveland CJ (ed) Encyclopedia of earth. URL: http://www.eoearth.org/article/Venezuelan_Andes_montane_forests
McGinley M (2008i) Moja desert. In: Cleveland CJ (ed) Encyclopedia of earth. URL: http://www.eoearth.org/article/Mojave_Desert
McGinley M (2009) Dong Phayayan Khao-Yai Forest Complex, Thailand. In: Cleveland CJ (ed) Encyclopedia of earth. URL: http://www.eoearth.org/article/Dong_Phayayan_Khao-Yai_Forest_Complex,_Thailand
McGlone MS, Duncan RP, Heenan PB (2001) Endemism, species selection and the origin and distribution of the vascular plant flora of New Zealand. J Biogeogr 28:199–216
Médail F, Diadema K (2009) Glacial refugia influence plant diversity patterns in the Mediterranean basin. J Biogeogr 36:1333–1345
Médail F, Verlaque R (1997) Ecological characteristics and rarity of endemic vascular plants from Southeastern France and Corsica: implications for biodiversity conservation. Biol Conserv 80:269–281
Meikle RD (1977/1985) Flora of Cyprus. The Bentham-Moxon Trust Royal Botanic Gardens, Kew
Meusel H, Jäger EJ (1992) Vergleichende Chorologie der zentraleuropäischen Flora. Text u. Karten. Bd. 3. Gustav Fischer Verlag, Stuttgart/New York
Meusel H, Jäger EJ, Weinert E (1965) Vergleichende Chorologie der zentraleuropäischen Flora. Text u. Karten. Bd. 1. VEB Fischer, Jena
Meusel H, Jäger EJ, Rauschert SW, Weinert E (1978) Vergleichende Chorologie der zentraleuropäischen Flora. Text u. Karten. Bd. 2. VEB Fischer, Jena
Miller AG, Morris M (2004) Ethnoflora of Soqotra Archipelago. Royal Botanic Garden, Edinburgh
Ministry for Natural Resources and the Environment (ed) (2007) National report on the status of biodiversity in S. Tomé and Principe. URL: www.cbd.int/doc/world/st/st-nr-03-en.pdf, 109pp
Mittermeier RA, Gil PR, Hoffman M, Pilgrim J, Brooks T, Mittermeier CG, Lamoreux J, da Fonseconda GAB (2005) Hotspots revisited: earth's biologically richest and most endangered terrestrial ecoregions. Cemex, Mexico City
Moat J, Smith P (eds) (2007) Atlas of the vegetation of Madagascar. Kew Publishing, Kew
Morat P (1993) Our knowledge of the flora of New Caledonia: endemism and diversity in relation to vegetation types and substrates. Biodivers Lett 1:72–81
Moustafa A, Zaghloul MS (1996) Environment and vegetation in the montane Saint Catherine area, south Sinai, Egypt. J Arid Environ 34:331–349
Mucina L, Rutherford MC (2006) The vegetation of South Africa, Lesotho and Swaziland. South African National Biodiversity Institute, Pretoria
Mueller-Dombois D, Fosberg FR (1998) Vegetation of the tropical Pacific Islands. Springer, New York
Myers N (1988) Threatened biotas: hotspots in tropical forests. Environmentalist 8:1–20
Myers N, Mittermeier RA, Mittermeier CG, da Fonseca GAB, Kent J (2000) Biodiversity hotspots for conservation priorities. Nature 403:853–858
Natori Y, Kohri M, Hayama S, De Silva N (2012) Key biodiversity areas identification in Japan Hotspot. J Threatened Taxa 4(8):2797–2805
Nau C (2003) Das Insel-Lexikon. Heel Verlag, Barcelona, 360S
Noroozi J, Akhani H, Breckle SW (2008) Biodiversity and phytogeography of alpine flora of Iran. Biodivers Conserv 17:493–521
Nowak A, Nobis M (2010) Tentative list of endemic vascular plants of the Zeravshan Mts in Tajikistan: distribution, habitat preferences and conservation status of species. Biodivers Res Conserv 19:65–80
Ojeda F, Simmons M, Arroyo J, Marañón T, Cowling RM (2001) Biodiversity in South African fynbos and Mediterranean heathland. J Veg Sci 12:867–874

Owiunji I, Nkuutu D, Kujirakwinja D, Liengola I, Plumptre AJ, Nsanzurwimo A, Fawcett K, Gray M, McNeilage A (2005) The biodiversity of the Virunga Volcanoes. URL: http://programs.wcs.org/portals/49/media/file/Volcanoes_Biodiv_survey.pdf. Downloaded 3/1/2012
Parks and Wildlife Service (2006) Macquarie island nature reserve and world heritage area management plan. Parks and Wildlife Service, Hobart
Parmentier I, Malhi Y, Senterre B, Whittaker RJ, Alonso A, Balinga MPB, Bakayoko A, Bongers FJJM, Chatelein C, Comiskey J et al (2007) The odd man out? Might climate explain the lower tree α-diversity of African rain forests relative to Amazonian rain forests? J Ecol 95:1058–1071
Parolly G (2004) The high mountain vegetation of Turkey – a state of the art report, including a first annotated conspectus of the major syntaxa. Turk J Bot 28:39–63
Pawar SS, Birand AC, Ahmed MF, Sengupta S, Shankar Raman TR (2007) Conservation biogeography in North-East India: hierarchical analysis of cross-taxon distributional congruence. Divers Distrib 13(1):53–65
Pawlowski B (1969) Der Endemismus in der Flora der Alpen, der Karpaten und der Balkanischen Gebirge im Verhältnis zu den Pflanzengesellschaften. Mitteilungen der ostalpin-dinarischen pflanzensoziologischen Arbeitsgemeinschaft 9:167–178
Pennington RT, Lewis GP, Ratter JA (eds) (2006) Neotropical savannas and seasonally dry forests. Plant diversity, biogeography, and conservation. CRC Press, Boca Raton
Pianka ER (1966) Latitudinal gradients in species diversity: a review of concepts. Am Nat 100:33–46
Pignatti S (1978) Evolutionary trends in Mediterranean flora and vegetation. Vegetatio 37:175–185
Pignatti S (1979) Plant geographical and morphological evidences in the evolution of the Mediterranean flora (with particular reference to the Italian representatives). Webbia 34:243–255
Plumptre AJ, Davenport TRB, Behangana M, Kityo R, Eilu G, Ssegawa P, Ewango C, Meirte D, Kahindo C, Herremans M et al (2007) The biodiversity of the Albertine Rift. Biol Conserv 134:178–194
Pott R (2010) Klimawandel im System Erde. Berichte der RTG 22:7–33
Pott R, Hüppe J, Wildpret de la Torre W (2003) Die Kanarischen Inseln. Ulmer, Stuttgart
Press JR, Short MJ (eds) (1994) Flora of Madeira. HMSO, London
Pryce M, Mitchell S, Burke A, McKenzie C, Stirling S, Ryan J, Simpson W, McGlashan D (2008) Jamaica: country report to the FAO international technical conference on plant genetic resources for food and agriculture. URL: www.moa.gov.jm/jam/jamaica2.pdf. Kingston, 59pp
Pyak AI, Shaw SC, Ebel AL, Zverev AA, Hodgson JG, Wheeler BD, Gaston KJ, Morenke MO, Revushkin AS, Kotukhov YA, Oyunchimeg D (2008) Endemic plants of the Altai mountain country. Wild Guides, Hampshire
Rabitsch W, Essl F (eds) (2009) Endemiten – Kostbarkeiten in Österreichs Pflanzen- und Tierwelt. Umweltbundesamt, Klagenfurt, Wien
Rapoport EH (1982) Areography: geographical strategies of species. Bergamon Press, New York
Rauer G, Ibisch PL, von den Driesch M, Lobin W, Barthlott W (2000) The convention on biodiversity and botanic gardens. In: Bundesamt für Naturschutz (ed) Botanic gardens and biodiversity. Landwirtschaftsverlag, Münster, pp 25–64
Raven PH, Axelrod DI (1978) Origin and relationships of the California flora. University of California Press, Berkeley
Republica de Cuba (ed) (2009) IV informe nacional al convenio sobre la diversidad biologica. URL: www.cbd.int/doc/world/cu/cu-nr-04-es.pdf, 162pp
Revenga C, Brunner J, Henninger N, Kassem K, Payne R (eds) (2000) Pilot analysis of global ecosystems: freshwater systems. World Resources Institute, Washington, DC
Ricklefs RE, Lovette IJ (1999) The roles of island area per se and habitat diversity in the species-area relationships of four Lesser Antiellean faunal groups. J Anim Ecol 68:1142–1160
Riemann H, Ezcurra E (2005) Plant endemism and natural protected areas in the Peninsula of Baja California, Mexico. Biol Conserv 122:141–150
Riemann H, Ezcurra E (2007) Endemic regions of the vascular flora of the peninsula of Baja California, Mexico. J Veg Sci 18:327–336

Roberts BA, Proctor J (eds) (1992) The ecology of areas with serpentinized rocks: a world view. Kluwer Academic Publishers, Dordrecht
Robertson SA (1989) Flowering plants of the Seychelles. Royal Botanic Gardens, Kew
Rohde K (1992) Latitudinal gradients in species diversity: the search for the primary cause. Oikos 65:514–527
Rosenzweig ML (1995a) Species diversity gradients: we know more and less than we thought. J Mammol 73:715–730
Rosenzweig ML (1995b) Species diversity in space and time. Cambridge University Press, Cambridge
Rull V (2004) Biogeography of the 'Lost World': a palaeoecological perspective. Earth Sci Rev 67:125–137
Schaefer H, Carine MA, Rumsey FJ (2011) From European priority species to invasive weed: *Marsilea azorica* (Marsileaceae) is a misidentified alien. Syst Bot 36(4):845–853
Schäfer H (2005a) Endemic vascular plants of the Azores: an updated list. Hoppea 66:275–283
Schäfer H (2005b) Flora of the Azores. Markgraf Publishers, Weikersheim
Schatz GE (2001) Generic tree flora of Madagascar. Royal Botanic Gardens/Missouri Botanical Garden, Kew/St. Louis
Scherrer D, Körner C (2011) Topographically controlled thermal-habitat differentiation buffers alpine plant diversity against climate warming. J Biogeogr 38:406–416
Schneckenburger S (1991) Neukaledonien. Palmengarten, Sonderheft 16: 78pp
Schroeder F-G (1998) Lehrbuch der Pflanzengeographie. Quelle & Meyer, Wiesbaden
Şekercioğlu ÇH, Anderson S, Akçay E, Bilgin R, Can ÖE, Semiz G, Tavşanoğlu Ç, Yokeş MB, Soyumert A, İpekdal K, Sağlam İK, Yücel M, Nüzhet Dalfes H (2011) Turkey's globally important biodiversity in crisis. Biol Conserv 144:2752–2769
Shou-qing Z (1988) The vulnerable and endangered plants of Xishuangbanna Prefecture, Yunnan Province, China. Arnoldiana 48(2):2–7
Song Y-C, Xu G-S (2003) A scheme of vegetation classification of Taiwan, China. Acta Botanica Sinica 45(8):883–895
Stebbins GL, Major J (1965) Endemism and speciation in the California flora. Ecol Monogr 35:2–35
Sterrer W (1998) How many species are there in Bermuda? Bull Mar Sci 62:809–840
Stevanović V, Tan K, Iatrou G (2003) Distribution of the endemic Balkan flora on serpentine I: obligate serpentine endemics. Plant Syst Evol 242(1):149–170
Stevens GC (1992) The elevation gradient in altitudinal range: an extension of Rapoport's latitudinal rule to altitude. Am Nat 140:893–911
Steyermark JA (1986) Speciation and endemism in the flora of the Venezuelan Tepuis. In: Vuilleumier F, Monasterio M (eds) High altitude tropical biogeography. Oxford University Press, New York, pp 317–373
Storch D (2000) Rapoport effect and speciation/extinction rates in the tropics. Trends Ecol Evol 15(12):514
Suttie JM, Reynolds SG, Batello C (2005) Grasslands of the world. Food and Agriculture Organisation of the United Nations, Rome
Talbot SS, Yurtsev BA, Murray DF, Argus GW, Bay C, Elvebakk A (1999) Atlas of rare endemic vascular plants of the arctic. In: Conservation of Arctic flora and fauna (CAFF). U.S. Fish and Wildlife Service, p. 73
Tokumaru H (2003) Nature conservation on Yakushima Island: Kagishima Prefecture's efforts. Global Environ Res 7(1):103–111
Tordorff AW, Baltzer MC, Fellowes JR, Pilgrim JD, Langhammer PF (2012) Key biodiversity areas in the Indo-Burma Hotspot: process, progress and future directions. J Threatened Taxa 4(8):2779–2787
Trinder-Smith TH, Cowling RM, Linder HP (1996) Profiling a besieged flora: endemic and threatened plants of the Cape Peninsula, South Africa. Biodivers Conserv 5:575–589

Tutin TG, Burges NA, Chater AO, Edmondson JR, Heywood VH, Moore DM, Valentine DH, Walters SM, Webb DA (1993 [1996a]). Flora Europaea, 2nd edn, vol 1, Psilotaceae-Platanaceae. Cambridge University Press, Cambridge

Tutin TG, Heywood VH, Burges NA, Moore DM, Valentine DH, Walters SM, Webb DA (1968 [1996b]) Flora Europaea, vol 2, Rosaceae-Umbelliferae. Cambridge University Press, Cambridge

Tutin TG, Heywood VH, Burges NA, Valentine DH, Walters SM, Webb DA (1968 [1996c]) Flora Europaea, vol 3, Diapensiaceae-Myoporaceae. Cambridge University Press, Cambridge

Tutin TG, Heywood VH, Burges NA, Valentine DH, Walters SM, Webb DA (1976 [1996d]) Flora Europaea, vol 4, Plantaginaceae-Compositae (and Rubiaceae). Cambridge University Press, Cambridge

Tutin TG, Heywood VH, Burges NA, Valentine DH, Walters SM, Webb DA (1980[1996e]) Flora Europaea, vol 5, Alismataceae-Orchidaceae. Cambridge University Press, Cambridge

United Nations Environment Programme, Clough LD (eds) (2008) Yakushima (Yaku Island), Japan. In: Cleveland CJ (ed) Encyclopedia of earth. URL: http://www.eoearth.org/article/Yakushima. Downloaded 20.10.2012

Valencia R, Pitman N, León-Yánez S, Jørgensen PM (eds) (2000) Libro Rojo de las Plantas Endémicas del Ecuador 2000. Publicaciones del Herbario QCA, Ponticicia Universidad Católica del Ecuador, Quito

Van der Maarel E, Van der Maarel-Versluys M (1996) Distribution and conservation status of littoral vascular plant species along the European coasts. J Coast Conserv 2:73–92

Van der Werff H, Consiglio T (2004) Distribution and conservation significance of endemic species of flowering plants in Peru. Biodivers Conserv 13:1699–1713

Van Wyk AE, Smith GF (2001) Regions of floristic endemism in southern Africa: a review with emphasis on succulents. Umdaus Press, Pretoria

Vetaas OR, Grytnes J-A (2002) Distribution of vascular species richness and endemic richness along the Himalayan elevation gradient in Nepal. Glob Ecol Biogeogr 11:291–301

Villarreal-Quintanilla JA, Encina-Dominguez JA (2005) Endemic vascular plants of Coahuila and adjacent areas, Mexico. Acta Botanica Mexicana 70:1–46

Vogiatsakis I (ed) (2012) Mediterranean mountain environments. Wiley-Blackwell, Oxford

Wagner WL, Lorence DH (1997) Studies of Marquesan vascular plants: introduction. Allertonia 7:221–225

Wagner WL, Herbst DR, Lorence DH (2005) Flora of the Hawaiian Islands – online. Smithsonia National Museum of History: http://ravenel.si.edu/botany/pacificislandbiodiversity/hawaiianflora/index.htm

Walter H (1954) Klimax und zonale Vegetation. Angewandte Pflanzensoziologie, Festschrift Aichinger 1:144–150

Werner U, Buszko J (2005) Detecting biodiversity hotspots using species-area and endemics-area relationships: the case of butterflies. Biodivers Conserv 14:1977–1988

Whistler A (2011) Rare plants of Samoa. Conservation International, Apia

White F (1983) The vegetation of Africa: a descriptive memoir to accompany the UNSECO/AEFTAT/UNSO vegetation map of Africa. United Nations Educational, Scientific and Cultural Organization, Paris

Wikramanayake E, Dinerstein E, Loucks CJ, Olson DM, Morrison J, Lamoreux J, McKnight M, Hedao P (2002) Terrestrial ecoregions of the Indo-Pacific. A conservation assessment. Island Press, Washington, DC

Wisheu IC, Rosenzweig ML, Olsvig-Whittaker L, Shmida A (2000) What makes nutrient-poor mediterranean heathlands so rich in plant diversity? Evol Ecol Res 2:935–955

Woinarski JCZ, Hempel C, Cowie I, Brennan K, Kerrigan R, Leach G, Russell-Smith J (2006) Distributional pattern of plant species endemic to the Northern Territory, Australia. Aust J Bot 54:627–640

World Wildlife Fund (ed) (2011) Tongan tropical moist forests. The Encyclopedia of Earth. URL: http://www.eoearth.org/article/Tongan_tropical_moist_forests. Downloaded 15 July 2012

WRI (World Resources Institute) (2003) Earth Trends. Country Profiles. URL: http://www.eoearth.org (downloaded for many regions; May-October 2012)
Yena AV (2007) Floristic endemism in the Crimea. Fritschiana 55:1–8
Zedler PH (1990) Life histories of vernal pool vascular plants. In: Ikeda DH, Schlising RA (eds) Vernal pool plants: their habitat and biology, Studies from the herbarium no. 8. California State University, Chico, pp 123–146

Chapter 6
Endemism on Islands – Case Studies

Andrés Moreira-Muñoz, Sergio Elórtegui Francioli, Carsten Hobohm, and Miguel Pinto da Silva Menezes de Sequeira

6.1 Juan Fernández Archipelago

Andrés Moreira-Muñoz (✉)
Instituto de Geografía, Pontificia Universidad Católica de Chile, Santiago, Chile

Sergio Elórtegui Francioli
Facultad de Ciencias de la Educación, Pontificia Universidad Católica de Chile, Santiago, Chile

6.1.1 Introduction

The Juan Fernández Archipelago is located around latitude 34°S offshore Chile, west of the Southern Cone of South America (Fig. 6.1). It comprises three small oceanic islands: Isla Robinson Crusoe (also known as Masatierra), situated at

A. Moreira-Muñoz (✉)
Instituto de Geografía, Pontificia Universidad Católica de Chile, Santiago, Chile
e-mail: asmoreir@uc.cl

S. Elórtegui Francioli
Facultad de Ciencias de la Educación, Pontificia Universidad Católica de Chile, Santiago, Chile
e-mail: saelorte@uc.cl

C. Hobohm (✉)
Ecology and Environmental Education Working Group, Interdisciplinary Institute of Environmental, Social and Human Studies, University of Flensburg, Flensburg, Germany
e-mail: hobohm@uni-flensburg.de

M.P. da S.M. de Sequeira (✉)
Centro de Ciências da Vida, Universidade da Madeira, Funchal, Portugal
e-mail: miguelmenezessequeira@gmail.com

C. Hobohm (ed.), *Endemism in Vascular Plants*, Plant and Vegetation 9,
DOI 10.1007/978-94-007-6913-7_6,

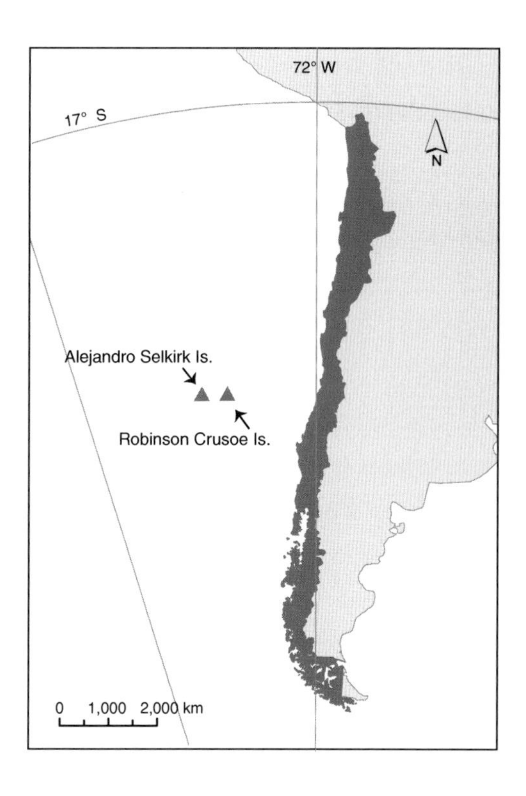

Fig. 6.1 The geographical position of the Juan Fernández Archipelago offshore Chile

670 km from the South American continent and encompassing 49 km^2; the little Isla Santa Clara (2.2 km^2), situated to the SSW of Robinson Crusoe; and Alejandro Selkirk (also known as Masafuera), situated 167 km to the west of the former two, and occupying an area of 53 km^2. The biotic and especially botanical value of these islands has long being recognized (Johow 1896; Skottsberg 1953; Muñoz Pizarro 1969; Marticorena et al. 1998; Danton 2006). Due to the steady suggestions from early botanists, the archipelago has been declared a National Park in 1935, also getting the status of Biosphere Reserve in 1977.

Sadly, the history of intensive human occupation and introduction of pests has transformed this territory in one of the most threatened all over the world. Only in recent times nine species extinctions have been documented (Danton and Perrier 2005, 2006), and most of the endemic species are in critical danger according to national assessment following IUCN criteria.

6.1.2 Overall Plant Endemism of the Archipelago

The archipelago's vascular flora comprises around 210 native species classified in 55 families and 110 native genera (Marticorena et al. 1998; Danton and Perrier 2006). Among these taxa, the islands contain 2 endemic families

Fig. 6.2 *Dendroseris litoralis* (Drawing from Sierra Ráfols in Muñoz Pizarro 1971)

(Lactoridaceae and Thyrsopteridaceae), 14 endemic plant genera, and 130 endemic taxa. This number of local endemic taxa is one of the highest levels of endemism per surface area reported globally for an island territory (Stuessy et al. 1992; Vargas et al. 2011). Most endemics pertain to the angiosperms, although many fern species also occur exclusively on these islands, including several species of Hymenophyllaceae, Blechnaceae and Dryopteridaceae among other endemic ferns such as the charismatic *Dicksonia* species. The endemic angiosperm species pertain mostly to the Asteraceae (28 species), including 22 species from the four endemic genera *Dendroseris*, *Robinsonia*, *Yunquea* and *Centaurodendron*. The species of these four genera have a remarkably growth form described as "rosette trees" (Skottsberg 1953; Hallé et al. 2004; Figs. 6.2, 6.3, and 6.4, Photo 6.1), which is rather uncommon among the Asteraceae.

The overall endemism for the archipelago is not homogenous across the three islands: Robinson Crusoe (RC) Island is the one that contains most endemic species, reaching a number of 90. From these, most pertain to the Asteraceae

Fig. 6.3 *Robinsonia gracilis* (Drawing from Sierra Ráfols in Muñoz Pizarro 1971)

(19 species), followed by the Hymehoplyllaceae (5 species). Conspicuous endemic species from the archipelago are restricted to this island, like *Lactoris fernandeziana* (Lactoridaceae), *Juania australis* (Arecaceae), three *Eryngium* species (Apiaceae), the two *Centaurodendron* species, the only one species from the genus *Yunquea* (Asteraceae), and most of the *Robinsonia* and *Dendroseris* species (Asteraceae).

Alejandro Selkirk (AS) Island contains less endemic species than Masatierra (Robinson Crusoe), but with 65 endemic species the number is considerable. They pertain mostly to the Asteraceae (6 *Erigeron* species, 3 *Dendroseris* species and 1 *Robinsonia*). From these endemic Asteraceae only *Erigeron fernandezianus* is also found in Robinson Crusoe Island. Conspicuous species not found elsewhere are the only orchid of the archipelago: *Gavilea insularis* (Photo 6.3), and also *Peperomia skottsbergii*, *Haloragis masafuerana* (including two varieties) and *Gunnera masafuerae*. Indeed, many species endemic to this island have been named with the specific epithet "*masafuerana*" or "*masafuerae*".

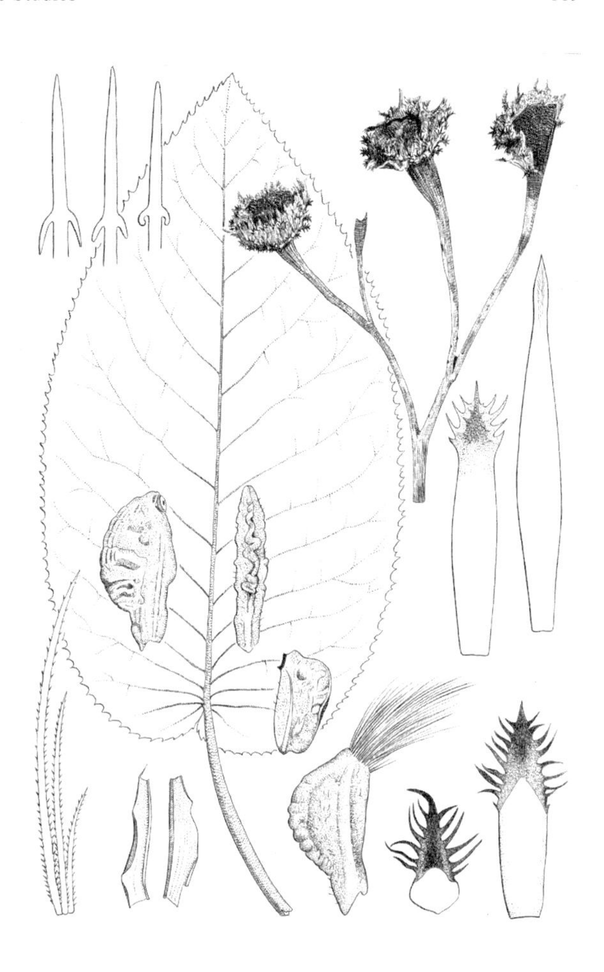

Fig. 6.4 *Yunquea tenzii* (Drawing from Sierra Ráfols in Muñoz Pizarro 1971)

The little island of Santa Clara harbours four endemic species, shared with Masatierra: *Wahlenbergia berteroi*, *Chenopodium sanctae-clarae*, *Dendroseris litoralis* and *D. pruinata*.

The islands Masatierra (RC) and Masafuera (AS) have 25 endemic taxa in common, and several of these are conspicuous and dominant elements of the vegetation, like the trees *Drimys confertifolia* (Winteraceae) and *Coprosma pyrifolia* (Rubiaceae) and the ferns *Blechnum cycadifolium* and *Rumohra berteroana*. Interestingly, several endemics show a vicariant pattern between both islands, having a closely related but different species occupying comparable habitats on each island: this is the case for *Fagara mayu*/*F. externa*; *Sophora fernandeziana*/*S. masafuerana*; *Berberis corymbosa*/*B. masafuerana*; *Nothomyrcia fernandeziana*/*Myrceugenia schulzei*. This is the case also for some ferns at species level (*Dicksonia berteroana*/*D. externa*) and at infraspecific rank: *Synammia intermedia* subsp. *intermedia*/*S. intermedia* subsp. *masafuerana*; *Megalastrum inaequalifolium* var. *inaequalifolium*/*M. inaequalifolium* var. *glabrior*.

Photo 6.1 *Robinsonia gayana* (Asteraceae) as representative of the 'rosette tree' growth form (Photographed by Sergio Elórtegui Francioli on Robinson Crusoe Island)

Nevertheless, the phylogenetic relationship of these vicariant species is not always as close as one could intuitively suspect, like the case of *Nothomyrcia fernandeziana* that seems to be more closely related to neotropical species of the genus *Blepharocalyx* than to *Myrceugenia schulzei* from Alejandro Selkirk Island (Murillo-Aldana and Ruiz 2011). Interestingly, both Myrtaceae species are the main component of the Fernandezian vegetation, to the point that a specific name for this forest has been proposed, namely the Fernandezian "myrtisylva" (Danton 2006). This vegetation unit seems to be the habitat for the highest number of endemic species, at least in Robinson Crusoe Island. This contribution seeks to analyze if this is really the case and whether there is a similar situation in Selkirk Island, which shows a higher diversity of habitats and apparently a lower level of human-induced impacts.

6.1.3 Material and Methods

The assessment of plant endemism for each different habitat in both main islands was based on available vegetation schemes. Both islands have an early vegetation map drawn at 1:50,000 scale (Skottsberg 1953) that has an immense historical value but a limited practical use. For the case of Robinson Crusoe, this map has been updated by Greimler et al. (2002) on the base of standard phytosociological methods. This last map, at a scale of 1:30,000, classifies the vegetation in 17 classes, including many anthropogenic units. For the purpose of the present study these classes have been reclassified in nine classes, nesting the units with minor areal importance. For the case of Alejandro Selkirk Island, a vegetation map at scale 1:12,000 based on aerial photographs and field work has been recently published by Greimler et al. (2013). Our own vegetation classification presented here on the base of field observations, puts the emphasis on the fern formations. The vegetation classification has been completed there on the base of field observations. Santa Clara Island has been omitted from the analysis since it harbours only an herbaceous carpet composed mainly of weeds. Vegetation units were superimposed to available species distribution information, by means of the software ArcGIS 9.3. A topological revision was undertaken to assure the consistency of the spatial units, avoiding overlapping and blank spaces, and granting a correct area calculation.

6.1.4 Results: Endemism Related to Habitats

The species-richest habitat, also containing the highest number of endemic species, is the montane forest of Robinson Crusoe Island. Physiognomically this forest resembles the Macaronesian laurisilva as well as the *Metrosideros* forest of the Hawaiian Islands (Skottsberg 1953). Danton (2006) proposed a proper name for the fernandezian forests, namely the Fernandezian Myrtisylva, reflecting both dominant species from the family Myrtaceae: *Nothomyrcia fernandeziana* in RC and *Myrceugenia schulzei* in AS. This myrtisylva has been divided since Skottsberg (1953) in "upper" and "lower" according to altitude, but the composition of both units is rather similar and the border between both units is very diffuse. Nevertheless, Greimler et al. (2002) recognized indeed a limit between both units, emphasizing the endemic character of their botanical components. One of the most remarkable aspects of this myrtisylva in its upper variant is the presence of numerous so called "rosette trees", the name Skottsberg (1953) gave to this remarkable growth-form of umbrella like little trees (Fig. 6.3).

Robinson Crusoe Island contains nine main habitat types or vegetation units. Figure 6.5 shows the vegetation map, and Table 6.1 the number of endemics present in every unit. Photo 6.2 shows several aspects of the vegetation of the island.

Rocks and eroded areas: these are highly eroded areas and rocky cliffs, barely vegetated, that occupy the largest area on the island, mostly at low altitude close to sea, also in Santa Clara Island. At several sites there are some plants, mostly weeds, that cover less than 10 % of the ground surface. The origin of these widespread eroded areas is a mixture of naturally exposed volcanic soil that is constantly being washed by the rain, and it also results from overexploitation of the environment by cattle and pests like rabbits. Nevertheless, some endemic species can be found in very low abundance, such as several *Dendroseris* and *Wahlenbergia* species, as

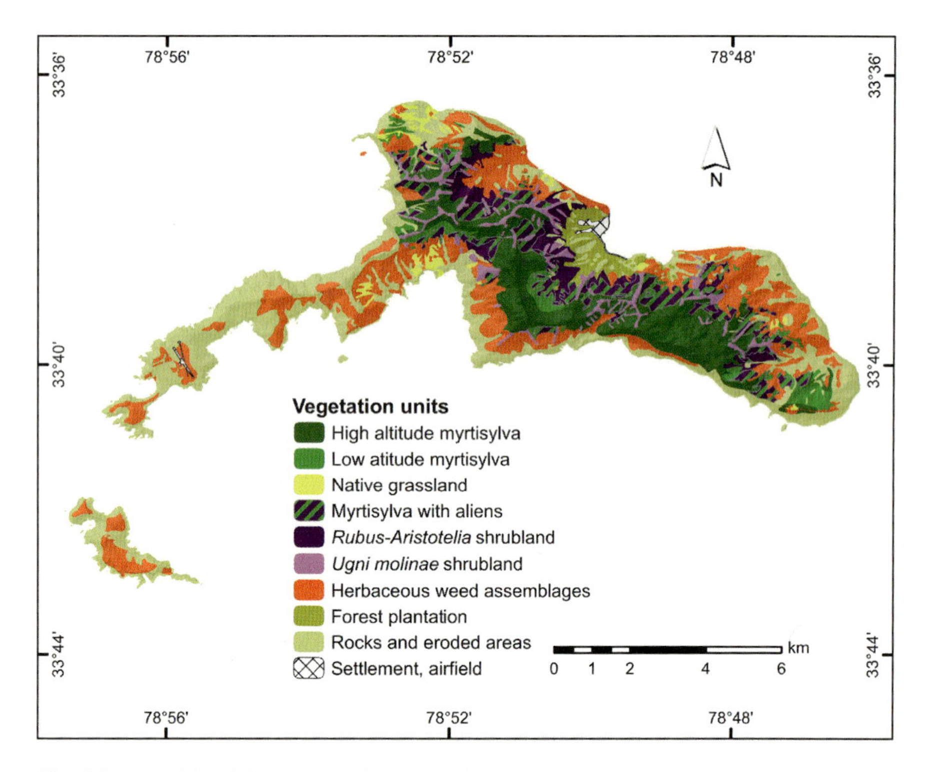

Fig. 6.5 Map of the different vegetation types of Robinson Crusoe Island

Table 6.1 Number of endemic species in each habitat of Robinson Crusoe Island

Vegetation unit	Area (hectares)	Endemic species
High altitude myrtisylva	835	69
Myrtisylva with aliens	499	48
Ugni molinae shrubland	232	38
Low altitude myrtisylva	191	31
Herbaceous weed assemblages	1,246	25
Rocks and eroded areas	1,473	24
Rubus-Aristotelia shrubland	219	15
Town and forest plantation	124	4
Native grassland	100	2

well as the more abundant *Ochagavia elegans*, *Eryngium inaccesum* and *Robinsonia gayana* on cliffs exposed to the sea.

Herbaceous weed assemblages: this unit is widespread at low to mid elevations (50–400 m asl) on both northeastern and southwestern slopes of the island, reaching the westernmost part and Santa Clara Island. One of the dominant species is the weed *Acaena argentea* that forms dense carpets leaving little space for other taxa. It was inadvertently introduced from the mainland during the nineteenth century and represents today "a main act in the tragedy of the island flora" (Greimler et al. 2002: 275). Several endemics occur in this unit; always in very scarce populations are the rare *Apium fernandezianum* and *Chenopodium crusoeanum.* It is highly possible that this unit represents different levels of ecosystem degradation, e.g. in Quebrada El Lápiz in the northeast, where the last exemplar of *Dendroseris neriifolia* serves as an indicator for better conditions for forest development in the past.

High altitude myrtisylva (= endemic upper montane forest, sensu Greimler et al. 2002) is situated mainly above 400 m, along the main ridge that divides the island, on deep and moist soils and also on vertical escarpments with less soil development. It is dominated by the endemic trees *Nothomyrcia fernadeziana*, *Drimys confertifolia*, *Fagara mayu*, *Coprosma pyrifolia* and *C. oliveri*, *Rhaphithamnus venustus*, together with the endemic palm *Juania australis* and the iconic species *Lactoris fernandeziana.* The unit is the most difficult to access on the island, and concentrates in ca. 800 ha the highest level of endemism in the whole archipelago (69 species). Most endemic species pertaining to the rosette-tree Asteraceae are to be found in this unit, such as both *Centaurodendron* species, *Dendroseris berteroana* and *D. pinnata*, different *Robinsonia* species, and the iconic *Yunquea tenzii*, known only from the summit of El Yunque massive.

Low altitude myrtisylva (= endemic lower montane forest. sensu Greimler et al. 2002) is situated mainly below 400 m, on the dryer northwestern and southeastern parts of RC. The dominant tree is *Nothomyrcia fernandeziana*, often accompanied by *Drimys confertifolia* and *Fagara mayu. Boehmeria excels*a is common on slopes adjacent to the bottoms of the "quebradas". This unit is widely affected by the aggressive intrusion of alien plants; therefore, most of it has been currently classified by Greimler et al. (2002) as "endemic forest with aliens".

Myrtisylva with aliens (endemic forest with aliens, sensu Greimler et al. 2002) is a unit that corresponds mainly to the lower altitude myrtisilva between 140 and 420 m, but is dominated by alien trees and shrubs from the species *Aristotelia chilensis* and *Rubus ulmifolius*. In spite of the high invasiveness of the mentioned species, a high number of endemics still survive under these conditions, the trees in a dramatic competition for light and space, and the herbaceous plants in very low abundances.

Ugni molinae shrubland is a unit that is dominated by the alien species *Ugni molinae*, a fleshy-fruit shrub that is native in the continent and has turned out to be the third worse pest in the island. It occupies mainly ridges and wind-exposed slopes between 100 and 700 m. Several endemic species that used to occupy this habitat still appear in the understory, but in low abundance, such as *Gaultheria racemulosa* at lower altitudes and *Blechnum cycadifolium*, *Robinsonia gracilis* and *Ugni selkirkii* above 400 m.

Rubus-Aristotelia shrubland is a unit that shows the extreme aspect of ecosystem degradation, located on low altitudes between 50 and 400 m. The native myrtisylva has been almost totally replaced by the alien species *Aristotelia chilensis* and *Rubus ulmifolius*. *Aristotelia*, which is native on the mainland, is supposed to have been taken to the archipelago in the middle of the nineteenth century, while the first report of *Rubus ulmifolius* as a weed was made by Looser (1927). During the second half of the twentieth century both species have turned out to be so invasive with the help of birds that disperse the fleshy fruits everywhere, that dramatic projections suggest that they could completely replace the native vegetation during the next 70 years (Dirnböck et al. 2003). Of course, the level of endemism in this unit is one of the lowest on the island.

Native grassland is a unit that comprises several patches in the lowest areas of the central part of the island between 20 and 460 m. It is dominated by tussock-forming native grasses, such as *Nassella laevissima*, *Piptochaetium bicolor* and occasionally *Nassella neesiana*, together with a high presence of alien herbs. This unit, would have been much more expanded towards the western side of the island under natural conditions. However, the area there is currently covered only by a weed assemblage. Several endemic grasses should have been found in this unit, such as *Podophorus bromoides*, but they got so rare that this last one has been classifies as "extinct".

Town and forest plantations: in and around the town San Juan Bautista, there are cultivated areas of *Eucalyptus globulus*, *Cupressus goveniana*, and *Pinus radiata* that were planted around the village to help stopping the erosion once the original forest was removed. Several other introduced trees, like *Acacia dealbata* and *A. melanoxylon*, are also occur in this unit, and the risk of invasion of these newly arrived species is still unknown. Very few endemic species are found in this unit. The number of introduced species on the island increases constantly due to the artificial "enrichment" of the town's gardens. Nevertheless, some species that are getting very rare in nature have been successfully cultivated in the gardens of the village, like *Dendroseris litoralis*, and many more of them in the greenhouses and gardens of the National Park's installations.

Alejandro Selkirk Island contains 12 main habitat types or vegetation units, plus the tiny occasional settlement and *Eucalyptus* plantation on the entrance to Quebrada Las Casas. Figure 6.6 shows the vegetation map, and Table 6.2 the number of endemics present in every unit (cf. Photos 6.2 and 6.3).

Grassland occupies the vastest extent on the island, especially above the quebradas of the eastern side between 0 and 800 m asl. It is covered by a carpet of native and introduced grasses, such as *Nassella laevissima*, *Aira praecox* and *Anthoxantum odoratum*. Presence of endemic species is rather high (30), including *Uncinia douglasii*, *Megalachne berteroana* and *Cardamine kruesselii.*

Fern Scrub occupies one of the largest amounts of area (1,500 ha) in the upper central and northern part of the island above 800 m. Dominant characteristic species is *Lophosoria quadripinnata*, with the continuous presence of *Gaultheria racemulosa* and the introduced grass *Anthoxantum odoratum*. The ample presence of this grass is an indicator of the high level of fragmentation of this unit due to the huge presence of goats, estimated at around a hundred individuals. Nevertheless, the unit shows the highest amount of endemic species (38) of the island, including *Thyrsopteris elegans*, *Gunnera masafuerae* (in locally humid conditions), *Berberis masafuerae*, *Erigeron ingae*, patches of *Peperomia fernandeziana* and the sole island orchid *Gavilea insularis* in a very restricted spot called La Cuchara-Morro El Pasto (Photo 6.3).

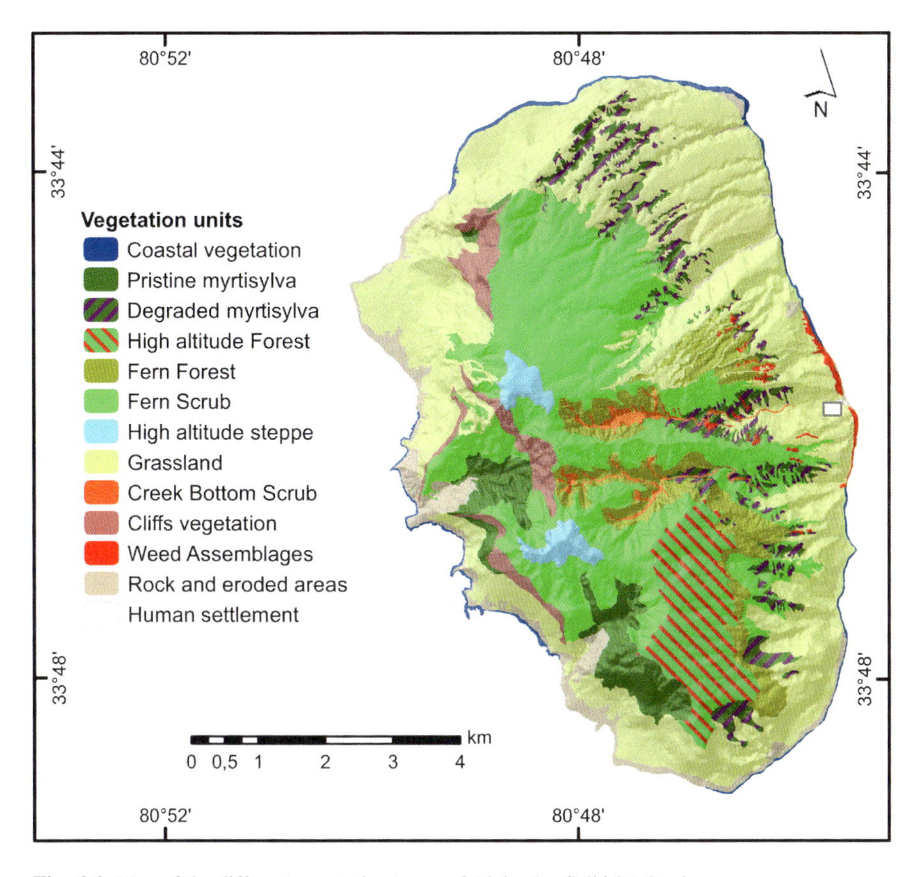

Fig. 6.6 Map of the different vegetation types of Alejandro Selkirk Island

Table 6.2 Number of endemic species in each habitat of Alejandro Selkirk Island

Vegetation unit	Area (hectares)	Endemic species
Fern scrub	1,361	38
Grassland	2,154	30
Fern forest	347	25
Degraded myrtisylva	310	24
Cliffs vegetation	155	23
Creek bottom vegetation	54	20
High altitude forest	320	18
Pristine myrtisylva	239	17
Rocks and eroded areas	225	8
Weed assemblages	33	6
Coastal vegetation	73	6
High altitude steppe	86	5
Human settlement	2	1

Photo 6.2 Representative photographs of the Juan Fernández Archipelago. (**a**) Santa Clara Island: grasslands and endemic species *Dendroseris pruinata*, (**b**) High altitude myrtisylva in Robinson Crusoe Island, (**c**) Myrtisylva in Alejandro Selkirk Island, (**d**) Grasslands on top of A. Selkirk Island, (**e**) Deep quebradas on the eastern side of A. Selkirk Island, (**f**) Fern scrub on top of the cliffs of the western side of A. Selkirk Island (Photographed by Sergio Elórtegui Francioli)

Fern forest is located above the deep valleys of the southern portion of the island, acting as a transition towards fern scrub. It is characterized by the tree fern *Dicksonia externa*, and other ferns such as *Blechnum cycadifolium*, *Blechnum chilensis*, *Rumohra berteroana*, *Pteris berteroana* and several *Hymenophyllum* species. Other species here are *Galium masafueranum*, *Gunnera masafuerae*, few exemplars of *Robinsonia masafuerae* and patches of *Drimys confertifolia.* On the border of this unit with the cliffs of the southeast there is the only known exemplar of *Dendroseris gigantea.*

High altitude forest is situated mainly between 700 and 1,000 m, in the upper part of the valleys and plateau on the southern portion of the island. Dominant endemic

Photo 6.3 A photographic selection of plant species endemic to the Juan Fernández Archipelago. (**a**) *Thyrsopteris elegans* (Thyrsopteridaceae), (**b**) *Drimys confertifolia* (Winteraceae), (**c**) *Rhaphithamnus venustus* (Verbenaceae), (**d**) *Wahlenbergia berteroi* (Campanulaceae) endemic to RC and Santa Clara Islands, (**e**) *Lactoris fernandeziana* (Lactoridaceae) endemic to RC Island, (**f**) *Plantago fernandezia* (Plantaginaceae) endemic to RC Island, (**g**) *Myrceugenia schulzei* (Myrtaceae) endemic to AS Island, (**h**) *Sophora masafuerana* (Fabaceae) endemic to AS Island, (**i**) *Gavilea insularis* (Orchidaceae) endemic to AS Island (Photographed by Sergio Elórtegui Francioli)

tree species are *Drimys confertifolia*, *Dicksonia externa*, *Blechnum cycadifolium* and many different *Hymenophyllum* species, together with a rich array of mosses and Hepaticae, including *Marchantia*.

Pristine myrtisylva occurs mainly between 100 and 650 m, in the valleys of the western part of the island. The unit is widely dominated by the endemic tree *Myrceugenia schulzei*, joined by *Coprosma pyrifolia*, *Rhaphithamnus venustus*, and *Fagara externa.* Endemic herbaceous species are *Peperomia skottsbergii* and *Haloragis masafuerana.* It seems to be one of the least degraded units of the island due to difficult access, but the overall level of endemism is relative low as it covers only limited area. It is also possible that the richness of this unit actually is higher but it has been rarely surveyed due to the aforementioned difficulties. In spite of the fact that *Santalum fernandezianum* is considered to have gone extinct long ago, there is a small possibility that a living exemplar of this species might be found in this unit.

Degraded myrtisylva seem to be the degraded version of the former one, situated mainly between 100 and 650 m, in the more accessible valleys of the eastern part of the island. The level of anthropogenic impacts is high, related to overgrazing and fires, to the point that physiognomically it resembles an open forest rather than a myrtisylva. Nevertheless, some very rare endemic species such as *Ranunculus caprarum* have been reported here, only with few individuals.

Cliff vegetation is situated mainly on the western side of the island, between 400 and 1,000 m. It shows a high number of endemic species (23 taxa). Characteristic endemic species are *Coprosma pyrifolia*, *Rhaphithamnus venustus*, *Acaena masafuerana* and *Galium masafueranum*, together with *Uncinia costata* and *Erigeron turricola.* From the endemic Asteraceae, *Dendroseris macrophylla* and *D. regia* are to be found at very few sites. It is possible that *Dendroseris gigantea*, and the putatively extinct *Chenopodium nesodendron* could have been common in this unit once.

Coastal vegetation occupies a thin band in contact with the sea, along the coast surrounding most of the island, characterized by a scanty herbaceous vegetation of *Nassella laevissima*, *Brassica oleracea*, *Juncus procerus*, together with the fern *Asplenium obtusatum.* Herbaceous endemic species typical of the unit are *Erigeron rupicola*, *Nicotiana cordifolia*, *Spergularia masafuerana* and *Spergularia confertiflora*.

Rocks and eroded areas occupy highly degraded sites above the coastal vegetation with introduced and native grasses. In spite of this, several endemic herbaceous species can be found here, such as *Haloragis masafuerana* or *Wahlenbergia masafuerae*.

Creek bottom vegetation, occurring in the deepest part of the quebradas of the eastern side, is adapted to extreme shady conditions. It is composed by many different ferns, such as *Blechnum longicauda*, and several *Peperomia* species, together with *Gunnera masafuerae*, *Urtica glumeruliflora*, *Erigeron fernandezianus*, *E. stuessyi*, *Wahlenbergia tuberosa* and scanty exemplars of *Sophora masafuerana*. In lower parts of this unit the invasive *Aristotelia* is rapidly gaining space.

High altitude steppe occurs above the "treeline" (fern-line), on the highest portion of the island between 1,160 and 1,380 m, with the characteristic presence of "magellanic" species of the shrubby cushions habit, such as *Abrotanella linearifolia*, *Lagenophora hariotii*, *Myrteola nummularia*, and *Rubus geoides*. Remarkably is the precipitation of snow at this altitude some times during the winter, which does not at all occur on Robinson Crusoe Island. The level of endemics is low, but endemic species characteristic to this unit are *Urtica masafuerae* (just few individuals) and *Agrostis masafuerana*.

Weed assemblages: Like in Robinson Crusoe Island, but somehow different, the tree aliens also occur on this island, but not as widespread as in the former island. Anyhow, the arboreal weed *Aristotelia chilensis* forms dense stands at middle altitudes on the eastern side of the island. *Rubus ulmifolius* is also entering in the quebradas, but still far not as dramatic as in Robinson Crusoe Island. The most widespread weed on the coast to the north and south of the settlement at the base of Quebrada Las Casas is *Bahia ambrosioides* (Asteraceae). The number of endemic species is drastically reduced in this unit to *Spergularia confertiflora* and very sparsely *Spergularia masafuerana* at the entrance to the quebradas, as well as *Uncinia douglassi* and *Megalachne berteroana* at middle altitudes.

Human settlement and plantation: this unit occupies a tiny portion at the base of Quebrada Las Casas, including a *Eucalyptus* plantation; the village is only sporadically occupied by fisherman. The only endemic species that has been reported there is the rare *Cardamine kruesselii* that occurs at very low presence at the entrance of several eastern quebradas.

6.1.5 Discussion and Notes on Conservation

Most of the units on Robinson Crusoe Island are characterized or dominated by alien species, with the high and low altitude myrtisylva being historically and systematically reduced due to invasive species. Both units plus the native grasslands occupy currently only a 23 % of Robinson Crusoe Island. Endemic species showing viable populations are today almost exclusively confined to the most inaccessible parts of the island: the slopes and summit around the El Yunque massive and the southeastern sector known as La Piña. This is why several conservation assessments suggest that at least 75 % of the endemic flora might be highly threatened (Stuessy et al. 1998; Cuevas and van Leersum 2001; Vargas et al. 2011). And physiognomically, it is highly possible that the myrtisilva as a whole might be replaced or severe invaded by alien species by the end of this twenty-first century (Dirnböck et al. 2003). Research leading to concrete actions towards ecological restoration is only recently being developed (Vargas et al. 2010).

In the case of Alejandro Selkirk Island the distribution of endemic species is more heterogeneous than in Robinson Crusoe, and many endemic species are not to be found in the myrtisylva but in the fern assemblages and in the vegetated cliffs and deep canyons. In fact, a relative pristine myrtisylva is restricted now to the most inaccessible parts on the western side. Even though the eastern myrtisylva are much more degraded, it seems that the intrusion of alien plant species is not as intensive as on Robinson Crusoe. But no assessment has been conducted to evaluate the potential spread velocity of these aliens, so the real threat for the endemic species remains uncertain.

The early vision of Skottsberg is still totally valid: ... "I wish to draw attention to the fact that so many of the endemic types are very scarce. [...]". Of species from Masatierra, *Podophorus bromoides* and *Phrygilanthus berteroi* have not been found by later collections and no exact locality is known. *Greigia berteroi*, *Chenopodium crusoeanum*, *Eryngium inaccesum*, *E. fernandezianum*, and *Dendroseris macrantha* were reported from one locality each: of the latter only two specimens could be found [...] other very rare plants are *Peperomia margaritifera*, *Selkirkia berteroi*, *Plantago fernandezia*, *Robinsonia thurifera* and *Centaurodenron*. One or more of these will probably share the fate of *Santalum fernandezianum*, which seems to be extinct.

Skottsberg continues regarding Masafuera Island: ... "also some species seem to be rare: *Peperomia skottsbergii*, *Ranunculus caprarum* and *Cardamine kruesselii* were found in one locality each; *Chenopodium nesodendron*, *Robinsonia masafuerae*, and *Dendroseris regia* are very scarce, and their recovery is checked by the too numerous wild goats. Many of the non-endemic species are also rare" (Skottsberg 1921: 212–213).

The conservation of the fernandezian flora is not just a romantic environmentalists' ideal, but an urgent need for better scientific understanding of the biogeography and evolution processes of island floras. The geographical relationships of the archipelago's flora show an important relationship to the American continent, as one could expect for these oceanic islands. Nevertheless, other important relationships with far territories like Australasia have intrigued biogeographers for a long time (Bernardello et al. 2006). Furthermore, the presence of many endemic genera and two endemic families suggests an old continental character of the flora, and traditional palaeogeographical reconstructions still do not fit well with this high degree of endemism. They suggest a much wider hypothetical territory that once could had better connections to Australasia or to ancient Gondwanaland (Skottsberg 1956; Moreira-Muñoz 2011). Under the current paradigm of molecular biology, oceanic islands are crucial for the calibration of phylogenies, despite the problems these calibrations might have (Heads 2011). Some species are so uncommon, that their systematic position is still problematic. In the words of Carlquist, referring to *Yunquea tenzii* restricted to the summit of El Yunque: "it may become extinct, as has so much of the autochthonous Juan Fernández vegetation, without having been collected in the flowering state" (Carlquist 1958: 20). Aside from botanical information, detailed ecological studies that consider variables such as soils, solar incidence, relief and micro-site characteristics are still lacking, in spite of being crucial for developing effective conservation actions (Vargas et al. 2010).

6.1.6 Conclusions

The main question of this section was the distribution of endemic species in relation to the different habitats present in the islands of the Juan Fernández archipelago. The answer seems to be very obvious in the case of Robinson Crusoe Island: the endemic species are today concentrated, or better said, are restricted due to human intervention, to the high altitude myrtisylva located at the most inaccessible parts of the island. Nevertheless, some species continue to exist in the low altitude myrtysilva and other human-disturbed units, but in critically low abundances. For many endemic species the population number does not exceed ten living exemplars or even just one remaining exemplar. At least nine species have gone extinct in historical times and two of them during the last decade (Danton and Perrier 2006). These species are *Podophorus bromoides*, *Santalum fernandezianum*, *Chenopodium nesodendron*, *Empetrum rubrum*, *Eryngium sarcophyllum*, *Notanthera heterophylla*, *Robinsonia berteroi*, *R. macrocephala*, and *Margyracaena skottsbergii.* This last species is still conserved ex-situ. The threats for the endemics in Alejandro Selkirk Island seem to be a little less dramatic; endemics still occupy the different habitats of the island, concentrated in the most inaccessible parts of the cliffs and deep valleys.

National Park rangers fight daily not only against pests and invaders, but also against the prevalent national carelessness and the local ignorance, dreaming of the day at which the big impulse for ecosystem restoration will generate again the conditions for the flourishing of this widely recognized botanical treasure. And who knows, maybe the future conditions are so favourable as to ensure that a hidden sandal-wood seed decides that it is time to show that life indeed can re-emerge when and where one does not believe it anymore.

Acknowledgements Compilation of the species data base and cartography was undertaken thanks to Fondecyt project N° 1120448. Juan Troncoso helped with the classification units, GIS calculations and map design. Rodrigo Vargas provided vegetation information for Masatierra. The park rangers and especially the park administrator don Iván Leiva have been for years of immense help in the field. Philippe Danton and Christophe Perrier provided always valuable information. Marcia Ricci also provided valuable information about the distribution of the species, while Mélica Muñoz-Schick and Alicia Marticorena from the SGO and CONC herbaria, provided information for checking the distribution of rare species. Patricio Novoa did a critical and much helpful revision on a first draft of the manuscript.

6.2 Madeira Islands

Miguel Pinto da Silva Menezes de Sequeira (✉)
Centro de Ciências da Vida, Universidade da Madeira, Funchal, Portugal

Carsten Hobohm
Ecology and Environmental Education Working Group, Interdisciplinary Institute of Environmental, Social and Human Studies, University of Flensburg, Flensburg, Germany

6.2.1 Physical Geography of the Madeira Archipelago

The Madeira archipelago (30–33°N) is a group of relatively old volcanic islands. The archipelago comprises Madeira as the largest island (737 km^2) and with the highest summit (Pico Ruivo, 1,861 m asl), Porto Santo (42 km^2), the Desertas (15 km^2), many islets, and, in political terms, the small Salvage Islands (4 km^2) which lie 300 km to the SE of Madeira. The islands and islets together cover an area of approximately 800 km^2.

The archipelago is located in the Atlantic Ocean SW of the Iberian Peninsula and W of North Africa. The distance from Madeira to the Iberian Peninsula is approximately 1,000 km and to Africa 600 km. The distance from the Selvage Islands to Africa is 300 km and to the Canary Islands 180 km.

The geological age differs from island to island and from one literature source to another. According to Press and Short (1994) the islands are 60–70 million years old, according to Jardim and Francisco (2000) the Madeira Islands are 18 million years old, whereas the Selvage Islands were formed about 27 million years ago (for details of Madeira's geology cf. Burton and MacDonald 2008).

Photo 6.4 *Aeonium glandulosum* is common on sea cliffs and rocks below 300 (700) m asl. in Madeira (Photographed by Miguel Sequeira)

Photo 6.5 *Isoplexis sceptrum* is a rare shrub (50–400 cm tall) of the Madeiran laurisiva (Photographed by Miguel Miguel Sequeira). On Madeira many individuals are planted along levadas, roadsides and in gardens

However, in contrast to the Canaries or Azores the islands have known no recent or historic volcanism. The terrain has suffered erosion for a very long time, is generally rugged and dissected by deep ravines. According to Mucina and Wardell-Johnson (2011) the landscapes of the Madeira Islands belong to the old stable landscapes of the world (OSLs) because of their old geological surface, relatively stable climate and the high predictability of the (non-) fire regime.

The climate is typical Mediterranean, with annual average temperatures between 6–10 °C at the summits and 15–20 °C on the southern coasts. Differences in precipitation and summer drought are the result of variation in altitude and geomorphology. High altitudes on Madeira Island are relatively wet (>1,600 mm/a), whereas Porto Santo and Desertas are relatively dry (<400 mm/a). The Salvage Islands can be prone to long periods of drought.

Press and Short (1994) counted 123 endemic vascular plant taxa (out of 1,226 indigenous and naturalized taxa), Borges et al. (2008) listed 154 species and subspecies (out of a total of 1,204 taxa). According to our own list (EvaplantE, latest version) the number is 134 (excluding endemic hybrids and varieties). Including the infraspecific categories varieties and natural hybrids, Jardim and Francisco (2000)

Photo 6.6 *Clethra arborea* is one of the major components of laurel forest on Madeira and the Azores at altitudes between 300 and 1,000 m (Photographed by Carsten Hobohm in Ribera de Janela, Madeira)

counted 165 endemics; in their book they list 143 endemics – species, several subspecies and a few hybrids. However, some 120 species or 170 taxa including species, subspecies and hybrids, or 10–12 % of the flora are endemic to the Madeira archipelago including the Salvage Islands.

Six endemic genera occur on Madeira Islands (de Sequeira et al. 2007).

6.2.2 Analysis of Floras and Databases

Important information about the endemics of the Madeira archipelago in EvaplantE (latest updated version) were extracted from Borges et al. (2008), de Sequeira et al. (2007), Jardim and Francisco (2000), Franquinho and Da Costa (1996), Press and Short (1994), Hansen and Sunding (1993), and Vieira (1992).

EvaplantE (latest updated version) and the literature cited above form the basis for the following analysis. During several excursions to Madeira and Porto Santo the second author had the opportunity to study the vegetation and to identify many of the endemic vascular plants in different landscapes and localities of the archipelago. Furthermore, we explored one of the most beautiful and impressive valleys in the world, the Janela valley, from the sea up to the highest parts. Many observations of species compositions in relation to their habitat types were noted and entered into the data base.

The importance of a specific habitat for endemics can be measured as the number of taxa which occur in this habitat – independent of the fact that most taxa occur in more than one habitat type.

6.2.3 Endemism Related to Landscapes and Habitats

The largest proportion of endemics on the Madeira archipelago occurs in rocky habitats and screes (82 taxa). The second largest group comprises forest endemics (41); these are followed by inhabitants of coastal and saline habitats (34), scrub and heath (21), habitats connected with running waters or waterfalls (12), and arable land, ruderal and settlement habitats (10). 4 Taxa (also) occur in grassland or grassy pastures, and 1 (also) in moorland (*Vaccinium padifolium*).

Examples of endemic inhabitants of rock and scree habitats are *Aeonium glutinosum*, *Agrostis obtusissima*, *Aichryson divaricatum* and *dumosum*, *Armeria maderensis*, *Asplenium trichomanes* ssp. *maderense*, *Berberis maderensis*, *Bunium brevifolium*, *Bystropogon punctatus*, *Ceterach lolegnamense*, *Chamaemeles coriacea* which inhabits both coastal and inland cliffs, *Cheirolophus massonianus*, *Crepis vesicaria* ssp. *andryaloides*, *Deschampsia maderensis*, *Erysimum arbuscula* and *maderense*, *Euphorbia desfoliata*, *Geranium maderense* and *palmatum*, *Helichrysum melaleucum* and *monizii*, *Hymenophyllum maderense*, *Jasminium azoricum*, *Lotus argyrodes*, *Luzula elegans*, *Melanoselinum decipiens*, *Monanthes lowei*, *Monizia edulis*, *Musschia aurea* and *wollastonii*, *Parafestuca albida*, *Rumex bucephalophorus* ssp. *frutescens*, *Saxifraga maderensis* and *portosanctana*, *Sideritis candicans*, *Sinapidendron frutescens*, *Sonchus pinnatus*, *Tolpis macrorhiza* and *Viola paradoxa*.

The following endemic taxa (also) occur in laurisilva, other forest types or woodland: *Arachniodes webbianum*, *Bystropogon maderensis*, *Carduus squarrosus* (also occurs on coastal cliffs), *Carex lowei* and *malato-belizii*, *Cirsium latifolium*, *Convolvulus massonii* which also occurs on low-altitude coastal rocks, *Dactylorhiza foliosa*, *Deschampsia argentea* on rocky slopes in damp woodland, *Dryopteris aitoniana* and *maderensis*, *Echium candicans*, *Festuca donax*, *Goodyera macrophylla*, *Hedera maderensis* ssp. *maderensis*, *Ilex perado* ssp. *perado*, *Isoplexis sceptrum*, *Luzula seubertii*, *Normania triphylla*, *Polystichum drepanum*, *Rosa mandonii*, *Rubus grandifolius* and *vahlii*, *Ruscus streptophyllus*, *Sambucus lanceolata* also occurs outside the forests and woodlands in thickets, just as *Sideroxylon mirmulans*, *Sonchus fruticosus*, *Teline maderensis*, *Teucrium abutiloides* and *betonicum*, *Vaccinium padifolium*, and *Vicia capreolata*. *Vicia capreolata* is found on Porto Santo and Desertas, also at low altitudes and independent of forest.

Coastal habitats such as coastal cliffs, maritime sands or organic layers on stony beaches are inhabited by e.g. *Andryala crithmifolia*, *Argyranthemum haematomma* and *thalassofilum*, *Betula patula*, *Calendula maderensis*, *Chamaemeles coriacea*, *Crambe fruticosa* (also some kilometers inland), *Echium nervosum* (also near roads), *Helichrysum devium* and *obconicum*, *Limonium ovalifolium* ssp. *pyramidatum*, *Lobularia canariensis* ssp. *pyramidatum* and ssp. *rosula-venti*, *Lotus loweanus* and *macranthus* which also occur in dry stony pastures or cornfields and other open and rocky places, respectively, *Matthiola maderensis*, *Maytenus umbellata*, *Olea maderensis*, *Senecio incrassatus*, *Sinapidendron angustifolium* which also grows a short distance inland, *Sonchus ustulatus*, and *Urtica portosanctana*.

Examples of endemics of heaths, thickets and scrub habitats are *Bystropogon maderensis* (also in laurisilva), *Erica maderensis* and *scoparia* ssp. *maderinicola*, *Genista tenera*, *Lavandula stoechas* ssp. *maderensis*, *Marcetella maderensis*, *Sorbus maderensis*, and again *Vaccinium padifolium*.

Some endemics are inhabitants of arable land, ruderal or urban habitats. Examples are *Crepis noronhaea* on roadsides, cultivated and waste ground, and *Delphinium maderense* in cornfields and on roadside banks.

Only few water plants occur on Madeira, and none of them are endemic to the archipelago. Several endemic taxa such as *Carex lowei*, *Ceterach lolegnamense*, *Dryopteris aitoniana*, *Dryopteris maderensis*, *Geranium palmatum*, *Geranium rubescens*, *Melanoselinum decipiens*, *Oenanthe divaricata*, *Scrophularia hirta*, *Scrophularia racemosa* and *Solanum patens* occur under wet and shady conditions in habitats which are connected to water ecosystems, e.g. river banks, dry river beds, or along levadas and beside streams. Some also occur in shady and wet forest, independent of the proximity to water bodies.

The average of the elevation minima is c. 300 m asl. and the maxima average at 1,000 m asl. Therefore, we assume that most endemics on Madeira occur between 300 and 1,000 m.

The high endemism of the archipelago as a whole can be explained by the age of the geological surface and the absence of historical or recent volcanism, by the rich environmental heterogeneity, isolation and – compared with the mainland of Europe or western North Africa – a relatively stable climate history. According to Mucina and Wardell-Johnson (2011) Madeira has old stable landscapes. Fossil records show that in the past laurisilva was part of the Tertiary flora, as it was in mainland Europe where this forest type was destroyed by climate change during the Pleistocene (see e.g. Kunkel 1993).

In general, differences in the endemism of different habitat types can be explained by the same arguments but smaller in time and space, e.g. the age of the current ecological conditions of a locality or habitat type. Ecological conditions, including meso-climate of coastal cliffs, of laurisilva, scrub and rocky habitats of the Madeira Islands, are much older than urban habitats, cornfields or roadsides. The area covered by a certain habitat type might also be relevant for the number of evolutionary processes and endemics. However, natural grassland and swamps are rare in Madeira's landscapes. The grassland area was extended by human activities such as cutting, burning and grazing by domestic animals. This grassland mostly is species-poor, and endemics of the island are seldom to be found here (Photos 6.7 and 6.8).

Photo 6.7 *Helichrysum obconicum*, endemic to the island of Madeira, inhabits coastal rocks and sea cliffs (Photographed by Carsten Hobohm on basaltic rock near Porto Moniz, Madeira)

Photo 6.8 *Plantago arborescens* ssp. *maderensis*, endemic to the Madeira archipelago, is common throughout Porto Santo and grows among rocks, on banks, on roadsides and in other open areas below 1,000 m (Photographed by Carsten Hobohm near the northwestern coast of Porto Santo)

6.3 Corsica, Mediterranean

Carsten Hobohm (✉)
Ecology and Environmental Education Working Group, Interdisciplinary Institute of Environmental, Social and Human Studies, University of Flensburg, Flensburg, Germany

6.3.1 *Physical Geography of Corsica*

The Mediterranean island of Corsica lies south of the mainland of France between latitudes 41 and 43°N. The highest mountains on Corsica consist predominantly of granites, gabbros and volcanic rocks which date back 280–250 Ma. At that time, the mountain ranges of Corsica, Sardinia and the Balearic Islands were part of the Iberian subplate west of the localities where they are now. Some 35 Ma ago, these mountain ranges began to move to the east. When sea water from the Atlantic Ocean reclaimed the Mediterranean Basin some 5 Ma ago the area around Corsica was flooded and Corsica became an island. Before that time, Corsica was part of the mainland and already covered by vegetation. However, since its birth the island has experienced different warm and cold climate periods. During the Pleistocene glaciation periods the highest mountain tops and shady slopes were most likely permanently covered with snow and ice. Climate change scenarios suggest that, as a consequence of increasing and decreasing temperatures during Pleistocene cycles, plant populations had to move up- and downwards, and a loss of species diversity might have been the result, not only in the northern parts of Europe but also in Mediterranean regions (Gillespie and Clague 2009; Gottfried et al. 2002).

The Mediterranean Island of Corsica (8,750 km^2) is situated between the coasts of mainland France to the North (160 km), Italy to the East (82 km) and the island of Sardinia to the South (12 km). More than 20 mountains reach a height of at least 2,000 m. The highest points are Monte Cintu (2,706 m), Monte Rotondo (2,622 m), and Monte Padro (2,393 m). Corsica is known as the most mountainous island in the Mediterranean Basin (Gillespie and Clague 2009; Gamisans and Marzocchi 1996).

Coastal regions are relatively dry (450–600 mm/a) in comparison with high mountain zones (>1,800 mm/a). Gamisans and Marzocchi (1996) distinguish three climate zones: a coastal zone of warm Mediterranean climate between the coastline and 600 m, with winter rain, summer drought and annual temperatures between 14 and 17 °C (1); a zone of cool Mediterranean climate between 600 and 1,200 m, with annual temperatures between 13 and 10 °C and precipitation rates of 800–1,500 mm/a (2); and a zone of relatively cold climate at elevations over 1,200 m, with high precipitation rates (>1,500 mm/a) and snowfall during winter (3).

A major part of the island is composed of crystalline rocks (e.g. granites, rhyolites), schists, ophiolites, serpentines and sedimentary limestone. Typical habitats comprise marine sands, dunes, saltmarshes and rocky shores at the coast. Large parts of the lowlands are cleared for agriculture. However, typical Mediterranean evergreen sclerophylleous forest, mixed forests and maquis remain in many parts of the island. The mountains are home to diverse mixed and broadleaf deciduous forests which are dominated by pine and oak species. The submontane zone is dominated by shrubs, the alpine zone by grassland, heaths with cushion plants, rocks and screes.

6.3.2 Analysis of Floras and Databases

The analysis of the composition of endemic vascular plants on the island of Corsica is based on Gamisans and Marzocchi (1996), Tutin et al. (1996a–e, 1964–1990), Bouchard (1968), EvaplantE (latest updated version which is based on diverse floras; cf. Bruchmann 2011; Hobohm and Bruchmann 2009) and Hobohm (2000).

According to the references cited it was possible to characterise most endemics in relation to habitat types and altitudes. In EvaplantE we use a relatively wide and conservative species concept in accordance with Tutin et al. (1996a–e). It is, therefore, possible to compare numbers of endemics from different regions on the basis of a consistent species concept.

6.3.3 Endemism Related to Landscapes and Habitats

According to Gamisans and Marzocchi (1996) c. 2,978 vascular plant taxa – 2,092 species, 264 subspecies, 89 varieties, 82 hybrids – are indigenous or naturalised on the island of Corsica. 131 taxa are endemic to the island (4.4 %). 165 taxa are subendemics. These are restricted to Corsica and Sardinia (75), Corsica together with Sardinia and the Balearic Islands (11), Corsica together with small neighbouring mainland regions, or together with mainland regions and neighbouring islands (86). Only 48 island-endemic taxa are listed for Corsica in EvaplantE. The discrepancy between these figures and those of Gamisans and Marzocchi (1996) has to do with the breadth of the species concept. In Gamisans and Marzocchi (1996: 28 ff.), for example, eight species of the genus *Limonium* which are endemic to the island are named and shown in photographs. They belong to the species groups of *Limonium articulatum*, *acutifolium* and *densiflorum*. None of these groups is endemic to Corsica. Bouchard (1968) distinguished six species of the genus *Limonium* in total. According to Tutin et al. (1996c) one is endemic to Corsica (and therefore listed in EvaplantE).

However, certain tendencies can be described independent of the interpretation of the species concept. Most endemics occur at altitudes between 1,000 and 2,200 m (average minimum altitude to average maximum altitude).

The highest proportion of endemics (c. 67 %) is associated with rocks and screes. The second largest group is found in grasslands (46 %); this is a relatively high proportion in comparison to other Mediterranean islands, and compared to the Canaries or Madeira (cf. Bruchmann 2011). A third (35 %) of the endemics are found in scrub and heaths, 19 % in freshwater habitats and wetlands, 10 % in forests, and only a few taxa occur in coastal habitats, mires or arable land. The sum of the percentages exceeds 100 % because the majority of the endemics are found in several habitat types. A few endemic taxa are rather habitat-specific, such as *Armeria leucocephala*, *Bupleurum falcatum* ssp. *corsicum* and *Phyteuma serratum* in rocky habitats or *Alnus viridis* ssp. *suaveolens* which dominates in subalpine scrub communities.

A comparison between the oceanic archipelago of Madeira and the continental island Corsica shows remarkable differences in the numbers of endemics, distribution patterns and habitat characteristics. Both regions are very mountainous and mainly influenced by a Mediterranean climate with winter rain and summer drought, at least at lower altitudes.

Madeira is much smaller than Corsica and richer in endemic vascular plants. This is most likely the result of a longer island history and a more continuous climate during the Pleistocene – stabilised by marine currents (cf. Mucina and Wardell-Johnson 2011). This interpretation is underlined by the fact that the ancient forest type laurisilva, which in the Tertiary also occurred in southern Europe, survived on Madeira, and a high percentage of endemics are found in these forest habitats.

A larger distance to the next landmass or mainland region clearly favours isolation. The Madeira archipelago is more isolated than Corsica. Genetic isolation favours speciation and speciation favours endemism. On the other hand, geographical separation successfully reduces dispersal events, and Madeira is poorer in vascular plant species overall; Corsica has more than twice the total number of vascular plant taxa in comparison with Madeira (cf. Borges et al. 2008; Gamisans and Marzocchi 1996), and a large species pool also favours speciation processes. However, the higher concentration of endemics on Madeira cannot be fully explained by the separation/isolation factor.

As in most regions of Europe, the highest proportion of endemics is found on rocks and screes. Grassland endemics constitute the second largest group on Corsica, whereas on Madeira the forest taxa are the second largest group. The high proportion of endemics associated with grasslands is astonishing because natural grassland on Corsica is rare, and most grassland habitats are the result of fertilisation, cutting, moving and/or grazing. No island endemic which occurs in grassland is found exclusively in this habitat type. Many of these taxa also occur in heathlands, scrub or rocky habitats. Scrub and heaths harbour a large percentage of the endemics on both islands, despite the fact that fire is a normal phenomenon on Corsica where scrub vegetation is a succession stage between fire and forest, whereas on the Madeira archipelago fire is fairly uncommon (favoured by plantations of pine and *Eucalyptus globulus*).

6.4 Madagascar

Carsten Hobohm (✉)
Ecology and Environmental Education Working Group, Interdisciplinary Institute of Environmental, Social and Human Studies, University of Flensburg, Flensburg, Germany

6.4.1 *Physical Geography of Madagascar*

Madagascar is the fourth largest island in the world (585,000–587,000 km^2; Table 5.4), situated >400 km off the SE coast of Africa between latitudes 12 and 25.5°S. With rocks dating back 3,200 million years and an origin as continental island some 160 million years ago it is most likely the oldest island in the world. Most landscapes are flattened, leached, and old in the sense of Mucina and Wardell-Johnson (2011). The highest mountain is Tsaratanana (2,876 m asl).

In most parts of the island the climate is tropical warm or hot with summer rain. Madagascar is influenced by mostly dry trade-wind conditions in winter and wet monsoon-driven tropical winds in summer. Only a few regions are influenced by a more subtropical (S Madagascar) or orogenic climate (central mountains). Eastern regions with humid forests receive precipitation throughout the year, without a distinct dry period, but also with higher precipitation in summer than in winter. Precipitation rates can be more than 6,500 mm/a on the northeastern coasts, whereas the driest parts in the SW receive less than 400 mm/a. Mean air temperatures range from 26–29 °C in the North to 20–27 °C in the South. Temperatures in the high mountain regions average between 19 and 12 °C, and can drop below zero in winter (minimum: −12 °C; Wells in Goodman and Benstead 2003).

Madagascar is part of the *"Madagascar and the Indian Ocean Islands Biodiversity Hotspot"* (Mittermeier et al. 1999, 2005). The outstanding plant and animal richness of Madagascar can be explained by the age of the geological surface (i), by the fact that Madagascar is of continental origin (ii) and by long and uniterrupted evolutionary histories (iii). The number of vascular plant species is estimated at 10,000–14,000 including c. 80–90 % endemics (Cribb and Hermans 2009; Gillespie and Clague 2009; Burga and Zanola 2007). A project called *Catalogue of the Vascular Plants of Madagascar* aims to enumerate and evaluate the native and naturalised flora of the island. Based on this list, Callmander et al. (2011) recently registered 11,220 species and 1,626 infraspecific taxa in total, including 8,884 endemic and 2,336 non-endemic species. The level of endemism of the indigenous vascular flora is calculated as 82 %, or 79 % of all vascular plants including 331 naturalised species.

Today, Madagascar is burning almost always and everywhere because of charcoal production, because of the tradition of burning pastures and arable land to produce nutrients from the ash and – in the case of pastures – give space for the young grass shoots which are eaten by zebus, and because charcoal and wood are used for cooking and heating.

The conversion of primary vegetation and forest to arable and pastural landscapes under the pressure of an overwhelming population still continues.

6.4.2 Analysis

Basic facts and numbers come from Callmander et al. (2011), Gillespie and Clague (2009), Gribb and Hermans (2009), Burga and Zanola (2007), Moat and Smith (2007), Goodman and Benstead (2003), Schatz (2001), Koechlin et al. (1974). However, the massive explosion of knowledge about the natural history of Madagascar and the description of plants and animals which are new to science at the beginning of the third millenium indicates a high potential for further investigations.

The author undertook several excursions to Madagascar with a couple of hypotheses in his baggage and with the plan to see and understand the natural units – landscapes, habitats, ecological conditions, biodiversity – just a little better.

6.4.3 Endemism Related to Landscapes and Habitats

> The flora of Madagascar is predominantly a woody flora. Despite the recent controversy surrounding the nature of the vegetation when humans first arrived, and the origin of the vast treeless areas on the Central High Plateu, the fact remains that even the driest and edaphically harshest areas in the southwest support woody vegetation. Moreover, the treeless expanses are especially depauperate in herbs, often consisting of only one or two species of grass widespread throughout Africa. With the exception of orchids - the majority of which are epiphytes on the woody vegetation - the Malagasy flora is relatively poor in herbaceous groups. Thus, trees provide the basic structure for most of the island's flora and fauna, as well as fuel wood and housing for the overwhelming majority of Malagasy people (Schatz 2001: 1).

Tree species, large and small shrubs and other woody species together form the majority of the flora on Madagascar. Species endemism for trees and shrubs together reaches a rate of 92 %, whereas the level for herbs is 72 % (Callmander et al. 2011). Before people inhabited the island, most of the regions of Madagascar were covered by forest, woodland and dry thickets.

Nowadays, the forest of Madagascar forms a near-coastal ring and includes a few inland forests which are concentrated in montane areas. Different processes and environmental circumstances have been discussed to explain this phenomenon.

Madagascar has a peripheral ring of lowland to montane forest around an upland that is covered by fire-simplified grassland and savanna. These largely degraded and depauperate open landscapes are dominated by a few cosmopolitan fire-adapted grasses.

The remaining primary vegetation, species richness and endemism are concentrated in near-coastal rather than inland regions.

Such a concentric order of richness can be found in various regions of the world (cf. e.g. Foster et al. 2012; Mittermeier et al. 2005). Furthermore, a peripheral forest ring can also be found in other parts of the world, e.g. Australia and Borneo.

In Australia, all species-rich regions are situated near the coast, whereas the central inland regions are dominated by species-poorer woodlands, thickets, grasslands and semi-desert. In the case of Australia, the low species diversity of the central parts can be explained by alternating dry and wet climate periods during the Pleistocene and Holocene. These cycles affected inland regions more than coastal regions. Unfortunately, and in contrast to Australia, the Cenozoic climate history and the development of the Malagasy biomes during this period remain largely unknown (cf. Wells in Goodman and Benstead 2003; Beard et al. 2000).

Raes et al. (2009) discussed and analysed distribution patterns of endemic vascular plants and centres of endemism on Borneo and compared these patterns with findings in other regions in the world and other groups of organisms. For Borneo they concluded that regional endemism hotspots represent near-coastal areas which are characterised by a small range in annual temperature and high environmental heterogeneity and which are least affected by El Niño drought events. Relatively stable climate conditions maintaining endemism have been found for various groups of organisms in different regions of the world (Jetz et al. 2004; Gathorne-Hardy et al. 2002, cf. also Araujo et al. 2008; ter Steege et al. 2003)

However, there are remarkable differences between the distribution patterns of Madagascar and e.g. Australia or Borneo. Whereas the species-poorer inner areas of Australia receive less than 250 mm precipitation per year, almost all parts of Madagascar have precipitation rates which doubtless would enable forest growth (>>800 mm/a, e.g. Antananarivo c. 1,300 mm, Fianarantsoa >1,200 mm). Furthermore, the driest parts of the island in the South, with precipitation rates <500 mm/a, are very rich in plants and endemics (dry spiny bush and forest). In contrast to Borneo, the near-coastal regions of Madagascar represent very different climates, from a wet tropical climate in the East (with precipitation rates >2,000 mm per year) to subarid in the South (<400 mm).

It is therefore curious that in Madagascar, under totally different climate conditions, there should be a peripheral ring of forest, rich biodiversity and endemism around a largely species- and endemic-poor inland area with grassland and savanna. If this peripheral ring reflected high environmental heterogeneity and steep mountain slopes, then the biodiversity would easily be explained by habitat diversity. However, maps show that this is only the case in a few parts of the island. In many parts forest occurs in lowland areas with obviously low environmental heterogeneity. It is possible that the higher endemism in coastal areas could be explained by unfavourable inland conditions rather than by a uniform marine-influenced coastal climate.

We assume that the risk of lightning fires corresponds with the length of the dry season. Therefore, the probability of natural fire increases from rainforest areas in the N and E to dry spiny bush in the SW. This view is underlined by the distribution patterns of fire-resistant plants which are concentrated in the western and southern parts of the island. Some plants have notably thick barks that allows them to survive fires (e.g. *Uapaca bojeri*, *Bismarckia nobilis*, *Hyphaene coriacea*, *Poupartia birrea* ssp. *caffra*, and several species of Sarcolaenaceae), others are able to resprout from their woody base or tuberlike roots after fire (Labat and Moat 2003; Schatz 2001; Rauh 1973). Furthermore, many of the succulent plants withstand fire simply because they do not burn well.

Because natural fire is a more or less normal event in the dry spiny bush in the S of Madagascar many of the related vegetation units are well (pre)adapted to fire. On the other hand, anthropogenic fires are less effective in terms of fire spread in areas without, or with only short, dry periods, as is the case in the northern and eastern parts with humid forest and high precipitation rates. In the central parts of Madagascar, the vegetation is not well fire-adapted on the one hand but on the other hand it burns easily in dry periods in wintertime. This might be a reason for the poorness of the inland grasslands and savannas, which in comparison with high mountainous and coastal or near-coastal regions are strongly influenced by man and fire.

Binggeli (in Goodman and Benstead 2003) discussed the impact of several neophytes on native ecosystems in Madagascar. Many exotic and invasive plants have been introduced in recent centuries. Alien species are omnipresent in almost all habitats and environments. Nonetheless, it is not easy to calculate the direct influence and to estimate the interrelationship between increasing neophyte populations and decreasing populations of native or endemic taxa. In many cases, human alteration of environmental conditions can promote both effects. Thus, it remains uncertain which neophytes directly threaten native or endemic vascular plants.

The impact of neophytes in several parts of the primary forest is astonishingly low. We saw untouched rainforest with a few individuals of e.g. *Clidemia hirta*, *Psidium guajava* or other introduced plant species – only along the footpaths. And we saw dry spiny bush and dry forest without any associated neophytes.

Three species that either have or have had an important impact on natural vegetation and humans alike (Binggeli 2003: 260) are *Eichhornia crassipes*, *Lantana camara* and *Opuntia monacantha*. However, these species do not occur everywhere either. *Eichornia* inhabits eutrophic, slow-running or standing waters, and many waterbodies in Madagascar are not covered by this plant. *Lantana camara* is named as widespread and an important pest in agricultural lands, secondary shrublands and succession stages after clearings. This can be seen in many cultivated areas and along the roads. Nonetheless, many other shrublands do not harbour a single individual of *Lantana camara*. The appearance of *Opuntia*, especially in dry regions, is extremely complex and its occurrence is scattered; as it is not infected by cochineal everywhere, it is widely used as resource and fodder, planted as an impenetrable hedge and so on (Binggeli 2003). Furthermore, this plant does not occur in every dry habitat.

However, the extent to which neophytes affect endemic vascular plant taxa remains unclear, whereas there is no doubt as to the constant threat created by the elimination and transformation of habitats by man.

The dominant vegetation types at the beginning of our millenium (Moat and Smith 2007: 19 ff.) are plateau grassland – wooded grassland mosaic (41.7 %) and wooded grassland-bushland (22.9 %), which together cover approximately two thirds of the island (65 %), and degraded humid forest (9.8 %) and primary and secondary humid forest (8.1 %) which together amount to c. 18 %. Western dry forest covers 5.4 %, south western dry spiny forest thicket 3.1 %, western subhumid forest 0.7 %, mangroves 0.4 %, tapia forest 0.2 %, littoral forest and western humid forest are the smallest forest types which together cover less than 1 % of the island's area. All forest types together make up c. 26 %. Approximately 4 % of the whole island is cultivated, 1 % is bare rock or soil, 1 % is wetlands.

Assuming that most parts of the island used to be covered by forest, the reduction in forest area can be estimated at c. 60–70 % of the whole area or more than two thirds of the island's forest. The majority of the remaining forest is highly impacted by various human activities.

One third of Madagascar's primary forest is estimated to have been lost since the 1970s. Less than 10 % of Madagascar's area is covered by natural vegetation and most of the open landscapes are grazed by zebu cattle and other livestock.

The dry spiny forest-thickets in SW Madagascar appear to be the vegetation type with the highest level of endemism (95 %; Wells in Goodman and Benstead 2003, Gautier & Goodman in Goodman and Benstead 2003). Endemism is also extremely high in primary forest (89 %), very high in rocky habitats (82 %), intermediate in swamps (56 %), lower in aquatic, semi-aquatic (38 %) and coastal habitats (21 %), and very low in savanna, grassland (6 %) and urban habitats (c. 0 %). However, information about absolute numbers of endemic taxa related to different vegetation types or information about species ranges is still scarce. As mentioned above, the majority of the endemics clearly inhabit forest. Because of the distribution patterns and area of the humid forest and because this is indicated by the floras (e.g. Schatz 2001; Dransfield et al. 2006; Cribb and Hermans 2009, and others) we assume that this vegetation unit harbours most of the endemic vascular plant taxa found in Madagascar.

According to Ranarijaona (in Goodman and Benstead 2003: 251) c. 128 aquatic and semiaquatic (of a total of 338 species and varieties, or 38 %) vascular plant taxa are endemic to Madagascar. This number is relatively low compared to that for terrestrial plants (81–82 % or 8,000–10,000; see Callmander et al. 2011; Gautier and Goodman 2003). Ferry et al. (1999, cited in Gautier and Goodman 2003) suggest that the low number of endemic aquatic plants on Madagascar may be a result of Quaternary climate fluctuations: during dry periods freshwater and semi-aquatic habitats were essentially dry, eliminating local floras which depend on wet conditions. However, the small total area of water bodies and the fact that lakes and running waters are normally relatively young and discrete habitats are not the best conditions for promoting endemism.

One of eight mangrove tree species (*Ceriops boiviniana*) is considered to be endemic or possibly a synonym of *Ceriops tagal* (Roger & Andrianasolo in Goodman and Benstead 2003: 209).

However, the absolute numbers and percentages of endemics in coastal habitats, freshwater habitats, wetlands, grasslands and savannas are relatively low in comparison with rocky habitats, dry thickets and different forest types.

Plant families on Madagascar rich in genera and/or species, are e.g. Orchidaceae, Rubiaceae, Fabaceae, Euphorbiaceae, Sapindaceae, Malvaceae, Arecaceae.

Ferns and orchids represent two very species-rich plant groups on the island and comprise c. 1,400–1,500 species (at least 862 species of orchids, 563 species of pteridophytes). Endemism in orchids is slightly lower than 90 % and in ferns c. 45–47 %.

Both groups are extremely species- and endemic-rich compared to every other region in continental Africa of a similar size. These groups are rich in epiphytes, and show a diversity gradient with increasing numbers from dry to wet conditions (cf. Cribb and Hermans 2009, Rakotondrainibe 2003 in Goodman and Benstead 2003). This fact confirms the significance of humid forest for the survival of very many endemic vascular plants on Madagascar.

The same diversity gradient from dry to wet and SW to NE can be found in several other systematic groups, e.g. palms. On a global scale and in comparison with the whole of continental Africa, Madagascar is extremely rich in palms (Arecaceae). Of the 170 or more species (continental Africa: approximately 60) all but 5 are endemic and most of them occur in the more humid eastern and Sambirano regions (Dransfield et al. 2006, Dransfield & Beentje in Goodman and Benstead 2003).

Other plant families and genera, such as Didieraceae (4 endemic genera, 11 endemic species) and *Aloe* (c. 120 endemic species, more than 50 of which were described after 1990), radiated in relatively dry environments (Castillon and Castillon 2010; Petignat and Cooke 2009; Schatz 2001).

Madagascar is a little larger than France, for example, and has more than twice the number of plant species. France is a mainland region with temperate to subtropical (Mediterranean) and orogenic (Alps, Pyrenees) climates, and was strongly affected by Pleistocene glaciation cycles. Madagascar is an island with winter-dry or wet tropical climates which was less impacted by Pleistocene glaciation cycles. We assume that the differences in the number of vascular plant taxa can be better explained by differences in the evolutionary histories than by differences in warmth or other aspects of contemporary climates (cf. Sandel et al. 2011; Normand et al. 2011). Nor can the differences be explained by environmental heterogeneity or habitat diversity.

6.5 Comparison of Endemism and Habitat Affinities on Juan Fernández Islands, Madeira Archipelago, Tenerife and Corsica

Carsten Hobohm (✉)
Ecology and Environmental Education Working Group, Interdisciplinary Institute of Environmental, Social and Human Studies, University of Flensburg, Flensburg, Germany

Andrés Moreira-Muñoz
Instituto de Geografía, Pontificia Universidad Católica de Chile, Santiago, Chile

The Juan Fernández archipelago (Chile), the Madeira archipelago (Portugal), and Tenerife (Canary Islands, Spain) are volcanic islands that originated in the ocean. Corsica (Mediterranean, France) is a continental island that became an island when the Mediterranean Basin was flooded. All islands are older than 1 million years, in most cases they originated a few or many million years ago (cf. Gillespie and Clague 2009, for the following comparison see also the literature cited above); the Juan Fernández Islands originated 1–4 Ma, Madeira Archipelago 18–70 Ma, Tenerife 7–50 Ma, and Corsica 5 Ma ago. According to the geological data, the Juan Fernández islands are younger than Corsica and Tenerife whereas the Madeira islands are the oldest. However, the exact numbers differ depending on the references (Table 6.3).

Juan Fernández (33°S), the Madeira islands (30–33°N) and Tenerife (28–29°N) lie c. 670, 600 or 275 km west of continents, respectively. Corsica (41–43°N) lies very close to the European mainland of France to the North (160 km), Italy to the East (58 km), and to the large continental island Sardinia to the South (12 km). The latter was also a mountain range before it became an island. Tenerife occupies a central position in the Canary Islands, and is c. 30–100 km from La Gomera, La Palma, El Hierro and Gran Canaria and 200–250 km from Fuerteventura and Lanzarote. The distance to the African mainland is approximately 275 km.

All islands are influenced by a Mediterranean climate with winter rain and summer drought. Mean temperatures from summits to warm coastal regions are c. 10–16 °C on Juan Fernández Islands, 6–20 °C on the Madeira archipelago, 9–21 °C on Tenerife, and 5–17 °C on Corsica. All islands can experience snow and frost on mountain tops in winter, irregularly on Juan Fernández Islands and Madeira, regularly on Tenerife and Corsica.

Precipitation rates differ from year to year but also from one to another locality on Juan Fernández Islands (c. 300–1,700 mm) with almost daily rain at high altitudes. Desertas, Porto Santo and the South of Madeira (Madeira archipelago) are relatively dry (<400 mm), whereas high mountain regions of Madeira Island receive more than 1,600 mm rain per year. Tenerife is very dry on southern coasts (100–200 mm), dry at high elevations (300–400 mm) and relatively wet at mid-altitudinal zones on northern slopes (c. 800 mm; Pott et al. 2003; Kämmer 1982). On Corsica the values

Table 6.3 Characteristics of four islands or archipelagos that are influenced by a Mediterranean climate, and have similar numbers of endemic vascular plant taxa (for references see text)

	Juan Fernandez Islands, Chile	Madeira Islands, Portugal	Tenerife, Spain	Corsica, France
Latitudes	34°S	30–33°N	28–29°N	41–43°N
Origin	Oceanic	Oceanic	Oceanic	Continental
Age	1–4 Ma	18–70 Ma	7–50 Ma	5 Ma
Area	104 km^2	800 km^2	2,034 km^2	8,750 km^2
Distance to mainland	670 km (to Chile)	600 km (to Morocco)	275 km (to Western Sahara)	58 km (to Italy)
Mean temperatures	10–16 °C	6–20 °C	9–21 °C	5–17 °C
Precipitation rates/a	300–1,700 mm	<400–>1,600 mm	<200–800 mm	<450–>1,800 mm
Endemism (rate)	62 %	10–12 %	8.4 %	4.4 %
No. of endemic plant families	2	0	0	0
No. of endemic genera	14	6	2	0
No. of endemic vascular plant species and subspecies	130	123–154	134	131
No. of vascular plant species and subspecies	c. 210	1,226	1,602	2,356
Density of endemics	c. 130/104 km^2	123–154/800 km^2	134/2,034 km^2	131/8,780 km^2
Endemism related to habitats, largest group	Forest (myrtisilva)	Rocks and screes	Rocks and screes	Rocks and screes
Endemism related to habitats, second largest group	Fern scrub	Forest (laurisilva)	Shrubland, scrub	Grassland

range from 450 mm in the driest parts to >1,800 mm in alpine regions (Gamisans and Marzocchi 1996). Thus, compared to the other islands Tenerife is relatively dry (cf. minima, maxima, averages).

Juan Fernández Islands have extremely high levels of endemism, c. 62 % on average. Furthermore, the islands harbour 14 endemic vascular plant genera and even two endemic plant families. The Madeira archipelago harbours 10–12 % endemics (including 6 endemic genera). According to our own counts based on Izquierdo et al. (2004) Tenerife has 8.4 % endemics, including two genera. The rate for Corsica is 4.4 % (no endemic genus). The rate of endemism itself is quite a good indicator for isolation. The lower rates of endemism on Tenerife and Corsica can easily be explained by the nearby mainland or island landmasses. Both islands are much less isolated compared to the other two archipelagos. As the much higher level of endemism of the Juan Fernández Islands shows, the degree of isolation of this archipelago is high compared to Madeira Archipelago. This is not easily explicable, as both archipelagos are located west of and at a similar distance to the continent and are not surrounded by other islands or archipelagos.

However, the absolute number of endemic vascular plant taxa on Juan Fernández Islands (104 km^2), the Madeira archipelago (800 km^2), Tenerife (2,034 km^2) and Corsica (8,750 km^2) is similar (c. 123–154 species and subspecies). Thus, the density of endemics is higher on Juan Fernández Islands than on Madeira Islands or Tenerife and much higher in comparison to Corsica.

We assume that these differences have to do with differences in ecological stability over long evolutionary periods. Climate and geological surface stability are important factors in explaining the differences; however, we can only speculate on the details.

Climate, in general, is buffered and stabilised by marine water bodies, especially by strong marine currents. As the oceanic archipelagos lie on the periphery of a circular current system the situation of the Juan Fernández, Madeira, and Canary Islands seems to be comparable. One clear difference, however, is the existence of a circumpolar current (Antarctic Circumpolar Current or West Wind Drift) in the southern hemisphere, which has no counterpart in the northern hemisphere. This strong current system which moves the cold southern oceans from West to East is today the largest ocean current and the major driver of exchange of water between the oceans. Estimates of the onset of this current range from 6 to 41 Ma (Barker et al. 2007). The discussion about the influence of such currents on climate or, the other way round, about the influence of climate change on the existence and stability of currents is still in progress (Hofmann and Rahmstorf 2009; Levermann et al. 2007; Zickfeld et al. 2007). However, it is possible that in the past the climate of Juan Fernández Islands was more stable than that of the Madeira archipelago and Canary Islands (Santelices 1992). Corsica, on the other hand, was more impacted by glacial cycles than the other archipelagos because of its position close to the European landmass.

There are also differences connected with the stability of the geological surface: for example, recent volcanism – i.e. during the past few centuries – is known from Tenerife but not from the other islands.

The habitat preferences of the endemics on these islands also differ remarkably. On Juan Fernández Islands, endemics are concentrated in montane forests (myrtisylva) and fern-dominated units.

In the Madeira archipelago, on Tenerife and Corsica most endemics occur in rocky habitats and screes. In most regions of Europe the highest proportion of endemics occur in this group of habitats.

Similar to the Azores, the second large group on the islands of Madeira comprises forest endemics. This fact is unusual compared to most regions of continental Europe or compared to Mediterranean islands and to the neighbouring Canaries (cf. Bruchmann 2011). The laurel forest of Madeira, the Canaries and Azores survived during glaciation cycles, whereas this forest type disappeared on the European mainland. Furthermore, in contrast to the Mediterranean, fire is very uncommon and mostly a man-made event on Madeira; also, the forest still covers large parts of the island. In contrast to the Canary Islands there has been no recent volcanism, and the main and maximum precipitation rate is much higher.

The second large group on the Canary Islands comprises endemics of shrub- and heathlands (Bruchmann 2011). This is more or less typical for Mediterranean regions. We assume that this relationship is also typical for the island Tenerife alone, as this is the largest island of the archipelago. Furthermore, Tenerife is richest in terms of different habitat types and species. Thus, we assume that the values for the whole archipelago are transferable to Tenerife, even if we do not have an analysis of the habitat affinities of Tenerife itself.

The second largest group of endemics on Corsica is found in grasslands. This is also unusual compared to other Mediterranean islands. For example, the second largest group on Sardinia, Sicily and the Balearic Islands consists of endemics of coastal and saline habitats. The second largest group following the endemics of rocks and screes on Crete and Cyprus comprises endemics with preference for scrub and heath habitats. The relatively large group of grassland endemics on Corsica might be explained by the large proportion of acidic granites in the central mountain regions which, additionally, receive relatively high rainfall. Also, summer rainfall in the high mountain regions of Corsica is a normal event. These climatic and edaphic conditions together promote the existence of grassland in alpine regions and might be advantageous over e.g. dwarf shrub or thorn cushion plant communities.

References

Araujo MB, Nogues-Bravo D, Diniz-Filho JAF, Haywood AM, Valdes PJ, Rahbek C (2008) Quaternary climate changes explain diversity among reptiles and amphibians. Ecography 31:8–15

Barker PF, Filppelli GM, Florindo F, Martin EE, Scher HD (2007) Onset and role of the Antarctic circumpolar current. Deep-Sea Res II 54:2388–2398

Beard JS, Chapman AR, Gioia P (2000) Species richness and endemism in the Western Australian flora. J Biogeogr 27:1257–1268

Bernardello G, Anderson GJ, Stuessy TF, Crawford DJ (2006) The angiosperm flora of the Archipelago Juan Fernández (Chile): origin and dispersal. Can J Bot 84:1266–1281
Binggeli I (2003) Introduced and invasive plants. In: Goodman SM, Benstedad JP (eds) The natural history of Madagascar. The University of Chicago Press, Chicago/London, pp 257–268
Borges PAV, Abreu C, Aguiar AMF, Carvalho P, Jardim R, Melo I, Oliveira P, Sergio C, Serrano ARM, Vieira P (eds) (2008) A list of the terrestrial fungi, flora and fauna of Madeira and Selvagens archipelagos. Direccao Regional do Ambiente da Madeira and Universidade dos Acores, Funchal und Angro do Heroismo, 440pp
Bouchard J (1968) Flore pratique de La Corse, 2nd edn. Société des Sciences Historiques et Naturelles de la Corse, Bastia
Bruchmann I (2011) Plant endemism in Europe: spatial distribution and habitat affinities of endemic veascular plants. Dissertation, University of Flensburg. Flensburg. URL: www.zhb-flensburg.de/dissert/bruchmann
Burga C, Zanola S (2007) Madagaskar – Hot Spot der Biodiversität. Exkursionsbericht und Landeskunde, Schriftenreihe Physische Geographie Bodenkunde und Biogeographie 55. Geographisches Institut der Universitat, Zurich, pp 1–201
Burton CJ, MacDonald JG (2008) A field guide to the geology of Madeira. Geological Society of Glasgow, Galsgow
Callmander MW, Phillipson PB, Schatz GE, Andriambololonera S, Rabarimanarivo M, Rakotonirina N, Raharimampionona J, Chatelain C, Gautier L, Lowry PP II (2011) The endemic and non-endemic vascular flora of Madagascar updated. Plant Ecol Evol 144(2):121–125
Carlquist S (1958) Anatomy and systematic position of *Centaurodendron* and *Yunquea* (Compositae). Brittonia 10(2):78–93
Castillon B, Castillon J-P (2010) Les Aloe de Madagascar. Published by the Authors, Reunion
Cribb P, Hermans J (2009) Field guide to the orchids of Madagascar. Kew Publishing, Kew
Cuevas JG, van Leersum G (2001) Proyecto conservación, restauración y desarrollo de las islas Juan Fernández, Chile. Rev Chil Hist Nat 74:899–910
Danton P (2006) La "myrtisylve" de l'archipel Juan Fernández (Chili), une forêt en voie de disparition rapide. Acta Bot Gallica 153(2):179–199
Danton P, Perrier C (2005) Notes sur la disparition d'une espèce emblématique: *Robinsonia berteroi* (DC.) Sanders, Stuessy & Martic. (Asteraceae), dans l'île Robinson Crusoe, archipel Juan Fernández (Chili). Le Journal de Botanique de la Société Botanique de France 31:3–8
Danton P, Perrier C (2006) Nouveau catalogue de la flore vasculaire de l'archipel Juan Fernández (Chili). Acta Bot Gallica 153:399–587
de Sequeira M, Jardim R, Silva M, Carvalho L (2007) Musschia isambertoi M. Seq., R. Jardim, M. Silva & L. Carvalho (Campanulaceae), a new species from the Madeira archipelago (Portugal). Anales del Jardim Botanico de Madeira 64(2):135–146
Dirnböck T, Greimler J, Lopez P, Stuessy TF (2003) Predicting future threats to the native vegetation of Robinson Crusoe Island, Juan Fernandez Archipelago, Chile. Cons Biol 17: 1650–1659
Dransfield J, Beentje H, Britt A, Ranarivelo T, Razafitsalama J (2006) Field guide to the palms of Madagascar. Kew Publishing, Kew
Foster MN, Brooks TM, Cuttelod A, De Silva N, Fishpool LDC, Radford EA, Woodley S (eds) (2012) Key biodiversity area special series. J Threat Taxa 4(8):2733–2844
Franquinho LO, Da Costa A (1996) Madeira: Plantas e flores. Francisco Ribeiro & Filhos, Funchal
Gamisans J, Marzocchi J-F (1996) La flore endémique de la Corse. Edisud, Aix-en-Provence
Gathorne-Hardy FJ, Syaukani, Davies RG, Eggleton P, Jones DT (2002) Quaternary rainforest refugia in south-east Asia: using termites (Isoptera) as indicators. Biol J Linn Soc 75(4):453–466
Gautier L, Goodman SM (2003) Introduction to the flora of Madagascar. In: Goodman SM, Benstedad JP (eds) The natural history of Madagascar. The University of Chicago Press, Chicago/London, pp 229–250
Gillespie RG, Clague DA (eds) (2009) Encyclopedia of islands. University Press of California, Berkeley

Goodman SM, Benstead JP (eds) (2003) The natural history of Madagascar. The University of Chicago Press, Chicago/London

Gottfried M, Pauli H, Reiter K, Grabherr G (2002) Potential effects of climate change on alpine and nival plants in the Alps. In: Körner C, Spehn EM (eds) Mountain biodiversity: a global assessment. Parthenion Publishing, Boca Raton et al, pp 213–223

Greimler J, López P, Stuessy TF, Dirnböck T (2002) The vegetation of Robinson Crusoe Island (Isla Masatierra), Juan Fernández archipelago, Chile. Pac Sci 56:263–284

Greimler J, López-Sepúlveda P, Reiter K, Baeza C, Peñailillo P, Ruiz E, Novoa P, Gatica A, Stuessy T (2013) Vegetation of Alejandro Selkirk Island (Isla Masafuera), Juan Fernández Archipelago. Chile Pac Sci 67(2):267–282

Gribb P, Hermans J (2009) Field guide to the orchids of Madagascar. Kew Publishing, Kew

Hallé F, Danton P, Perrier C (2004) Architectures de plantes de l'Ile Robinson Crusoe, Archipel Juan Fernández, Chili. In: Vester H, Romeijn P, van der Wal H (eds) 25 Jaar een "Boom der vrijheid" – Liber Amicorum, Prof Dr Ir R.A.A Oldeman, Treebooks 8. Treemail Publishers, Heelsum, pp 43–55

Hansen A, Sunding P (1993) Flora of Macaronesia. Checklist of vascular plants, Sommerfeltia 17. Botanical Garden and Museum, University of Oslo, Oslo, pp 1–295

Heads M (2011) Using island age to estimate the age of island-endemic clades and calibrate molecular clocks. Syst Biol 60:204–218

Hobohm C (2000) Biodiversität. Quelle & Meyer, Wiebelsheim

Hobohm C, Bruchmann I (2009) Endemische Gefäßpflanzen und ihre Habitate in Europa – Plädoyer für den Schutz der Grasland-Ökosysteme. RTG-Berichte 21:142–161

Hofmann M, Rahmstorf S (2009) On the stability of the Atlantic meridional overturning circulation. PNAS. www.pnas.org_cgi_doi_10.1073_pnas.0909146106. Downloaded 21 Sept 2012

Izquierdo I, Martín JL, Zurita N, Arechavaleta M (eds) (2004) Lista de especies silvestres de Canarias (hongos, plantas y animales terrestres) 2004. Consejeria de Medio Ambiente y Ordención Territorial. Gobierno de Canarias, Tenerife

Jardim R, Francisco D (2000) Flora Endémica da Madeira. Corlito/Setubal, Funchal

Jetz W, Rahbek C, Colwell RK (2004) The coincidence of rarity and richness as the potential signature of history in centres of endemism. Ecol Lett 7:1180–1191

Johow F (1896) Estudios sobre la flora de las Islas de Juan Fernández. Imprenta Cervantes, Santiago

Kämmer F (1982) Beiträge zu einer kritischen Interpretation der rezenten und fossilen Gefässpflanzenflora und Wirbeltierfauna der Azoren, des Madeira-Archipels, der Ilhas Selvagens, der Kanarischen Inseln und der Kapverdischen Inseln, mit einem Ausblick auf Probleme des Artenschwundes in Makaronesien. Polycopy, Freiburg i. Br, 179pp

Koechlin J, Guillaumet J-L, Morat P (1974 [1997]) Flore et végétation de Madagascar. Reprint. A.R.G. Gantner Verlag, Vaduz

Kunkel G (1993) Die Kanarischen Inseln und ihre Pflanzenwelt, 3rd edn. Gustav Fischer, Stuttgart

Labat JN, Moat J (2003) Leguminosae (Fabaceae). In: Goodman SM, Benstead JP (eds) The natural history of Madagascar. The University of Chicago Press, Chicago/London, pp 346–373

Levermann A, Mignot J, Nawrath S, Rahmstorf S (2007) The role of Northern sea ice cover for the weakening of the thermohaline circulation under global warming. J Clim 20:4160–4171

Looser G (1927) La zarzamora (*Rubus ulmifolius*) en Juan Fernández. Rev Chil Hist Nat 31:84–85, Santiago

Marticorena C, Stuessy TF, Baeza CM (1998) Catalogue of the vascular flora of the Robinson Crusoe or Juan Fernández Islands, Chile. Gayana Bot 55:187–211

Mittermeier RA, Myers N, Mittermeier CG, Gil PR (1999) Hotspots: earth's biologically richest and most endangered terrestrial ecoregions. Cemex, Mexico City

Mittermeier RA, Gil PR, Hoffman M, Pilgrim J, Brooks T, Mittermeier CG, Lamoreux J, da Fonseconda GAB (2005) Hotspots revisited: earth's biologically richest and most endangered terrestrial ecoregions. Cemex, Mexico City

Moat J, Smith P (eds) (2007) Atlas of the vegetation of Madagascar. Kew Publishing, Kew

Moreira-Muñoz A (2011) Plant geography of Chile, vol 5, Plant and vegetation. Springer, Dordrecht

Mucina L, Wardell-Johnson GW (2011) Landscape age and soil fertility, climate stability, and fire regime predictability: beyond the OCBIL framework. Plant Soil 341:1–23. doi:10.1007/s11104-011-0734-x

Muñoz Pizarro C (1969) El Archipiélago de Juan Fernández y la conservación de sus recursos naturales renovables. Museo Nacional de Historia Natural (Santiago) Serie Educativa 9:17–47

Muñoz Pizarro C (1971) Chile: Plantas en Extinción. Editorial Universitaria, Santiago

Murillo-Aldana J, Ruiz E (2011) Revalidación de *Nothomyrcia* (Myrtaceae), un género endémico del Archipiélago de Juan Fernández. Gayana Bot 68(2):129–134

Normand S, Ricklefs RE, Skov F, Bladt J, Tackenberg O, Svenning J-C (2011) Postglacial migration supplements climate in determining plant species ranges in Europe. Proc R Soc B 278:3644–3653

Petignat A, Cooke B (2009) Guide des plantes succulentes du Sud-Ouest de Madagascar. Arboretum d'Antsokay, Toliara

Pott R, Hüppe J, Wildpret de la Torre W (2003) Die Kanarischen Inseln: Natur- und Kulturlandschaften. Ulmer, Stuttgart

Press JR, Short MJ (1994) Flora of Madeira. The Natural History Museum, London

Raes N, Roos MC, Slik JWF, van Loon E, ter Steege H (2009) Botanical richness and endemicity patterns of Borneo derived from species distribution models. Ecography 32(1):180–192

Rauh W (1973) Über die Zonierung und Differenzierung der Vegetation Madagaskars, Tropische und Subtropische Pflanzenwelt 1. Steiner, Wiesbaden, pp 1–146

Sandel B, Arge L, Dalsgaard B, Davies RG, Gaston KJ, Sutherland WJ, Svenning J-C (2011) The influence of late quaternary climate-change velocity on species endemism. Science 334:660–664

Santelices B (1992) Marine phytogeography of the Juan Fernández archipelago: a new assessment. Pac Sci 46(4):438–452

Schatz GE (2001) Generic tree flora. Royal Botanic Gardens/Missouri, Botanical Garden, Kew Louis

Skottsberg C (1921) The phanerogams of the Juan Fernández Islands. In: Skottsberg C (ed) The natural history of the Juan Fernandez and Easter Islands, vol 2 (Botany). Almqvist & Wiksell, Uppsala, pp 95–241

Skottsberg C (1953) The vegetation of the Juan Fernández Islands. In: Skottsberg C (ed) The natural history of the Juan Fernandez and Easter Islands, vol 2 (Botany). Almqvist & Wiksell, Uppsala, pp 793–960

Skottsberg C (1956) Derivation of the flora and fauna of Juan Fernández and Easter Island. In: Skottsberg C (ed) The natural history of the Juan Fernández and Easter Islands, vol 1. Almqvist & Wiksell, Uppsala, pp 193–405

Stuessy TF, Grau J, Zizka G (1992) Diversidad de plantas en las Islas Robinson Crusoe. In: Grau J, Ziska G (eds) Flora silvestre de Chile, Palmengarten Sonderheft 19, Frankfurt am Main, pp 54–66

Stuessy TF, Swenson U, Marticorena C, Matthei O, Crawford DJ (1998) Loss of plant diversity and extinction on Robinson Crusoe Islands, Chile. In: Peng CI, Lowry PP (eds) Rare, threatened and endangered floras of Asia and the Pacific Rim, Academia Sinica monographs 16. Institute of Botany, Academia Sinica, Taipei, pp 243–257

ter Steege H, Pitman N, Sabatier D, Castellanos H, van der Hout P, Daly DC, Silveira M, Phillips O, Vsquez R, van Andel T et al (2003) A spatial model of tree alpha-diversity and tree density for the Amazon. Biodivers Conserv 12:2255–2277

Tutin TG, Burges NA, Chater AO, Edmondson JR, Heywood VH, Moore DM, Valentine DH, Walters SM, Webb DA (1996a) Flora Europaea, vol 1: Psilotaceae-Platanaceae, 2nd edn (Reprint, first published 1993). Cambridge University Press, Cambridge

Tutin TG, Heywood VH, Burges NA, Moore DM, Valentine DH, Walters SM, Webb DA (1996b) Flora Europaea, vol 2: Rosaceae-Umbelliferae (Reprint, first published 1968). Cambridge University Press, Cambridge

Tutin TG, Heywood VH, Burges NA, Valentine DH, Walters SM, Webb DA (1996c) Flora Europaea, vol 3: Diapensiaceae-Myoporaceae (Reprint, first published 1968). Cambridge University Press, Cambridge

Tutin TG, Heywood VH, Burges NA, Valentine DH, Walters SM, Webb DA (1996d) Flora Europaea, vol 4: Plantaginaceae-Compositae (and Rubiaceae) (Reprint, first published 1976). Cambridge University Press, Cambridge

Tutin TG, Heywood VH, Burges NA, Valentine DH, Walters SM, Webb DA (1996e) Flora Europaea, vol 5: Alismataceae-Orchidaceae (Reprint, first published 1980). Cambridge University Press, Cambridge

Vargas R, Cuevas JG, Le-Quesne C, Reif A, Bannister J (2010) Spatial distribution and regeneration strategies of the main forest species on Robinson Crusoe Island. Rev Chil Hist Nat 83:349–363

Vargas R, Reif A, Faúndez MJ (2011) The forest of the Robinson Crusoe Island, Chile: an endemism hotspot in danger. Bosque 32(2):61–70

Vieira R (1992) Flora da Madeira: o interesse das plantas endémicas Macaronesicas, Coleccao Natureza e Paisagem 11. Serviço Nacional de Parques, Lisboa, 155pp

Zickfeld K, Lvermann A, Morgan MG, Kuhlbrodt T, Rahmstorf S, Keith DW (2007) Expert judgement on the response of the Atlantic meridional overturning circulation to climate change. Clim Chang 82:235–265

Chapter 7
Endemism in Mainland Regions – Case Studies

Sula E. Vanderplank, Andrés Moreira-Muñoz, Carsten Hobohm, Gerhard Pils, Jalil Noroozi, V. Ralph Clark, Nigel P. Barker, Wenjing Yang, Jihong Huang, Keping Ma, Cindy Q. Tang, Marinus J.A. Werger, Masahiko Ohsawa, and Yongchuan Yang

7.1 Endemism in an Ecotone: From Chaparral to Desert in Baja California, Mexico

Sula E. Vanderplank (✉)
Department of Botany & Plant Sciences, University of California, Riverside, CA, USA
e-mail: sula.vanderplank@gmail.com

S.E. Vanderplank (✉)
Department of Botany & Plant Sciences, University of California, Riverside, CA, USA
e-mail: sula.vanderplank@gmail.com

A. Moreira-Muñoz (✉)
Instituto de Geografía, Pontificia Universidad Católica de Chile, Santiago, Chile
e-mail: asmoreir@uc.cl

C. Hobohm (✉)
Ecology and Environmental Education Working Group, Interdisciplinary Institute of Environmental, Social and Human Studies, University of Flensburg, Flensburg, Germany
e-mail: hobohm@uni-flensburg.de

G. Pils (✉)
HAK Spittal/Drau, Kärnten, Austria
e-mail: gerhardpils@yahoo.de

J. Noroozi (✉)
Department of Conservation Biology, Vegetation and Landscape Ecology, Faculty Centre of Biodiversity, University of Vienna, Vienna, Austria

Plant Science Department, University of Tabriz, 51666 Tabriz, Iran
e-mail: noroozi.jalil@gmail.com

V.R. Clark • N.P. Barker (✉)
Department of Botany, Rhodes University, Grahamstown, South Africa
e-mail: vincentralph.clark@gmail.com

C. Hobohm (ed.), *Endemism in Vascular Plants*, Plant and Vegetation 9,
DOI 10.1007/978-94-007-6913-7_7, © Springer Science+Business Media Dordrecht 2014

7.1.1 Introduction

Vegetation patterns on a global scale are primarily determined by climate. The Mediterranean climate of coastal California and the first true desert conditions in western North America are thought to have originated during the late Miocene and the Pliocene, 5–10 million years ago, with modern warm-desert vegetation becoming extensive approximately 12,000 years ago, after the end of the last glacial period (Axelrod 1978; Frenzel 2005; Raven and Axelrod 1978). Regional geography also has a significant impact on plant distributions; when the Baja California peninsula broke away from mainland Mexico new barriers to plant migrations were formed. The peninsula is recognized as a center of unique biodiversity, largely due to its isolation from the mainland (Garcillán et al. 2010). The northwestern region of the peninsula has a Mediterranean climate and chaparral vegetation, but to the south and east lies the Sonoran Desert, with vegetation adapted to some of the hottest conditions in the Americas.

This case-study is focused in the region between parallels 30 and 33 in Baja California, Mexico and SW California (see Fig. 7.1). The size of the region is approximately 14,000 km^2. Along the Pacific Coast of North America a Mediterranean climate prevails, resulting in a phytogeographic region known as the California Floristic Province (CFP) stretching from southern Oregon to northern Baja California. The Mediterranean climate is characterized by hot dry summers and cool wet winters. Differences in the air temperatures in the Tropozone cause jet stream winds to sweep down along the Pacific Coast. In North America, the Coriolis effect causes the jet stream to hit the Pacific Coast, and water precipitates out as the moisture-laden jet stream passes over land. The strength of the jet stream has a positive correlation with the latitudinal temperature gradient (i.e., a greater temperature difference between the North Pole and the equator results in a stronger jet stream). A rainfall gradient from winter to summer precipitation is

W. Yang • J. Huang • K. Ma (✉)
Institute of Botany, Chinese Academy of Sciences, Beijing, China
e-mail: kpma@ibcas.ac.cn

C.Q. Tang
Institute of Ecology and Geobotany, Yunnan University, Kunming, China
e-mail: cindytang@ynu.edu.cn

M.J.A. Werger
Department of Plant Ecology, University of Utrecht, Utrecht, The Netherlands

M. Ohsawa (✉)
Institute of Ecology and Geobotany, Kunming University in China, Kunming, China

Y. Yang
Faculty of Urban Construction and Environmental Engineering, Chongqing University, Chongqing, China

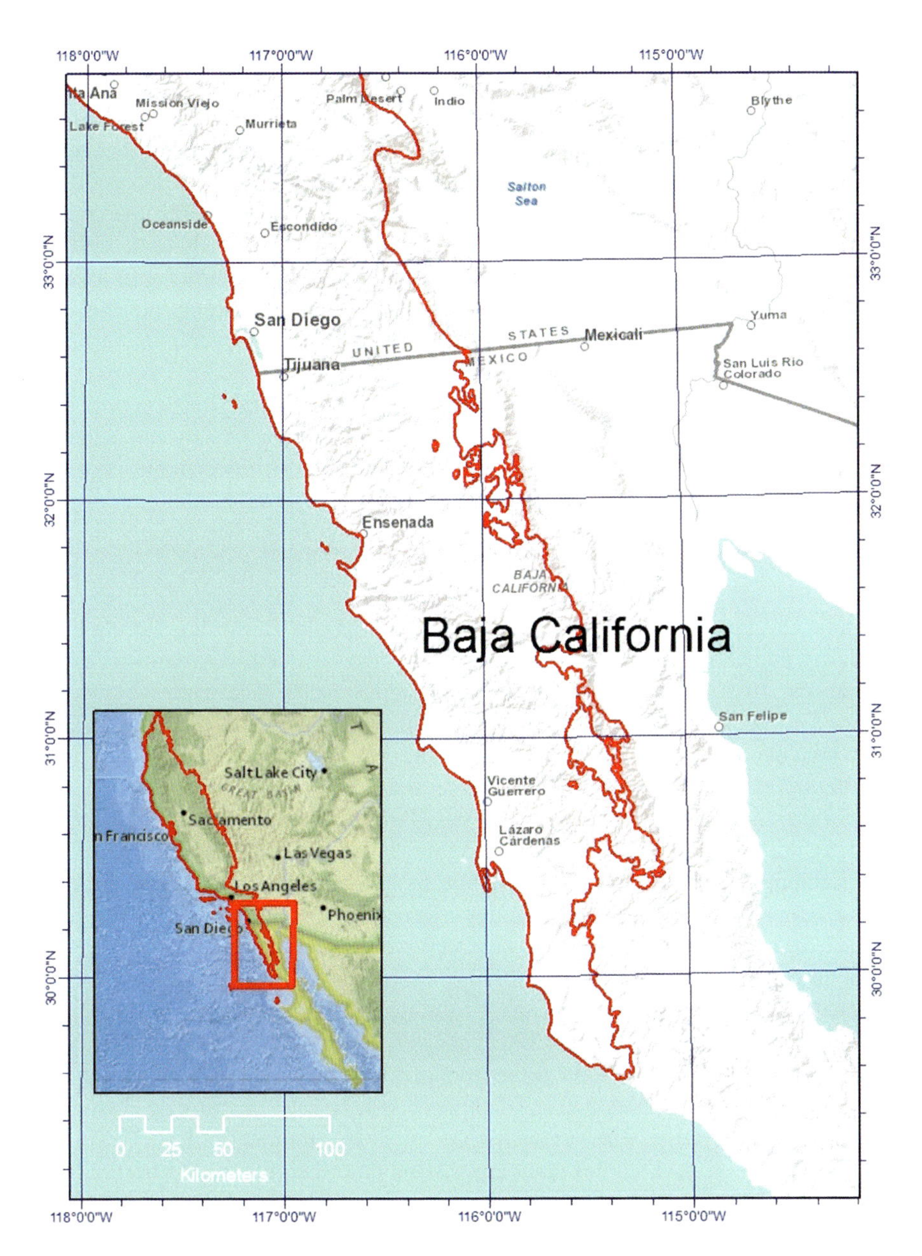

Fig. 7.1 California Floristic Province (*red shape*), Baja California and region of case study (*red shape* south of latitude 33° N)

clearly observed as one moves southward along the peninsula of Baja California (Aschmann 1959; Caso et al. 2007; Hastings and Turner 1965; Shreve 1936). Mean monthly temperatures vary little across northwestern Baja California (13 °C, or 55 °F, in winter and 23 °C, or 73 °F, in summer in the north at Tijuana, and

14 °C, or 57 °F, in winter and 24 °C, or 75 °F, in summer to the south at El Rosario). However, rainfall is not exactly predictable in the ecotone and the evapotranspirative balance varies significantly between seasons and between years. The massive 3,000-m tall Sierra San Pedro Mártir forms the spine of the peninsular ranges that act as a barrier to westerly winds and cause a rainshadow effect in the east. The southern end of the Sierra San Pedro Mártir meets the relatively low relief of the desert plain and that region coincides with the shift from Mediterranean climate to desert near the 30th parallel. That break in the topography allows north-westerly winds to blow across the peninsula moving moisture quickly across the landscape into the Gulf of California and the Mexican mainland, resulting in reduced precipitation in the peninsula.

The California Floristic Province (CFP) has been designated as a global biodiversity hotspot, an area of high endemism, that has been heavily impacted by human activity (Myers et al. 2000). The CFP, home to 2,125 endemic vascular plants, originally occupied 324,000 km^2; however, today only 80,000 km^2 – less than 25 %, remain naturally vegetated (Myers et al. 2000). The Baja California portion of the CFP is home to around 1,800 native vascular plant species, with almost half being rare, threatened, or locally endemic in the region. O'Brien et al. (in prep.) evaluated all plant taxa within the CFP of Baja California, scoring them for rarity and endemism, and documenting 172 plant taxa entirely endemic to the region (ca. 10 % of the native flora) and an additional 67 near-endemic species. These 239 species represent more than 11 % of total endemism for the entire CFP in an area that is less than 5 % of the size of the total province.

There has been debate concerning the southern limit of the CFP in Baja California, with some authors excluding the Maritime Succulent Scrub. The debate is partly due to the difficulty of classifying the vegetation as an increasing number of succulent plants are seen southward and more mesic plants become increasing scarce. However, the botanical community generally recognizes the southern limit of the CFP reaching the 30th parallel near El Rosario (Garcillán et al. 2010; Minnich and Franco-Vizcaíno 1998; Thorne 1993). Studies on the flora of San Quintín and Colonet also highlight the predominant contingent of the CFP flora in this region (Harper et al. 2010; Vanderplank 2011a, b).

Cowling et al. (1996) looked at the flora of California and other Mediterranean regions and found speciation from disruptive selection to be strongly driven by fire and climate change. Fire suppression is much less frequent in Baja California than in California, and studies comparing areas on either side of the border have shown dramatic changes in fire dynamics as a result of fire suppression in California (Minnich 1985, 2006). As a result of reduced fire suppression (stable fire interval and more frequent disturbance regime), the chaparral plants of Baja California may be more actively undergoing speciation processes, and perhaps more resistant to extinction. Within northwest Baja California there are two primary areas of high plant endemism: the mountain ranges and the coastal plain between Ensenada and El Rosario (Riemann and Ezcurra 2007). The influence of the relatively cold California current along the coast results in a heavy marine layer of low coastal fog that is highly stable and may sit over the land for days at a time (Vanderplank 2011a).

The effect of the cold current is strongest inland during the warmer months of the year and therefore fogs are most prevalent during those months, buffering the evapotranspiration balance until the winter months. Unique habitats have been identified within this region – namely the Maritime Succulent Scrub, noted for its high numbers of rosette-forming taxa that harvest moisture from the heavy coastal fogs (Rundel et al. 1972; Martorell and Ezcurra 2002). There are also small areas of Maritime Chaparral that favor non-sprouting chaparral species adapted to longer intervals between fire cycles and increased moisture from fog than inland areas.

The low relief coastal plain between Ensenada and El Rosario is an area of exceptionally high endemism in the state of Baja California (Garcillán et al. 2010; Riemann and Ezcurra 2007). As an area with a relatively strong climate gradient, this region may be key to species migrations under changing climate regimes. The area seems to be a hotbed of speciation and an important species refugium since the coastal fog mitigates some of the changing precipitation and temperature patterns (Minnich 2007). El Niño events typically favor higher winter rainfall and dense spring flowering events; but many years are very dry. The flora responds rapidly to local weather, which in turn affects all higher trophic levels, causing pulses in the availability of resources (Minnich 1985). The coastal plain has relatively little variation in elevation, but minor topographical features and underlying edaphic conditions have a strong effect on floristic composition. A rich patchwork of species assemblages (within the same broad vegetation belt) is also observed along the California coast. Microclimatic conditions appear to have a strong influence on plant distributions and putatively relict vegetation associations are fragmented throughout the Californias, and in particular in northwestern Baja California (Minnich and Franco-Vizcaíno 1998; Peinado et al. 1994; Delgadillo 1998; Raven and Axelrod 1978). The occurrence of micro-endemics in northwest Baja California suggests that plant distributions are not homogeneous and thus unlikely to be entirely controlled by broad climatic and latitudinal effects.

Most endemics inhabit succulent scrub and chaparral, at higher elevations also rocky habitats. There are very few endemics of riparian habitats because almost all riparian areas are arroyos that are dry most of the year (with underground water) and therefore the sandy soil tends to attract (semi-)desert plants. Some endemics like the blue palm (*Brahea armata*) are associated with permanent oases but they are few (Photos 7.1 and 7.2).

7.1.2 Primary Hypothesis

The coastal area between Ensenada and El Rosario appears to have been a plant species refugium with a more stable climate than the adjacent regions to the north and south. As an area that sits between two climate regimes, it presently has a more variable climate than either of the adjacent areas. Historically, however, climatic change was probably greater either side of this transitional area than within it. This hypothesis assumes that the area has had a climatic transition near the 30th parallel

Photo 7.1 Pristine Succulent Maritime Scrub, near Cerro Solo, on the coast of northwestern Baja California (Photographed by Sean Lahmeyer)

for most of the Quaternary, which seems reasonable in light of the Coriolis Effect on the jet stream. These weather patterns have resulted in a flora that is rich in both paleotaxa (maintained by the relatively stable climate through time) and neo-endemic taxa that appear to have radiated in response to changing climate not severe enough to cause extinction. The most significant refugium is seen along the coast where fog buffers variation in radiant loadings from the sun.

7.1.3 How Has Water Availability Shaped the Plant Communities?

Northwestern Baja California currently experiences relatively extreme short-term weather variation, ranging from severe droughts to major flooding events that may cause bridges to collapse and scour the arroyos of vegetation. Heavy rainfall events also create vernal pools, and snowmelt from winter storms in nearby mountains can augment water resources in the aquifer for long periods of time. Historical data indicate that in this region the moist glacial (Pleistocene) climates gradually transitioned into moist Holocene climates, with shifts in the source of precipitation

Photo 7.2 Endemic palm *Brahea armata* (Arecaceae) in the southern Sierra San Pedro Martir of Baja California. This area experiences light grazing and fires are occasionally started to keep mountain lions away from cattle and mules (Photographed by Sula E. Vanderplank)

and the radiant loadings (Bartlein et al. 1998). The predominant rainfall regimes north and south of the 30th parallel varied in intensity with the ecotone receiving some precipitation from each regime.

The degree of endemism in northwestern Baja California suggests that the amplitude of historic climatic change has not been as large as that in the adjacent regions. The plant communities of northwest Baja California are dynamic and, while some species have migrated, others have remained in situ and evolved in a locally changed climate regime. Ability to respond to climatic change is probably more limited in long-lived slow-growing species such as certain tree species as compared to shorter-lived herbaceous species due to generation times. The impact of climate on phenology may be the actual driver of plant evolution and migration in many cases (i.e., in short growing seasons flowering times are strongly correlated to temperatures and day lengths).

There are numerous putatively ancient paleoendemic taxa in northwestern Baja California (Raven and Axelrod 1978). The putative paleotaxa are often assumed to have undergone niche conservatism through time, suggesting that they occupy microclimates that are remnants of historical weather patterns. The southern side of the transverse range just south of Ensenada houses many of these species.

Small-scale topography and microclimates have a significant effect on species distributions. For example, conditions in a canyon (e.g., one about 20 m deep) will greatly alter the vegetation found on either side of the canyon walls and exposed ridges (which also differs from the wetter areas at the canyon bottom). Such canyons often support populations of narrow endemics and disjunct taxa. This distributional pattern suggests that this diversity of microhabitats and niches offers greater opportunity for refuge to the micro-endemic species of the region, hence the paleo-nature of many of the plants in the ecotone. Similarly, the large islands of Guadalupe and Cedros, offshore west Baja California, have altered evapotranspirative potentials from the fog that allow relict populations of pines (*Pinus*; Axelrod 1980) and cypress (*Cupressus*) to persist. The relative absence of fossil packrat middens (due to a humid climate; see Betancourt et al. 1990) in the region makes the reliable identification of paleoendemics challenging. However, the absence of close relatives suggests a long lineage in taxa such as *Adenothamnus*, and less obviously in monotypic genera such as *Xylococcus* and *Ornithostaphylos* (Francisco 2001).

In the same region there are also several genera with relatively numerous closely related species occupying small, often allopatric, ranges (e.g., *Arctostaphylos, Astragalus, Ceanothus, Dudleya*), which might be considered neoendemics or species undergoing adaptive radiations. Several species of *Hazardia* in northwest Baja California form a "patchwork quilt" when their ranges are mapped. Although largely not sympatric, each species borders the other, occupying a small geographic region. This pattern alludes to a common origin and perhaps an adaptive radiation event in their evolutionary history. These occurrences provide supportive evidence that the ecotone has been a refugium through times of global climate change, fostering both ancient lineages and more recent species diversifications. The buffering effect of the fog may have been a significant factor in the provision of climatic refugia for plants under changing climate.

7.1.4 Drought-Avoidance Strategies and Plant Physiognomies

Throughout the peninsula of Baja California there is great variation in rainfall both spatially and temporally (e.g., Hastings and Turner 1965). Studies by Franco-Vizcaíno (1994) show that even within the arid desert province to the south, floristic composition continues to vary southward along the gradient from winter to summer rainfall. The frequency and intensity of rainfall events affects soil properties and plant species composition (Shreve 1951). The unpredictability of water in this region has resulted in a suite of different drought-adaptations in the flora. Throughout this precipitation gradient a variety of life-strategies are observed.

Forrest Shreve (1936) published a seminal paper observing the precipitation gradient and the transition in the vegetation of the ecotone region between Ensenada and El Rosario in NW Baja California. In particular he noted the increased number of succulent species as one moves south, combined with decreasing numbers of chaparral species, and the increase in locally endemic taxa that are adapted to this small area with its unique climate. Shreve also commented on the nature of this ecotone and the increased heterogeneity of vegetation and life-forms, and therefore of the landscape.

Many of the drought-tolerant physiognomies of the northwest Baja California ecotonal plants have phenological phases that correlate directly to precipitation, with varied responses depending on the timing, intensity and duration of rainfall events. The phenological plasticity of the ecotone plants correlates with rainfall and fog moisture, often producing pulses that vary in magnitude with the availability of water.

Annual plants: This strategy represents the ultimate in opportunism for plants in an area of uncertain rainfall. After sufficient rainfall these species germinate rapidly, often flowering and fruiting very quickly without necessarily becoming large (Felger 2000). In northwestern Baja California these species are often showy with flowers and/or reproductive organs that may be larger than the vegetative structures. In areas of summer rainfall a contingent of C4 summer annuals appear that are well adapted to a very rapid growth and high temperatures following rains (Mulroy and Rundel 1977), however, the ecotone region has mostly winter annuals given the stochasticity of summer rainfall in modern times.

Deep-rooted trees: Several broad-leaved deciduous tree species in northwestern Baja California survive the dry warm late-summer–early fall weather by forming roots deep into perennial water sources. During Pleistocene times these trees species were generally more widespread in what is now arid North America. Currently they are largely restricted to permanent streams, and areas where bedrock pushes the aquifer into the reach of the roots. The role of mycorrhizal fungi also enables some of these species to survive in bedrock that roots cannot penetrate but does not prevent the fungal hyphae reaching the water table (Allen 2009). These tree species are generally wind pollinated and dispersed, with predictable flowering times. Only putative paleo-endemics, such as *Brahea armata*, and *Pinus muricata*, fall into this category in northwestern Baja California.

Evergreen chaparral: The evergreen sclerophyllous shrubs of California's chaparral exhibit a suite of characteristics that make them highly tolerant of seasonal (late summer and early fall) drought. Small, tough, leaves with short internodes have strong stomatal controls and high cuticular resistance to reduce water-loss (Minnich 1985). Their evergreen habit allows them to respond quickly after rainfall events and flower/fruit profusely. The large amounts of organic matter that these plants accumulate make them highly fire-prone in dry weather. Eventually the evapotranspirative ratio is so high that fire weather results in regular burns. As such, the distribution of the chaparral and its evolutionary history are tightly linked to the history of fire (Minnich 2006). Chaparral ecotone endemics include various parapatric species of *Arctostaphylos* and *Ceanothus*, genera which appear to still have high potential for future speciation. There are also several of the putatively paleo-endemic large woody shrubs (e.g., *Ornithostaphylos, Xylococcus*, and *Arctostaphylos* species such as *A. australis*).

Succulent plants: Succulence is a well-known strategy for dealing with the pressures of increasingly arid conditions and drought. In combination with CAM (crassulacean acid metabolism) photosynthesis, this strategy is particularly effective for resisting high radiant loadings and irregular precipitation. There are an increasing number of species in the Cactaceae and Crassulaceae at the southern end of the CFP, many of which are locally endemic (e.g., *Ferocactus fordii, Echinocereus maritimus, Mammillaria louisae*). Special adaptations are seen in *Mammillaria brandegeei*, a locally endemic species that actually has its vegetative body underground, exposing on the top of the plant at the soil surface. During drought it actually shrinks down into the soil, where the microclimate is much more equitable.

Drought deciduousness: Particularly common in the coastal scrub is a drought-deciduous habit, which is the condition of many narrowly endemic habitat dominants (e.g., *Ambrosia chenopodifolia, Aesculus parryi, Bahiopsis laciniata*). The number of months a plant is without leaves varies depending on local weather conditions. Drought deciduousness is often combined with other drought-tolerant strategies, e.g., succulent stems with drought-deciduous leaves (*Euphorbia misera*).

Rosettes and clumping: The Maritime Succulent Scrub has a dominant element of rosette-forming plants, e.g., the near-endemic *Agave shawii* subsp. *shawii* and various locally endemic *Dudleya* species. In Spanish it is known as the 'Matorral Costero Rosetofilo' or rosetophyllous coastal scrub. These rosette-forming plants have been shown to be efficient fog harvesters (Martorell and Ezcurra 2002).

Geophytes/underground storage organs (culms, bulbs, tubers, etc.): Although there are few true bulbs in the region (excepting a few species of wild onion (*Allium*) and mariposa lily (*Calochortus*)), there are several species that have some kind of under-ground storage organ (e.g., *Dichelostemma pulchellum, Marah macrocarpa, Jepsonia parryi*). These species avoid hot dry spells, remaining underground until soil moisture reaches levels appropriate to stimulate growth, yet there a very few, if any, endemic taxa with underground storage organs as their primary drought adaptation.

Some taxa have combined multiples of the above strategies. Of particular note are some of the most narrowly restricted taxa – three *Dudleya* species that grow in areas with different lithologies on Colonet mesa (Harper et al. 2010). All three are *Hasseanthus*-complex *Dudleya* species that are drought-deciduous with underground storage organs (rhizomes), and above-ground leaves are succulent and pseudo-rosetophyllous.

7.1.5 Plant Distributions: Rarity and Endemism

In reviewing the work of Grime (1977), Kruckeberg and Rabinowitz (1985) note that many plants that are adapted to come with an environmental extreme (e.g., serpentine endemism, halophytes) appear to be restricted to these extreme habitats only by competition from other species in more favorable habitats (i.e., many halophytes can grow in non-saline conditions, but appear to be out-competed in natural habitats). These processes highlight the role that repeated disturbances (e.g., fire) or stress (e.g., basic soils) will have on selecting for the composition of a vegetation type. There is evidence that species in the lower stages of plant succession are much less likely to go extinct; however, endemic species in climax communities are much more susceptible to catastrophic events and human activities (Kruckeberg and Rabinowitz 1985).

Vanderplank (2011a, b) found no correlation between the local and global abundance of taxa in the ecotone (i.e., a plant that was globally scarce was equally likely to be rare, frequent or abundant locally, as was a cosmopolitan species). This puts most of the endemics of this region in the 'locally abundant but restricted geographically' or 'constantly spare and geographically restricted' categories (Rabinowitz 1981). Stebbins (1980) states that the primary cause of localized endemism is adaptation to localized ecological factors. Kruckeberg and Rabinowitz (1985) note that narrow endemics are most often members of distinctive communities or singular habitats, but this statement is difficult to verify globally. The number of narrow or micro-endemics in the ecotone that are seemingly not restricted to a harsh environment or highly specific micro-habitat is noteworthy.

Qian (1998) showed that globally, along the latitudinal gradient, generic richness (in terms of the number of genera) shows a striking increase with decreasing latitude. However, recent research from Jansson (2003) indicates that global patterns in locally endemic taxa are caused by the amplitude of climatic change during peaks of Milankovitch oscillations (every 10,000–100,000 years). Smaller climatic shifts allow the survival of paleoendemics and diverging gene pools (neoendemics) are able to persist. Using change in mean annual temperature since the last glacial maximum Jansson showed that areas that have experienced higher temperature changes have lower endemism in mammals, birds, reptiles, amphibians and vascular plants (robust to area, latitude, extent of former glaciation and oceanic island syndrome). This research suggests that Rapoport's rule (species range increases with latitude) is a product of the increase in the amplitude of climatic oscillations towards the poles.

Consistent with the patterns seen in northwest Baja California, Sorrie and Weakley (2010) show that topography may be a minor consideration in endemism; for example, the coastal plain of Florida is flat (less than 250 m (800 ft) change in relief in the whole coastal plain) yet the state is second in endemism only to California. They suggest that this may be partially the result of the climatic transition between ecotones but also involves edaphic factors. Endemism in ecotones is not well-studied but often a peak in endemism is seen in ecotones. The forest-savannah ecotone in Africa has been shown to be important to divergence and speciation (Smith et al. 1997, 2005). If one compares the ecoregions of Baja California with the hotspots of endemism and species richness for the peninsula (Garcillán et al. 2010; Riemann and Ezcurra 2007), we see that where each ecotone occurs there is almost always a peak in species richness (presumably from the overlapping ranges of many species from two distinct biomes). In contrast; however, only sometimes is a peak in local endemism also seen in these ecotones, suggesting that the presence of an ecotone alone is not responsible for the elevated levels of endemism.

Recent research on the biotic interactions between desert plants has revealed that facilitation as a mutualism evolved between taxa from the Quaternary (nurse plants) and taxa from the Tertiary that were adapted to wetter conditions, but persist as a result of the nursing effects of more recently derived lineages (Valiente-Banuet et al. 2006). This facilitation saved many species from extinction during times of climatic change at the end of the Tertiary, and facilitation is often facilitated by a few key-stone 'nurse' species (Verdú and Valiente-Banuet 2008). Surprisingly the generalist nurses are often the most abundant species in the community, providing strong resistance to extinction for the dependent species. This facilitation also allows niche conservatism in ancient lineages and increases phylogenetic diversity in plant communities (Valiente-Banuet and Verdú 2007). However, in closely related species, these facilitative mutualisms can turn into competition in times of stress of changing climate (Valiente-Banuet and Verdú 2007; Verdú et al. 2003).

The amount of literature on the flora of California far exceeds that available for Baja California, yet it gives some insight into the origins of the present endemism and diversity (Stebbins and Major 1965; Richerson and Lum 2008; Thorne and Viers 2009; Vandergast et al. 2008; Viers et al. 2006). Most recently, Kraft et al. (2010) looked at the distribution of neoendemics in California, which correlated poorly to climate and topography. They found the endemics of the western edge of deserts to be very young, with most endemism in habitats that have undergone post-Pleistocene isolation or climatic change; with sky islands having wetter climates and the greatest temporal diversity of endemics.

Jansson (2009) stresses the relevance of emerging information on historical climate change that can be used to study microrefugia of the Pleistocene climate oscillations. The heterogeneous environment of the ecotone in Baja California lends itself to the concept of microrefugia and studies of the likelihood of microrefugia in this zone should be pursued. Médail and Diadema (2009) found that glacial refugia are climatically stable areas that are determined by complex historical and environmental factors. Refugia are priorities for the long-term conservation of species and genetic diversity, representing 'phylogeographical hotspots' especially under changing climate regimes.

Ackerly (2009) reviews the factors affecting the age and origin of California and Mediterranean vegetation. He introduces the concepts of synclimatic (with climate) migration, resulting in niche conservatism (as documented by Kelly and Goulden 2008); and anticlimatic (not following climate) migration which often results in adaptive evolution. As such, species that moved with climate have conserved niches, species that didn't move fast enough had to adapt. Ackerly (2009) points out that survival may be heavily dependent on biotic contexts and for those species not migrating with the climate it may be the presence or absence of competition from new species arriving (or not) that dictates whether a species survives the new regime (rather than it being wholly necessary to 'adapt' to the new climate). In light of this, successful 'adaptive' response may simply be factors that are barriers to dispersal of other species. This may be even a depleted gene pool in the potentially competing taxa. Ackerly states "By this logic, the greatest opportunity for adaptive evolution will occur on the trailing edge of species ranges during episodes of climate change, as changing conditions kill off the existing vegetation". Assuming there will always be a 'trailing edge' near the 30th parallel due to the position of the jet stream and its physical limits, one can tentatively hypothesize that the ecotone will always be a hotspot of potential for adaptive evolution under any climate change regime.

7.1.6 Conclusions

Since the southern extent of the jet stream does not pass further south than the 30th parallel, it is likely that there has long been a transition in the vegetation near the 30th parallel in northwestern Baja California, corresponding to the southern limit of winter precipitation originating from the west. It follows that plant taxa ranging through this region would be unable to follow synclimatic migration routes, and thus migration in the ecotone is likely to have been anticlimatic (favoring adaptive radiations). Assuming the climate change theories of Jansson (2003), the southern end of the CFP may hold more relict diversity and more neo-endemic lineages than other regions of the CFP. However, it is likely that the increased temperatures during the Pleistocene shifted some species farther north during the Pleistocene.

There are many putatively ancient 'paleoendemic' taxa in the ecotone region (Raven and Axelrod 1978); however, the absence of fossil packrat middens and a dearth of detailed fossil and molecular information on the region makes identification of paleoendemics challenging. There are also several genera in the ecotone that have many closely related species occupying small allopatric ranges (e.g., *Astragalus*, *Dudleya*, *Hazardia*), which might be considered neoendemics. I hypothesize that the 30th parallel ecotone has been a refugium through times of global climate change, fostering both ancient lineages and neoendemics. Predicting the severity of climate change as part of future Milankovitch cycles may be particularly valuable in assessing whether the ecotone will remain a refugium, or be likely to experience a high rate of extinction. Future research should look at range sizes of endemics in Baja California to allow mapping and phylogenetic distance analyses. The effects of species interactions remain unstudied, and nurse plant effects are only reported from the arid regions to the south. It seems probable that similar processes may be active in the ecotone, and perhaps other undiscovered interspecific processes are contributing to the distributions of the narrow endemics.

Water plays a vital role in the distribution of plants in the ecotone of northwest Baja California, and other dryland systems. Life-strategies, distributions, trophic interactions, phenological timings and evolutionary histories are strongly linked to patterns in water availability, and systems for drought-tolerance. Future research should address the resource-pulses caused by sporadic rainfall events and their effects on the entire ecosystem.

Acknowledgements I am very grateful to Richard Felger and John Vanderplank for their editorial assistance and long term support, also to Rich Minnich, Lucinda McDade, and Exequiel Ezcurra for their ongoing mentoring, and to my friends and colleagues Alan Harper, Ben Wilder, Chris DiVittorio, and Naomi Fraga for inspiring conversations. Many thanks to Luis Barragan and Terra Peninsular A.C. for Figure 1.

7.2 Ecuador

Carsten Hobohm (✉)
Ecology and Environmental Education Working Group, Interdisciplinary Institute of Environmental, Social and Human Studies, University of Flensburg, Flensburg, Germany
e-mail: hobohm@uni-flensburg.de

7.2.1 Physical Geography of Continental Ecuador

Continental Ecuador, situated in northwestern South America (latitudes 2°N to 5°S) and bordering the Pacific Ocean at the Equator, is divided into three broad biogeographical zones, the western coastal zone, the central Andes mountains (Sierra or highlands), and parts of the Amazon Basin (Oriente) to the east. Almost all rivers rise in the Andes region and flow eastwards to the Amazon River or westwards to the Pacific Ocean. The highest mountains are Chimborazo (6,310 m) and several others over 4,000 and 5,000 m.

The main temperatures range from 23 to 26 °C at the coast and 28 °C in the Amazon Basin to c. 0–3 °C at the highest summits where some parts are constantly covered with ice and snow. Pleistocene moraines occur in paramo zones. These indicate that many parts of the high mountain regions were glaciated during cold periods.

Extremely small differences between the average of the warmest and coldest months can be measured e.g. in Izobamba, Province Pichincha (3,058 m), and in Otavalo, Province Imbabura (2,555 m), where the difference is only 0.5 °C. In Quito (2,818 m) an amplitude of 0.6 °C and e.g. in Esmeraldas (6 m) an amplitude of 0.8 °C represent extremely low amplitudes in a global context (Deutscher Wetterdienst 2002, 2008, 2010). However, dependent on altitude, humidity and geomorphology diurnal temperatures can vary considerably, from cold mornings to hot afternoons.

Precipitation per year ranges from less than 300 mm in the driest parts to more than 5,000 mm in the wettest. Arid regions are located along the southern coast and in valleys of the Andes mountains, whereas northern coastal regions and western or eastern foothills of the mountains and the Upper Amazon Basin show high precipitation values. Precipitation in general decreases with increasing altitude and decreasing temperature in regions of the Andes.

According to Groombridge and Jenkins (2002) Ecuador is one of the endemic- and species-richest countries in the world. Large parts of two Biodiversity Hotspots, the Tropical Andes Hotspot and the Tumbes-Chocó-Magdalena Hotspot, together cover two thirds of the country's area (Mittermeier et al. 2005). The main vegetation belts from the coast to the highest mountains are mangroves, rain forest, wet to dry and deciduous forest, scrub paramo, other paramo types, rock and scree habitats.

7.2.2 Red List Analysis

Fortunately a very comprehensive Red List of endemic plants in Ecuador was published in 2000 (Valencia et al. 2000). This book is the basis and reason for the following review and analysis. If no other reference is given in this chapter, Valencia et al. (2000) is the source.

We counted numbers of endemics per vegetation type and altitude, for the list as a whole and specifically for certain plant families. Because a lot of species occur in more than one habitat type many of the following numbers represent overlapping species compositions and not necessarily different taxa.

13 habitat groups were defined by Valencia et al. (2000: 2 f.). These are:

Páramo húmedo or wet paramo (1), Páramo seco or dry paramo (2), páramo arbustivo or scrub paramo (3), bosque andino alto or high Andean forest above 2,000 m (4), Bosque andino bajo or Andean forest at mid-altitudes between 1,000 and 2,000 m (5), Vegetación interandina seca o húmeda or dry or moist Andean vegetation in the central Andean valleys normally at elevations between (1,000) 1,500 and 3,000 m (6), Bosque litoral piemontano or forest of western Ecuador in the transition zone between coast and mountain range at elevations from 500 to 1,000 m (7), Bosque litoral humédo or wet coastal forest at the northern coast (8), Bosque (o vegetación) litoral seco or dry coastal vegetation at the southern coast (9), Bosque amazónico piemontano or forest between 500 and 1,000 m in the transition zone between Andes and Amazonia (10), Bosque amazónico periodicamente inundato or inundated Amazonian forest below 500 m (11), Bosque amazónico de pantano or palm forest below 500 m normally with the palm *Mauritia flexuosa* (12), and finally (13) Bosque amazónico de tierra firma or Amazonian forest below 500 m on well drained soil.

Here we classify altitudes below 1,000 m as lowland, elevations between 1,000 and 2,000 m as mid-altitudes, and forests over 2,000 m as high-mountain forests.

7.2.3 Endemic Vascular Plants of Ecuador and Their Relationship to Ecological Conditions and Habitats

Of the 4,011 endemic vascular plant species in the whole of Ecuador (including 177 taxa from the Galapagos archipelago) a third (33 %) belongs to the Orchidaceae,

9 % to the Asteraceae, 5 % to the Melastomataceae, 4 % to Araceae, 4 % to various fern families, 4 % to Bromeliaceae, 3 % to Piperaceae, 3 % to Ericaceae, 2 % to Rubiaceae, 2 % to Campanulaceae, and 31 % to other plant families.

Most endemics are epiphytes (36 %), 26 % are shrubs or dwarf shrubs, 22 % are herbs, 8 % are small or high trees, 6 % are lianas and vines, and 2 % are hemi-epiphytes or others.

In the regions of continental Ecuador most endemics (more than three quarters) occur in forest or woodland and only a minority occurs in paramo vegetation, including scrub paramo in the subalpine belt and other habitat types such as scrub or thicket in the dry parts of the Andes or of SW Ecuador. Some opportunists occur in different habitat types. The orchid *Elleanthus ecuadoriensis*, for example, tolerates disturbed areas such as roadsides.

In the continental regions of Ecuador (without Galapagos) 2,965 endemic species occur in the Andes, 743 in the coastal zone (western Ecuador), and 453 in Upper Amazonia (eastern Ecuador).

Endemism peaks between 1,300 and 2,500 m. This finding is roughly consistent with results of the analysis of van der Werff and Consiglio (2004; their Fig. 2A) for Peru. When comparing Ecuador and Peru it must be remembered that the coastal zone in Peru is much norrower and extremely dry and the average elevation in the country as a whole is higher than in Ecuador. van der Werff and Consiglio (2004) found the richest belt in Peru between 2,500 and 3,000 m. They also found different elevational patterns for different functional types. Herbs and shrubs showed maximum numbers between 2,500 and 3,000 m, the majority of epiphytes was found at altitudes between 500 and 2,000 m, and trees, lianas and vines showed highest absolute numbers in the lowlands below 500 m.

However, different plant families also prefer different altitudinal zones and habitats. The two species-richest plant families in the world – Orchidaceae and Asteraceae – are also the two richest families in Ecuador. 1,318 orchid species (of 3,013 species in total) and 318 composite species (of 863 in total) are listed as endemic to Ecuador. Unlike orchids, ferns or bromeliad taxa, asteraceous species do not normally live as epiphytes, but, in Ecuador, are mostly shrubs or perennial herbs which root in the ground.

The endemic species diversity of the family Asteraceae peaks at higher altitudes in Ecuador than the orchids do. In paramo habitats there are the same number of endemic Asteraceae as orchids (c. 115). 22 endemic Asteraceae are catalogued for the dry inner parts of the Andes, whereas only a single endemic orchid occurs there. In high mountain forests the number of endemic orchids (536) is more than twice the number of endemic Asteraceae species (207), whereas below 2,000 m the number of endemic orchids explodes in relationship to Asteraceae (c. ten times more orchids).

Endemic ferns are also concentrated at relatively high altitudes e.g. with far more species in high mountain forest of the Andes (81 species in bosque andino alto) in comparison to mid-altitude forests (49 species in bosque andino bajo).

In contrast, the Bromeliaceae, for example, are represented by 70 endemic species at mid-altitudes and 40 species in high mountain forest.

Obviously, the high number of endemics in such a small country must be discussed in the context of high spatial heterogeneity combined with relatively stable climate conditions. The neighbourhood of old stable landscapes sensu Mucina and Wardell-Johnson (2011), as in the western and eastern parts of Ecuador, to high younger landscapes, such as the Andes, might have favoured total species richness and the development and survival of endemic vascular plants. The relatively low endemism in the Amazon Basin is most likely related to reduced topographical and geological heterogeneity.

Many endemics are threatened by the same human activities as in many other (wet) tropical parts of the world, such as deforestation and destruction of habitats as a result of the expansion of arable lands, settlements, roads, extraction of oil, and so on. There is no indication of threats caused by global warming.

7.3 Central Chile Ecoregion

Andrés Moreira-Muñoz (✉)
Instituto de Geografía, Pontificia Universidad Católica de Chile, Santiago, Chile
e-mail: asmoreir@uc.cl

7.3.1 Introduction

Chile's biota, trapped at the southern cone of South America by immense geographical barriers, has evolved in isolation, gaining a unique character. The limits of this "biogeographic island" are the Pacific Ocean to the west, the Andean cordillera to the East, the Peruvian Desert to the north and Cape Horn to the south. The condition of isolation dates at least from the Pliocene, when the Southern Andes gained their maximal altitude, setting the geographical (climatic, geomorphic, edaphic) and biological conditions for the evolution of a unique flora with high levels of endemism; nevertheless, the origin of several lineages has much deeper roots, associated with ancient Earth processes like the fragmentation of Gondwana (Moreira-Muñoz 2011). According to this interplay between ancient and newer conditions for the development of endemism, the vascular native flora of Chile shows high levels of endemism especially at high taxonomic levels: while having around 1,930 endemic species, it also has 80 endemic genera and 3–4 endemic families. The endemism at genus and family level is higher than in other territories widely recognized due to their high level of endemism, like Peru or Ecuador, that base their endemism at the species level. Only the Cape Flora of South Africa shows higher levels of endemism at genus and family level (Moreira-Muñoz 2011).

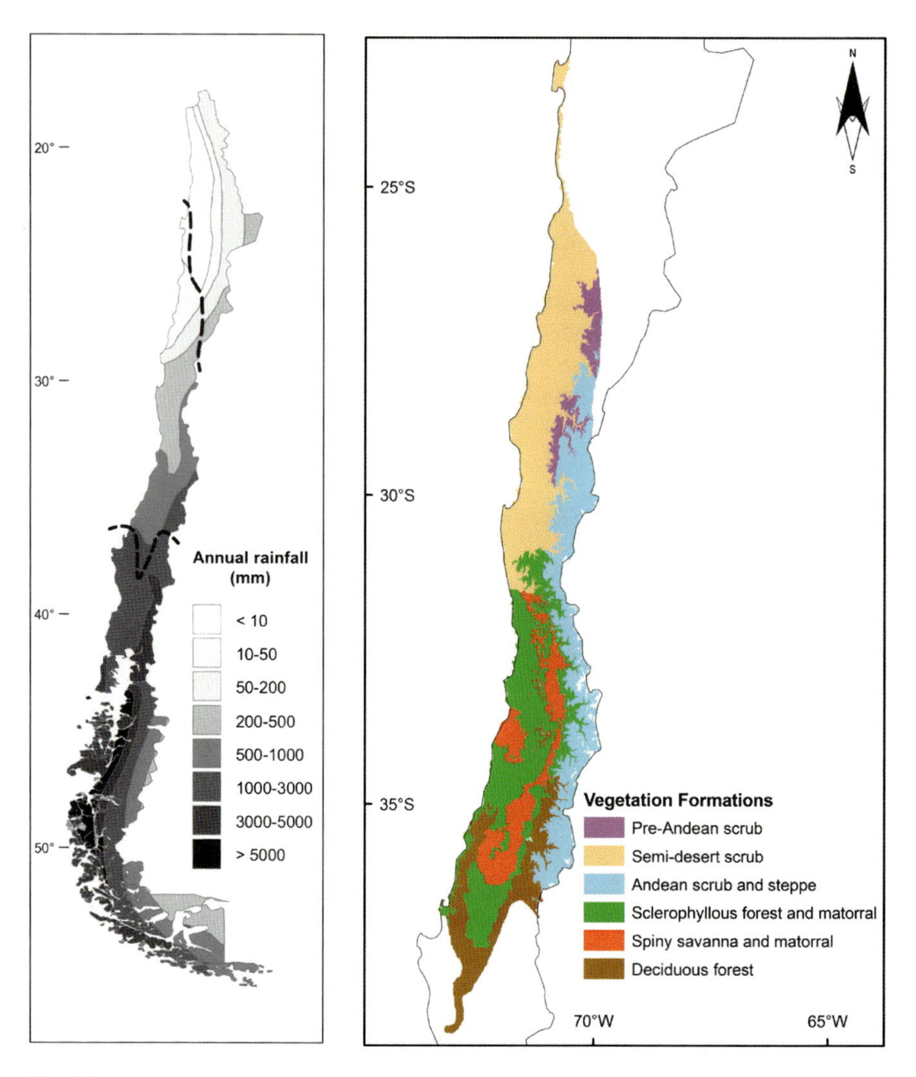

Fig. 7.2 *Left side*: Mediterranean Central Chile (*segmented line*) upon the annual precipitation gradient scheme; *right side*: Vegetation formations in Mediterranean Chile (Adapted from Schmithüsen 1956; Luebert and Pliscoff 2006)

7.3.2 Plant Endemism in Central Chile

The core of this notable floristic unique condition is Mediterranean Central Chile south of the Atacama Desert at latitude 23 °S (coast) and extending to the limit with temperate Chile at 38 °S (Fig. 7.2). The ecoregion shows a climatic gradient characterized by an increase in annual rainfall from less than 10 mm to more than 1,000 mm towards the South, generating many different habitats for plant growth and evolution (Fig. 7.2). This is one of five Mediterranean-type climate ecoregions of the world that have been recognized as main global centres of plant diversity (Cowling et al. 1996; Davis et al. 1997; Dallman 1998), and as biodiversity global hotspots. This last concept explicitly searches for global territories that show high species richness and endemism and are the subject of ample threats that currently affects their biota (Myers et al. 2000).

The original conception of the Chilean hotspot was restricted to the Mediterranean-type climate of Central Chile (Myers 1990); while nowadays it has been expanded towards temperate Chile including also the oceanic islands (Mittermeier et al. 2005). This broader conception of the Chilean hotspot increased its area from 180,000 km^2 to almost 400,000 km^2. The quantity of habitats also increases when expanding the hotspot towards the south. Several remarkable biotic characteristics justify this wider conception, like several faunistic groups and also floristic communities dominated by iconic native tree species such as araucaria (*Araucaria araucana*) and alerce (*Fitzroya cupressoides*). But the core of plant endemism is still to be found in Mediterranean Chile, more related to semiarid climatic conditions than to temperate ones.

Mediterranean Chile contains about 2,500 native species, i.e. 58 % of the country's native flora in 19 % of the territory. Endemics comprise around 1,200 species, related to an exceptional high habitat diversity (Myers 1990). From those Chilean endemics around 900 are strictly endemic to Central Chile (Arroyo et al. 2003).

The objective of this contribution is to address spatial patterns of plant endemism in main habitats (vegetation types) of Mediterranean Central Chile. Such studies are still lacking (see Beard et al. 2000), or have been developed partially for specific taxonomic groups, such as the genera *Adesmia* (Mihoc et al. 2006), and *Valeriana* (Kutschker and Morrone 2012), or the families Cactaceae (Guerrero et al. 2011) or Asteraceae (Moreira-Muñoz and Muñoz-Schick 2007).

7.3.3 Material and Methods

The checklist of species endemic to Chile, together with their distribution, has been compiled from different sources for the project "Plant Geography of Chile" (Moreira-Muñoz 2011). A main source has been the so far available volumes of the *Flora de Chile* (Marticorena and Rodríguez 1995 onwards) complemented and updated with available monographs and recent checklists (Zuloaga et al. 2008). When groups/families appear taxonomically inflated, a conservative approach has been preferred, e.g. for Chilean Alliaceae and Alstroemeriaceae (Muñoz-Schick and Moreira-Muñoz 2000, 2003b). Subspecies and varieties were considered as occupying part of the distribution range of the typical species. Some highly diverse groups within the Asteraceae have been recently revised by Moreira-Muñoz et al. (2012).

Table 7.1 Families rich in Chilean endemic species and occuring in Mediterranean Chile

Family	Endemic Chilean species
Asteraceae	222
Fabaceae	99
Solanaceae	61
Cactaceae	59
Calceolariaceae	43
Boraginaceae	38
Poaceae	37
Brassicaceae	34
Alstroemeriaceae	28
Apiaceae	24

The habitats at a meso-scale are the vegetation formations defined by means of bioclimatic methods by Luebert and Pliscoff (2006). These units are being considered as main ecosystems and as the appropriate units for conservation planning in Chile (Pliscoff and Luebert 2008). The distribution ranges of endemic species and vegetation formations were compared by means of ArcGis 9.3. To complement the analysis, a comparison with the different altitudinal levels was undertaken, on the base of the GTOPO30 elevation data set (www.usgs.gov). Species presence was compared with altitude ranges of 250–4,250 m asl.

7.3.4 Results: Endemism in Central Chile

Mediterranean Central Chile ranges from latitude 23 °S (coast) to 39 °S according to bioclimatic classifications (Amigo and Ramírez 1998; Luebert and Pliscoff 2006) (Fig. 7.2). The vegetation formations that compose this biodiversity hotspot show a north–south gradient of increasing vegetation cover from a semiarid sparse scrub to the sclerophyllous matorral and deciduous forests, reaching the northern limit of the austral conifer and laurifolious forests at 39 °S. The west–east gradient is a coast-Andean gradient, and ranges from the coastal scrubs through thicker matorral and forests towards the Andean scrub and the sparse treeline around 2,200 m depending on latitudinal position.

Mediterranean Chile harbours 1,164 Chilean endemic species. Most of them pertain to the Asteraceae (222), the Fabaceae (99), the Solanaceae (61, including genus *Nolana*), Cactaceae (59), and Calceolariaceae (43) (Table 7.1). These families are among the species-richest families in the Chilean flora (Moreira-Muñoz 2011, p. 59). Within these families we find also the species-richest genera (Table 7.2).

The overlap in distribution ranges with the vegetation formations shows a preliminary picture of the distribution of endemism within the Central Chilean hotspot. Table 7.3 shows the number of endemic species for each vegetation

Table 7.2 Genera rich in Chilean endemic species and occuring in Mediterranean Chile

Genus	Endemic Chilean species
Adesmia (Fabaceae)	61
Senecio (Asteraceae)	60
Calceolaria (Calceolariaceae)	41
Nolana (Solanaceae)	34
Haplopappus (Asteraceae)	27
Alstroemeria (Alstroemeriaceae)	26
Copiapoa (Cactaceae)	20
Oxalis (Oxalidaceae)	20
Viola (Violaceae)	20
Heliotropium (Boraginaceae)	15

Table 7.3 Vegetation formations from Central Chile and their number of endemic species

Vegetation formation	Area (km^2)	Chilean endemic species
Semi-desert scrub	54,738	677
Sclerophyllous forest and matorral	45,340	655
Spiny savanna and matorral	19,803	362
Andean scrub and steppe	37,000	355
Deciduous forest	22,019	301
Pre-andean scrub	8,230	63

formation. The highest number of Chilean endemic species is found in the Semi-desert scrub (677 endemics), (Photos 7.3, 7.4) followed by the Sclerophyllous forests and matorral (655 endemics), (Photos 7.5, 7.6) both reaching almost twice the numbers than the Spiny savanna (362), the Andean scrub and steppe (355), (Photos 7.7, 7.8) and the Deciduous forest (301). The Pre-Andean scrub, situated towards the north of the ecoregion, shows the lowest number of endemics, since it is also has the smallest area of the units analysed here.

In the altitudinal profile, there is a marked decrease in endemics towards higher belts, ranging from almost 750 endemic species at the coast between 0 and 250 m, decreasing to 500 at 1.000 m, 170 at 2.000 m, 140 at 3.000 m, and 50 at 4.000 m (Fig. 7.2). The profile shows little increases, e.g. around 2.250 m, associated with the increased richness in the Andean scrub around latitude 33° (Muñoz-Schick et al. 2000). The lower level of endemism at higher altitudes has to do with two facts: the overall decrease in species with altitude, and the presence of more native species that are also distributed in adjacent Argentina (Muñoz-Schick et al. 2000).

The overall tendency of decreasing numbers with increasing altitude will surely vary in each vegetation formation. Figure 7.3 shows that this tendency is emphasized by the two richer formations, the semi-desert scrub and the sclerophyllous forest and

Photo 7.3 Representative photographs of the Central Chilean Semi-desert scrub. (**a**) Coastal scrub North of Huasco, (**b**) *Cristaria cyanea and Zephyra elegans* fields at Aguada Tongoy, (**c**) *Rhodophiala bagnoldii* at Isla Damas, (**d**) Scrub with *Schizanthus litoralis* North of La Serena, (**e**) Inland scrub with *Bahia ambrosioides*, (**f**) *Copiapoa dealbata* fields at Carrizal coast, with Marinus Werger (Photographed by Andrés Moreira-Muñoz)

matorral. The spiny savanna and the deciduous forest show an opposite trend around 750 m, which has to do with the macro-relief morphology: the deciduous forest is practically absent from the Central Depression, while the Spiny savanna occupies a big area in this orographic unit (see Fig. 7.4). Contrary to the overall tendency, the Pre-Andean scrub and Andean scrub show different endemism peaks, more marked in the second one. This is due to overlap in formations at 1,500 m, a peak at 2,750 m, and a slow decrease towards the highest altitudes (Fig. 7.5).

Photo 7.4 A photographic selection of plant species endemic to the Central Chilean Semi-desert scrub. (**a**) *Dinemagonum gayanum* (Malpighiaceae), (**b**) *Calceolaria picta* (Calceolariaceae), (**c**) *Schizanthus candidus* (Solanaceae), (**d**) *Adesmia argentea* (Fabaceae), (**e**) *Pleurophora polyandra* (Lythraceae), (**f**) *Cistanthe longiscapa* (Montiaceae), (**g**) *Leucocoryne vittata* (Alliaceae), (**h**) *Centaurea chilensis* (Asteraceae), (**i**) *Cyphocarpus rigescens* (Campanulaceae) (Photographed by Andrés Moreira-Muñoz)

Photo 7.5 Sclerophyllous forest and matorral, Central Chile (Photographed by Andrés Moreira-Muñoz)

7.3.5 Discussion and Notes on Conservation

Mediterranean Chile, considered a global biodiversity hotspot, has a long history of anthropogenic impacts, including native forest and scrub substitution for urbanization, grazing and agriculture, habitat fragmentation, and direct exploitation of natural resources. This situation is similar to the other Mediterrranean regions in the world (Underwood et al. 2009). No wonder that most Chilean threatened species are located within the ecoregion (Muñoz-Schick and Moreira-Muñoz 2003a). A high amount of endemic species is found in the Semi-desertic scrub, which is mainly composed of shrubs and herbaceous species. Indeed, the two genera that are rich in endemic species, *Adesmia* and *Senecio*, do show the herbaceous to shrubby growth form. Coincidently, many Semi-desertic endemic species tend to show restricted geographical ranges and are of concern for conservation purposes (Squeo et al. 2001, 2008). This is also the case for endemic-rich genera from the Cactaceae, such as *Pyrrhocactus* and *Copiapoa*.

But not only genera rich in endemics are of interest for conservation planning. Many endemic species pertain to monotypic genera that are supposed to have also a high interest for biogeographers due to their phylogenetic uniqueness. Relative novel approaches like phylogenetic diversity should account for this so far cryptic

Photo 7.6 A photographic selection of plant species endemic to the Central Chilean Sclerophyllous forest and matorral. (**a**) *Peumus boldus* (Monimiaceae), (**b**) *Placea ornata* (Amaryllidaceae), (**c**) *Miersia chilensis* (Alliaceae), (**d**) *Bomarea salsilla* (Alstroemeriaceae), (**e**) *Adenopeltis serrata* (Euphorbiaceae), (**f**) *Lathyrus subandinus* (Fabaceae), (**g**) *Lithrea caustica* (Anacardiaceae), (**h**) *Schinus latifolius* (Anacardiaceae), (**i**) *Escallonia illinita* (Escalloniaceae) (Photographed by Andrés Moreira-Muñoz)

Photo 7.7 Spiny savanna, Central Chile (Photographed by Andrés Moreira-Muñoz)

diversity. Some of these genera are concentrated in the Semi-desertic scrub, such as *Balsamocarpon* (Fabaceae), *Dinemandra* (Malpighiaceae), *Leontochir* (Alstroemeriaceae), or *Pintoa* (Zygophyllaceae). Others have their main distribution in the Sclerophyllous forests and matorral, such as *Avellanita*, *Adenopeltis* (Euphorbiacea), and *Miersia* (Alliaceae). The Spiny savannah and the Andean scrub show less endemic monotypic genera, e.g. *Calopappus* (Asteraceae), occurring in the Andean scrub (Photo 7.8).

A constant question in assessments of endemism is whether the patterns are the result of rapid diversification or the remaining of old lineages, or "plant cradles" versus "plant museums" (López-Pujol et al. 2011). This has also to do with the latitudinal diversity gradient of increasing diversity towards the tropics, a global biogeographical pattern that has been mentioned recurrently but also contested. It is in part explained by a combination of processes, including large scale climate changes, tectonic and geological events, interactions among species, and differential rates of speciation and extinction of lineages (Ackerly 2009; Lomolino et al. 2010). In this ecoregion, plant richness and endemism seems to be related to climatic conditions and soils, but also to relief and morphology, i.e. to its heterogeneous geodiversity. In fact, research on other Mediterranean biodiversity hotspots like the California Floristic Province shows that diversity is not evenly distributed, mainly due to geographic complexity (Crain and White 2011, and this chapter).

Photo 7.8 Andean scrub, Central Chile (Photographed by Andrés Moreira-Muñoz)

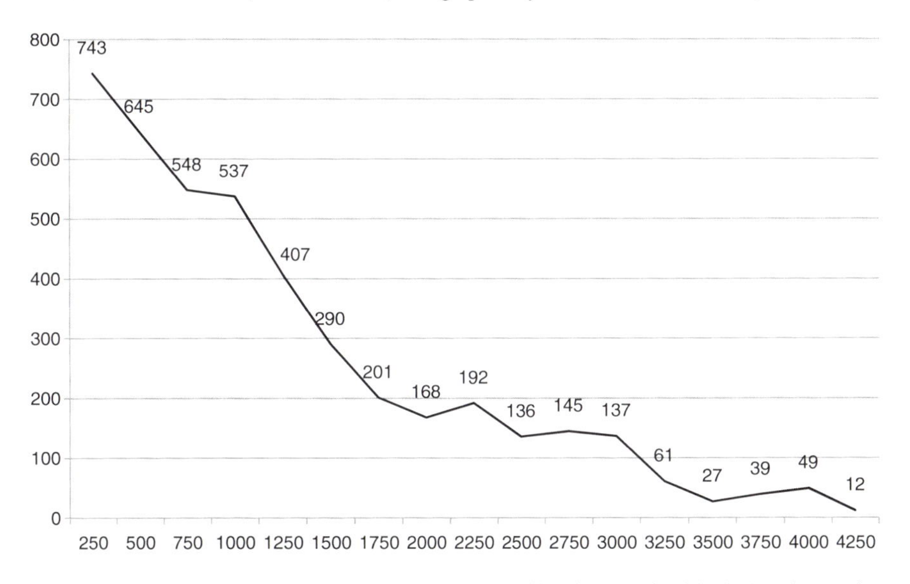

Fig. 7.3 Number of endemic species in each stretch of 250 m increase in altitude (y-axis: number of endemic taxa, x-axis: altitude in m)

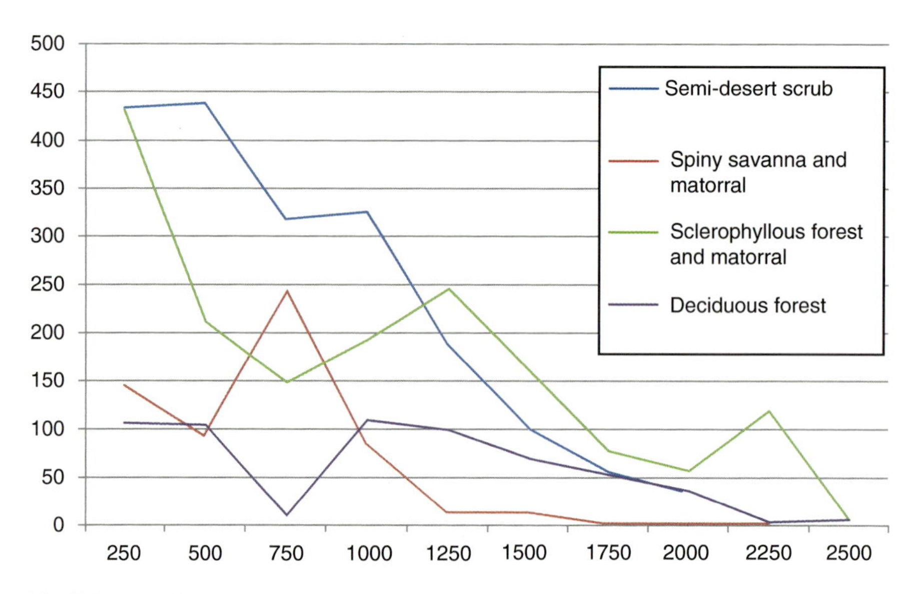

Fig. 7.4 Endemic species in different low altitude formations (y-axis: number of endemic taxa, x-axis: altitude in m asl.)

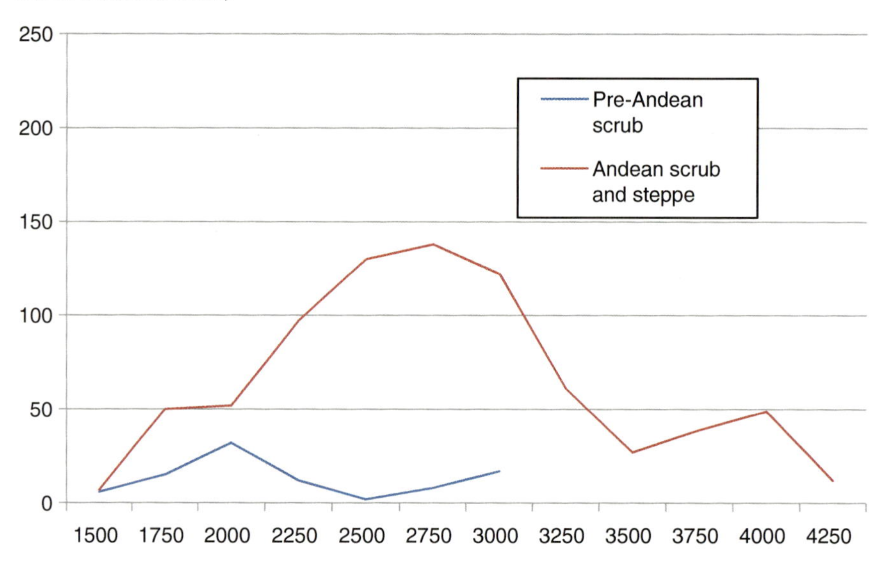

Fig. 7.5 Endemic species at different altitudes for high altitude formations (y-axis: number of endemic taxa, x-axis: altitude in m asl.)

Anyhow, endemism seems to be, at least in Mediterranean-type ecoregions, a consequence of a long interaction between relative dynamic speciation processes for several groups and species persistence for others. This last process seems to usually encompass extinctions and can be the reason for the high presence of monotypic genera. In this sense, centres of endemism can be considered as past centres of cladogenesis (Croizat et al. 1974; Jetz et al. 2004; Heads 2009). Indeed, Central Chile has been recognized as a contact zone between the Austral realm and the Neotropical floristic realms (Moreira-Muñoz 2007), composed by pre-Mediterranean lineages (Verdú et al. 2003), and with deep roots in old Gondwanic connections.

In the contact to the tropics, palaeoclimatical trends towards increasing aridity can explain the floristic and ecomorphological patterns detected in the Mediterranean ecoregions (Axelrod 1975). This seems to be the case of the Semi-arid desert scrub endemics. Regional broad geological-climatic events like the Middle-Miocene to Pliocene Andean uplift seem to be highly related to the increasing aridity of the Atacama Desert, having crucial effects on the relative rapid evolution and consequently high level of endemism of several groups. This is potentially one of the main causes for the origination of endemics via vicariant speciation, in unrelated genera such as *Heliotropium* (Luebert and Wen 2008), *Nolana* (Dillon et al. 2009), or *Copiapoa* (Guerrero et al. 2011). This might be also the case for disjunct distribution patterns in Desert-scrub endemics (Viruel et al. 2012).

It is highly necessary to improve the knowledge of the ecological, evolutionary and landscape-level processes that are responsible for the high level of endemism, integrating range size data with molecular-based estimates of taxon age (Kraft et al. 2010). In the near future a better understanding of endemism patterns and processes at different scales will be crucial for conservation planning and reserves design (Crain et al. 2011). Taxonomic knowledge is still incomplete and constantly changing, and endemic species are being revised, renamed and resurrected (e.g. Muñoz-Schick and Moreira-Muñoz 2008; Muñoz-Schick et al. 2011). Geographic distribution knowledge is usually even more elusive (e.g. Muñoz-Schick et al. 2010). These are the most important challenges for biodiversity information and decision-making within the novel approach of conservation biogeography (Whittaker et al. 2005).

Acknowledgements This contribution has the financial support of Fondecyt project N° 1120448. Mélica Muñoz-Schick from SGO herbarium helped with information regarding the endemic species' taxonomy and distribution. Patricio Pliscoff (University of Lausanne) provided the GIS data of the vegetation formations. Miguel Verdú (CSIC-UV-GV) kindly sent important papers.

7.4 Europe's Mainland

Carsten Hobohm (✉)
Ecology and Environmental Education Working Group, Interdisciplinary Institute of Environmental, Social and Human Studies, University of Flensburg, Flensburg, Germany
e-mail: hobohm@uni-flensburg.de

7.4.1 Physical Geography of Europe's Mainland

The climate in the North of Europe is arctic or subarctic, in Central Europe temperate, in South Europe Mediterranean with winter rain and summer drought. Mean temperatures range from some degrees below 0 °C in northern Scandinavia, northern Russia and high mountain zones of the Alps to c. 19 °C in some parts of the Mediterranean.

Annual precipitation totals range from <200 mm at Cabo de Gata, SE Spain, to over 2,000 mm in high mountain zones of the Alps and in oceanic regions of South Norway (e.g. Bergen: 2,250 mm/a).

Continentality increases from the West to the East. The difference between the average temperature of the coldest and the warmest month is less than 15 °C in some western parts (e.g. W France), but also in the highest mountain zones of the Alps. The difference is much larger in eastern parts of European Russia, with values >30 °C (e.g. Astrakhan: –5.5 °C in January, 25.2 °C in July, Syktyvkar –16.7 °C in January, 17.2 °C in July; cf. Deutscher Wetterdienst 2002, 2008).

Most substrates (granite, gravel, sand, organic layer) in the North are acidic, whereas the amount of calcareous and/or alkaline substrates increases to the South (Ewald 2003).

Vascular plant endemism and species richness at regional scales is higher in regions of the Pyrenees, Alps, Carpathians, mountains of the Balkan Peninsula and in countries bordering the Mediterranean Sea than in northern regions which were flattened by glaciers or affected by extremely cold arctic climate during the Pleistocene.

7.4.2 Analysis of EvaplantE

The goal of this chapter is to analyse mainland endemism of the vascular plants in Europe in relation to habitat and to compare the results with findings from islands, other mainland regions, or the whole of Europe, i.e. mainland and island regions combined (Bruchmann 2011).

Therefore, all data and information about island regions in EvaplantE (latest updated version) were eliminated. Data of taxa which occur on both mainland and islands of Europe were also eliminated.

Mainland Europe as defined here excludes the larger islands of the Mediterranean Sea and the Atlantic Ocean: Cyprus, Crete, Sicily, Corsica, Sardinia, the Balearic Islands, the Canary Islands, Madeira Islands, the Azores, Great Britain, Ireland, Faroe Islands, Iceland and Svalbard (with some smaller islands and islets in their vicinity; cf. Tutin et al. 1996a, b, c, d, e; Fontaine et al. 2007), but includes e.g. the Channel Islands and many Greek islands.

We predefined Mainland Europe into 28 regions. *Mainland endemics* are restricted to one or more regions We define an endemic which occurs in a single region out of the 28 predefined regions of mainland Europe as *1-region-endemic* (*local-endemics* sensu Bruchmann 2011). *European endemics* are all taxa which are restricted to the continent (42 predefined regions), independent of the fact as to whether these occur on islands or the mainland, in a single region or in many regions of Europe.

The percentage values in the table are standardised to a sum of 100 % (see Bruchmann 2011 or Hobohm and Bruchmann 2009). This is important because many endemics are listed in more than one habitat category. Thus, the non-standardised percentage-values would be higher and add up to more than 100 %.

7.4.3 Endemism of Mainland Europe

The flora of Europe is relatively well known. However, the number of vascular plant taxa and endemics has often been seriously underestimated. Davis et al. (1994), for example, estimate the number of endemic vascular plant species at 3,500, whereas Hobohm and Bruchmann (2009) estimate that at least 6,500 vascular plant taxa (species, subspecies and groups of microspecies) are endemic to Europe. This number would be even higher if a narrower concept of what constitutes a species would be applied. We estimate that about 20–30 % of all European vascular plant taxa are endemics.

According to EvaplantE (latest updated version), Europe is inhabited by 6,242 European endemics (of which 4,827 are characteristic for specific habitats), including 4,243 mainland endemics (3,084 characteristic). 1,805 mainland endemics are restricted to one of the 28 regions (1,761 1-region-endemics characteristic).

Table 7.4 shows numbers of European mainland endemics and 1-region-endemics of the European mainland characterised by habitats.

Rocky habitats harbour the highest number of mainland endemics (>1,800). The second largest group is grassland endemics (>1,100) followed by scrub and heath (>600), forest (>400), arable lands, horticultural and artificial habitats (>250), coastal/saline habitats (200), freshwater habitats (>150), and finally mires and swamps (>50).

Obviously, the mean altitudes of endemics occurring in rocky habitats or grassland, and most other habitats, are higher than the mean altitudes of endemics in coastal or anthropogenic habitats.

The proportion of basiphytic endemics in Mainland Europe is higher than that of acidophytes. Only the inhabitants of swamps and mires show comparable numbers of basiphytes and acidophytes.

Large regions in southern Europe, such as Spain, France or former Yugoslavia, show high total numbers of endemics.

Table 7.4 Numbers of European mainland- and 1-region-endemics in relation to habitat (EvaplantE; latest updated version, numbers of taxa are minimum values)

	Coastal/saline	Freshwater	Mires and swamps	Forest	Scrub and heath	Grassland	Rock and scree habitats	Arable, horticultural, artificial
Proportion of the whole area (Hobohm and Bruchmann 2009)	<1 %	A few %	A few %	<30 %	>10 %	<10 %	A few %	>>30 %
No. of European mainland endemics in total	200	178	58	422	639	1,134	1,813	277
In %	4.2	3.8	1.2	8.9	13.5	24.0	38.4	5.9
Min. altitude (m asl., median)	0	510	350	300	300	700	800	50
Max. altitude (m asl., median)	400	2,000	2,100	1,780	1,750	2,100	2,100	1,230
No. of basiphytes	46	30	12	83	167	243	694	50
No. of acidophytes	4	17	10	23	56	84	127	14
No. of 1-region-endemics	111	46	10	100	212	300	845	137
In %	6.3	2.6	0.6	5.7	12.0	17.0	48.0	7.8
No. of European mainland endemics by region								
Al (Albania)	3	12	5	71	93	179	266	24
Au (Austria)	6	62	19	143	164	363	310	16
Be (Belgium, Luxembourg)	2	1	1	15	16	23	9	7
Bu (Bulgaria)	10	18	7	98	116	248	252	29
Cz (Czech Republik, Slovakia)	7	34	13	112	108	220	131	17
Da (Denmark)	9	2	2	9	9	13	3	3

(continued)

Table 7.4 (continued)

	Coastal/saline	Freshwater	Mires and swamps	Forest	Scrub and heath	Grassland	Rock and scree habitats	Arable, horticultural, artificial
Fe (Finland)	11	1	5	16	7	19	5	2
Ga (mainland France)	31	71	18	127	178	379	444	48
Ge (Germany)	13	54	19	100	102	232	177	18
Gr (Greece, except Crete)	17	17	4	75	112	181	459	74
He (Switzerland)	1	50	14	91	109	270	239	19
Ho (The Netherlands)	4	2	1	7	8	10	1	6
Hs (mainland Spain)	94	72	21	121	292	302	566	129
Hu (Hungary)	10	8	8	82	73	114	53	16
It (mainland Italy)	15	55	19	149	184	436	522	44
Ju (former Yugoslavia)	25	41	20	179	204	455	519	51
Lu (mainland Portugal)	48	31	10	45	115	68	103	60
No (Norway, without Svalb.)	5	7	4	13	8	18	11	2
Po (Poland)	9	27	11	77	75	159	103	14
Rm (Romania)	13	32	16	134	131	279	216	22
RsB (f. Eur. SU, Baltic)	10	5	6	20	12	23	5	3
RsC (f. Eur. SU, Central)	7	9	4	42	40	52	27	15
RsE (f. Eur. SU, East)	2	6	0	18	16	34	22	6
RsK (f. Eur. SU, Krim)	8	1	0	20	17	32	45	10
RsN (f. Eur. SU, North)	4	1	4	11	4	16	6	2
RsW (f. Eur. SU, West)	17	34	9	99	95	181	106	13
Su (Sweden)	11	7	7	18	12	30	20	2
Tu (European part of Turkey)	3	3	1	14	12	25	21	6

7.4.3.1 Coastal and Saline Habitats

This group includes 4.2 % of the endemic vascular plants that are restricted to the mainland of Europe as defined in this chapter. Endemism of coastal and saline habitats shows a southwestern gradient. The Iberian Peninsula and western part of the Mediterranean is richer in endemic plants that occur in salt marshes, sand dunes or coastal rocks than the eastern part of the Mediterranean or northern parts of the Atlantic coasts. This might be due to climate conditions in the past, the age of water bodies, marine currents and the constancy of local ecological conditions including age of the geological surface.

7.4.3.2 Freshwater Habitats, Mires and Swamps

3.8 % of the European mainland endemics are associated with freshwater habitats, 1.2 % with mires and swamps. Plant endemism of water bodies, swamps and mires, or seasonally inundated habitats, is relatively low, as in most wetlands of the world. We explain this fact by the young age of most aquatic habitats: swamps, mires, lakes, rivers, or seasonally inundated habitats are almost all much younger than 10,000 years old. Additionally, these habitats are very often isolated. Isolation combined with young age restricts endemism.

7.4.3.3 Forest

According to numbers published by FAO (2010) forest coverage is 34 % for the whole of Europe excluding the Russian Federation and 45 % including the Russian Federation. These relatively high percentages depend on the definition of *forest*. The definition of the FAO covers much more than what biologists would normally include. Hobohm and Bruchmann (2009) estimate that the percentage of forest (defined as vegetation type which is dominated by trees, crown cover >50 %) for the whole of Europe (including the European part of the Russian Federation) is less than 30 %. However, the area is much larger than the area of e.g. rocks and screes, coastal and saline habitats, grassland or scrub and heath.

Compared to endemics of other habitat types the number of forest endemics (8.9 %) is relatively low. We assume that this is due to Pleistocene glaciation periods which destroyed most of the forests. Many forest plants which nowadays occur in temperate or boreal climate zones of Europe survived during the cold periods in refugia of southern Europe (cf. e.g. Médail and Diadema 2009; Krebs et al. 2004).

7.4.3.4 Scrub and Heath

About 13.5 % of the European mainland endemics are associated with scrub and heath. We do not know much about distribution patterns of scrub and heath habitats or about the size of these habitats in Europe. They probably cover a little more than 10 % of the surface. However, in Europe more endemics are related to scrub and heath than e.g. to forest. Additionally, the mean ranges seem to be smaller than those of forest endemics; the mean range size (median) covers a single region in the case of endemics related to scrub or heath and two regions for forest endemics.

Portugal has more endemics in scrub and heath communities than in any other habitat category (Table 7.4).

7.4.3.5 Grassland

In comparison to the whole of Europe including the islands (Bruchmann 2011) the percentage values for grassland endemics that are restricted to the mainland is higher (24 % as opposed to 18.3 % for the whole of Europe). This trend is also reflected by a higher number of 1-region-endemics (17 % grassland-endemics on the mainland and 10 % for the whole of Europe).

Grassland endemics which only occur in a single region of the mainland are the largest group in 5 regions, whereas endemics of rocky habitats are the largest group in 16 regions (cf. Bruchmann 2011).

The situation is the opposite if one compares numbers of all mainland endemics. In this case, most regions have higher numbers of endemics associated with grassland rather than rocky habitats. Grassland endemics are the largest group in 19 regions; in 8 regions, endemics of rocks and screes represent the largest group.

This is in accordance with the finding (Hobohm and Bruchmann 2009) that shows that the mean ranges of rock endemics in Europe are smaller (median 1) than the ranges of grassland endemics (median 3). Thus, many grassland endemics of mainland Europe are distributed over two, three or more regions (countries). This may result from the connectivity of grassland habitats and transhumance in the past.

Furthermore, grassland endemics are stronger concentrated in temperate regions, in Central and eastern Europe, than endemics of other habitat types. Almost all northern and eastern regions as well as regions in Central Europe (Switzerland, Austria) harbour more grassland taxa that are endemic to mainland Europe than endemics of other habitat types (see Table 7.4).

7.4.3.6 Rock and Scree Habitats

Rocky habitats including rock outcrops, screes and caves, in general, can be found all over the mainland of Europe and in all altitudinal zones. At the moment we do not know the exact coverage of this habitat category in Europe, but we assume that rocks and screes cover a few percent of the total surface (Hobohm and Bruchmann 2009).

More than a third (38.4 %) of the European mainland endemics inhabits rocky habitats. In 16 regions of continental Europe rocky habitats harbour the highest number of 1-region-endemics (Bruchmann 2011).

The large number of endemics in rocky habitats of Europe might be explained by *stability* of ecological conditions (high diversity of microclimates over short distances) and short-distance *vertical displacement* of the inhabitants during climate change (cf. Rull 2004). The large number cannot simply be explained by area because other habitat types cover larger areas and have lower endemism.

7.4.3.7 Arable, Horticultural and Artificial Habitats

Compared to other regions in the world, a relatively large proportion of endemics (5.9 %) occurs regularly in anthropogenic habitats of European mainland. A few taxa, such as *Anthemis lithuanica, Bromus secalinus ssp. multiflorus, Bromus interruptus, Carduus litigiosus, Centaurea polymorpha, Erucastrum gallicum* and *Urtica atrovirens,* are relatively strongly associated with arable lands or ruderal habitats and do not normally occur in semi-natural or natural habitats (Hobohm and Bruchmann 2009; Hobohm 2008). Gams (1938) and Pignatti (1978, 1979) discussed the possibility of the coevolution between plant and man.

7.5 Turkey

Gerhard Pils (✉)
HAK Spittal/Drau, Kärnten, Austria
e-mail: gerhardpils@yahoo.de

7.5.1 Introduction: Regional Overview

The territory of the Turkish Republic covers c. 783.000 km^2, surpassing France as the largest EU-country (with c. 540.000 km^2). Only 3 % of this area belongs to Europe (Eastern Thrace), the other 97 % forms the westernmost protrusion of Asia, and is named Anatolia or Asia Minor. This region forms part of the Alpide Belt, and was covered by the Tethys during the Mesozoic. The final uplift of Anatolia did not begin until the end of the Neogene. The resulting highland resembles a ramp, emerging from the Mediterranean Sea in the west, reaching maximum elevations in the east and enclosed by impressive mountain chains in the north and south. Turkey is one of the most mountainous countries in the world, with an average elevation of 1,130 m asl. (Swiss: 1,307 m, Spain: 660 m). Due to this rugged surface and its large area Turkey has a highly diversified climate, resulting in the occurrence of three major phytogeographical regions.

The Euxine Province of the Euro Siberian Region extends along N. Anatolia, where the climate is dominated by humid westerly winds all the year round. Climax vegetation on the seaward slopes of the Pontic Mountains consists of lush deciduous forests, often dominated by *Fagus orientalis* and – at high altitudes – *Abies nordmanniana*. Annual precipitation in this region oscillates mostly between 500 and 1,300 mm, but rises sharply in the Colchic sector east of Ordu to a maximum of 2,241 mm in Rize. Leeward of the coastal ranges humidity drops rapidly to <500 mm, paralleled by a change in climax vegetation to pure stands of hardy and drought-resistant *Pinus sylvestris*, which in the west is partially substituted by *Pinus nigra*.

The Mediterranean region is confined to the seaward parts of W. and S. Anatolia. Annual precipitation is about the same as in the Euro Siberian sector (from 448 mm in Mut to 1,197 mm in Manavgat), but the seasonal distribution is completely different. Wet winters are contrasted by a pronounced summer drought, which is longest in the south and also includes the high mountains. What is left of climax vegetation is sclerophyllous forest, dominated by *Quercus coccifera* and *Pinus brutia*. Nowadays, however, the fertile alluvial plains are almost exclusively dedicated to agricultural use, whilst the remainder has mostly been replaced by Mediterranean maquis or even further degraded phrygana. The timberline in the Taurus is made up of *Pinus nigra* ssp. *pallasiana, Abies cilicica* und *Cedrus libani*, on the much drier northern side partly also of *Juniperus excelsa* and *J. foetidissima*.

The Irano-Turanian Region is by far the largest of the three zones, comprising Central and E. Anatolia. Due to its position behind the coastal ranges, precipitation is low and winters are cold. The area is subdivided by the Taurus mountains which cross it diagonally ("Anatolian Diagonal") into two ecologically distinct territories.

The western part consists of a system of sedimentary basins (ovas) und hills formed of basement rocks. Annual precipitation oscillates mostly between 320 and 420 mm, but drops to values of 270–300 mm in the central ovas of the Great Salt Lake (Tuzgölü), Konya and Yeşilhizar. It is generally agreed that recent steppe or salt-steppe vegetation there is of natural origin, whereas climax vegetation on the hillsides and mountain slopes is dry forest of a submediterranean type, with *Quercus pubescens, Q. cerris, Pinus nigra* and peripherally also *Quercus coccifera*. But after 4,500 years of forest destruction it became difficult to distinguish between natural steppe, semi natural-steppe (due to forest destruction) und subalpine thorn-cushion formations. Moreover this issue is nowadays increasingly of theoretical interest only, as huge steppe areas have been transformed into monotonous fields over the last 40 years.

East of the Anatolian Diagonal altitudes increase considerably, with Van Lake basin (1,650 m) lying more than 700 m higher than Tüzgölü basin (925 m). A lot of mountains surpass 3,000 m, some even reach 4,000 m (Mt. Ararat, Süphan Daği, Cilodağ), and huge areas are covered by young volcanic material. Climate is very

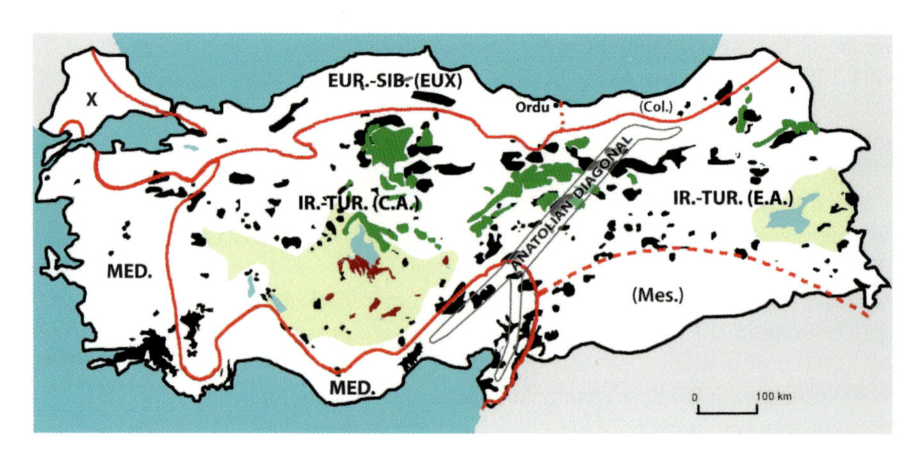

Fig. 7.6 Climatic and edaphic diversity in Turkey. *Red lines*: approximate limits of the phytogeographical regions in Turkey after Davis (1971). EUR.-SIB (EUX): Euro-Siberian Region (Euxine province); *Col.*: Colchic sector of Euxine province; *MED.*: Mediterranean Region (east Mediterranean province); *IR.-TUR.*: Irano-Turanian Region; *C.A.*: central Anatolia; *E.A.*: E. Anatolia; (*Mes.*: Mesopotamia); *X*: Probably central European/Balkan province of Euro-Siberian Region. – *Light yellow* shaded areas: endorheic basins (Elektrik İşleri 2012); *black*: ophiolites (Billor and Gibb 2002); *green*: evaporites (Doğan and Özel 2005); *red*: salt steppe (Atlas Harita Servisi 2004)

severe, with average temperatures in January corresponding to those of northern Turkestan (Erzurum: −8.9°, Kars: −18.3°). Large areas are treeless and most of the remaining forests are degraded to humble scrub, consisting mainly of deciduous oaks (*Quercus brandtii, Q. infectoria, Q. libani, Q. robur* ssp. *pedunculiflora, Q. petraea* ssp. *pinnatiloba*) .

The driest region of Turkey is the Aras valley between Kağızman and the Armenian border, with an average precipitation in Iğdır of only 252 mm (in dry years only 114 mm, Atalay and Mortan 2003). Drought is so severe there that snow-like patches of salt outlets are leaking out of the barren hills. A considerable number of more easterly distributed (semi) desert plants enter Turkish territory only in this remote corner, e.g. *Limonium meyeri* (Plumbaginaceae), *Calligonum polygonoides* (Polygonaceae), *Halanthium rarifolium, H. roseum, Salsola verrucosa, S. dendroides, Halogeton glomeratus, Suaeda microphylla, S. linifolia* (Chenopodiaceae), *Tamarix octandra* (Tamaricaceae).

The genesis of the recent Turkish flora began at the end of the Palaeogene, when the Anatolian mass had just formed but still remained partly submerged by transgressions of the Tethys. Subtropic conditions at that time can be deduced from widespread brown coal deposits in the Neogene basins, with remains of *Taxodium, Cinnamomum, Glyptostrobus, Sequioa* etc. (Gemici and Akgün 2001). East of Ankara, sediments of this period contain huge masses of evaporits, which nowadays form the impressive gypsum hills around Çankırı and Sivas (Fig. 7.6,

Günay 2002). This suggests higher aridity towards the east, a supposition which is also supported by palynological evidence, showing an increase of Compositae, Chenopodiaceae and Poaceae from Middle to Late Miocene (Akgün et al. 2007). Towards the end of the Neogene, temperatures and humidity sank to recent levels and the final uplift of the Pontus and Taurus chains began. Parts of the Palaeotropic flora disappeared, others formed the basis for the evolution of the present day Mediterranean (sclerophyllous) flora, and some survived as relicts, mainly in the Colchis. Dry central Anatolia was invaded by the ancestors of the recent Irano-Turanian genera, which experienced explosive radiative evolution, some of them also penetrating into the Mediterranean zone (e.g. *Verbascum, Onosma*).

Glaciations during the subsequent Pleistocene period were of much less importance in Anatolia than in the Alps. Local mountain glaciers were mainly restricted to the northern slopes of the highest mountains. The most extensive glaciations occurred in the Kaçkar Mountains (3,932 m) in the northeast and on Çilo Dağları (4,135 m) in the southeast, where the longest of these local glaciers reached 10 km and descended to 1,600 m (Güldali 1979). Pollen profiles from the lower regions of central Anatolia (Eski Acigöl, 1,270 m, Nevşehir Province) show an Artemisia-Chenopod steppe towards the end of the last glacial period (16.000 BP., Roberts et al. 2001). There is evidence that Irano-Turanian genera like *Cousinia* not only easily survived glacial periods, but in fact were more abundant and more widespread than in interglacials (Djamali et al. 2012). In the Mediterranean zone, thermophilous plants disappeared and vegetation consisted of a dry forest steppe with *Prunus amygdalus* s.l, *Pinus* sp., deciduous oaks and again much *Artemisia* and Chenopodiaceae (Emery-Barbier and Thiébault 2005). As to the Euro-Siberian Flora of northern Anatolia, there is general agreement that temperate forests were able to survive glacial periods at least in the Colchic sector, as well as in the adjacent Caucasus and Hyrcanian regions (DellaSala 2010).

During the Holocene, temperature and humidity increased, causing the development of a mosaic of woodland with deciduous oaks, shrubs (e.g. *Corylus*) and open grassland about 8,000 BP. After 6,500 BP the more mesophilous trees such as *Ulmus* and *Corylus* disappeared, probably due to an increase in aridity. Human impact becomes noticeable in the Bronze Age (c. 4,500–4,000 BP), with a marked decline in oak pollen paralleled by a spread of anthropogenic species (Roberts et al. 2001). Since then huge areas of Central Anatolia have been cleared of any forest cover. The same is true for the Mediterranean region, where all types of degraded vegetation, from maquis and phrygana to open rock- and scree vegetation, experienced dramatic increases since prehistoric times.

Today the flora of Turkey numbers about 9,000 species of vascular plants, with an endemism ratio of 1/3 (33,3 %). Both values are astonishingly high and unique for the western Palaearctic. Even when distorting effects of different territory size on endemism ratios are reduced by comparing Bykov's index instead of mere species numbers, Turkey (2,29; using data from Güner et al. 2000) is far ahead of Greece (1,18) and continental Spain (−1,08), and comes second only after isolated islands like the Canary Islands (10,68), Madeira (10,36) or Crete (3,08; Bruchman 2011).

7.5.2 Material and Methods

If not stated otherwise, data on climate, geology or geomorphology are taken from Atalay and Mortan (2003) and Hütteroth and Höhfeld (1982), data on Turkish plants from Flora of Turkey (Davis 1965–1988) and its supplements (Güner et al. 2000; Özhatay and Kültür 2006; Özhatay et al. 2009, 2011). Important additional information about the distribution of regional endemics in Turkey was extracted from Boulos et al. (1994) and Özhatay et al. (2003).

Of very limited use are existing maps showing the frequencies of Turkish endemics within Turkey (Kutluk and Aytuğ 2001). Such maps seriously underestimate regional endemism in all borderline grids. If such a methodology was applied to Europe a whole, Switzerland lying in the centre of Europe and therefore containing 30 % European endemics would range far ahead of Greece with only 21.9 % European endemics (calculations based on data from Bruchmann 2011).

Knowledge about the ecology of Turkish plants is still rather rudimentary, leading to such unspecific descriptions as "rocky slopes", "mountainous districts", "pastures". Clear ecological distinction between the different kinds of steppes, pastures or rock associations entails a profound knowledge of the Turkish flora. For any single researcher this was virtually impossible to achieve before completion of the "Flora of Turkey". Unfortunately it is not much easier today, due to bureaucratic restrictions on research and collecting samples for non-Turkish citizens (Pils 2006).

7.5.3 The Main Components of the Turkish Endemic Flora

Table 7.5 shows important genera in the Turkish endemic flora. The first 10 genera on the list are those with the highest numbers of endemics in Turkey. Excluded is the mostly apomictic genus *Hieracium*. Turkish values compiled from Davis (1965–1988), Güner et al. (2000), Özhatay and Kultür (2006) Özhatay et al. (2009, 2011), European values from Flora Europaea.

The most important share in the Turkish endemic flora falls upon genera with global distribution centres in the Irano-Turanian region, i.e. the dry regions of Turkey and eastwards to Mongolia and Afghanistan. Such Irano-Turanian genera are by definition poor in European species, as is shown in Table 7.5 for the Flora Europaea area. In this sense *Astragalus* and *Verbascum*, the two most diversified genera in Turkey are perfectly Irano-Turanian. The same is true for *Salvia, Onosma* and 11 other examples listed in Table 7.5 Some of these are represented in the Flora Iranica region with far more species than in Turkey, e.g. *Astragalus* (965 species!), *Acantholimon* (>160) or *Cousinia* (c. 350), others such as *Verbascum* and, to a lesser extent, *Onosma* and *Salvia* are clearly "Turkish".

As stated already by Hedge (1986), the highest numbers of species in Anatolia do not necessarily indicate highest level of morphological variation here. Hedge illustrated this with *Salvia*; Turkey has by far the most species (95) of this genus, but in Afghanistan which has only 23 species there is a greater range of morphological

Table 7.5 Endemic-rich genera of vascular plants in Turkey (for further information see text)

	No. of endemics in Turkey	No. of species in total	Level of endemics (%)	No. of species in Europe
Astragalus	277	437	63	133
Verbascum	194	242	80	87
Centaurea	122	193	63	221
Allium	66	163	40	110
Silene	66	147	45	194
Campanula	63	112	56	144
Galium	55	109	50	145
Alyssum	59	96	61	70
Salvia	50	95	53	36
Onosma	49	96	51	33
Further examples for important irano-turanic genera in Turkey				
Gypsophila	31	84	37	27
Cousinia	26	38	68	1
Alkanna	26	35	74	17
Acantholimon	25	39	64	1
Fritillaria	23	40	58	23
Phlomis	22	36	61	12
Aethionema	21	41	51	9
Muscari	20	31	65	13
Paracaryum	19	28	68	0
Delphinium	17	32	53	25
Consolida	14	28	50	13
Ebenus	14	14	100	2

variation. It seems probable, therefore, that Anatolia is not the primary centre of evolution for most of its Irano-Turanian genera, but a centre for secondary adaptive radiation, giving rise to numerous knots of closely related, often vicarious, neoendemic species. Such a point of view is also in accordance with the relatively recent genesis of the Anatolian micro-continent as outlined above.

All of these successful Irano-Turanian genera are well adapted to treeless, dry and heavily grazed habitats, in some cases also to soils with high contents of heavy metals (e.g. Brassicaceae like *Alyssum*, *Aethionema*) or gypsum soils (*Gypsophila*). *Astragalus* as the most successful genus in terms of species numbers is also the most plastic one with regard to niche exploitation in such habitats. There are annuals (9 sections), acaulous perennials with their leaves pressed to the ground (sect. *Myobroma*), high-growing but unarmed perennials relying on chemical defence (sect. *Alopecias*) and several sections of heavily armed thorn-cushions, which are very characteristic for the subalpine regions of Irano-Turanic Anatolia. Such thorn-cushions evolved convergently in different *Astragalus* sections, but also in *Onobrychis cornuta, Acantholimon* (Plumbaginaceae), some Asteraceae (e.g. *Centaurea urvillei, C. iberica, Lactuca intricata*) and Caryophyllaceae (e.g. *Minuartia juniperina*).

Very successful under Irano-Turanian conditions are thistles (e.g. *Onopordum* with 17 species, *Cousinia* with 38 species) or prickly herbs such as many *Onosma* species. The latter are covered by clusters of stiff setulae or spinules, which in some cases may be as annoying as the hair-like glochides of *Opuntia*. A rather singular Anatolian success story are the mostly pannose, tap-rooted candelabras of *Verbascum* spp., which are highly unpalatable to cattle and sheep. Turkey alone has c. three times as many species of *Verbascum* (242) as the whole Flora Europaea area (87).

An outline of the Turkish endemic flora would be incomplete without mentioning the numerous bulbous Monocotyledons, which are preferential "objects of desire" to rock gardeners all over the world. Species formation also seems to be in full progress in some of these genera. An instructive example is *Crocus*. At the moment about 37 *Crocus* species are accepted for Turkey (depending on the species concept applied); 24 of these are endemic. A closer look reveals that some of these "species" are better referred to as "superspecies". This is true especially for *Crocus biflorus* s. latiss., which is distributed all over Turkey in a broad spectrum of habitats and at altitudes from 200 to 3,000 m. Today 17 Anatolian subspecies are accepted within *C. biflorus* (Euro+Med 2006-), excluding some better feasible forms, that are accepted as good species (e.g. *C. nerimaniae* and *C. wattiorum*). However, still more local populations exist, whose allocation to one of the already described subspecies is unresolved (Kerndorf and Pasche 2006). Typically each of these shows a certain degree of morphological differentiation and occupies its own territory, often a slightly isolated mountain range. The ecological niches of some of these geographical races show broad overlap, but areas of distribution as a rule are neatly separated. Delimitation of such "microspecies in statu nascendi" probably works by competitive exclusion, and in the case of strictly allogamous species also through positively frequency-correlated fitness functions (Pils 1995).

Situations are similar in other actively evolving species clusters in Turkey, e.g. *Ranunculus dissectus* s.latiss. At present 7 subspecies are distinguished, but 3 taxa have already gained greater evolutionary distance and are therefore treated as good species (*R. fenzlii, R. crateris, R. anatolicus*). They all have mutually excluding distribution areas covering the major part of Anatolia in the form of a somewhat irregular mosaic. One of the main reasons for evolving towards such a high degree of differentiation obviously comes from the highly patchy structure of Anatolia (Fig. 7.6), possibly in combination with a temporary increase of separation between populations due to changing forest cover, as recently suggested for *Cousinia* (Djamali et al. 2012).

Crocus biflorus is a nice example for the mutual dependence of progress in systematics and a better understanding of endemism. Still more striking is the case of *Ornithogalum* s.lat., a traditionally rather sidelined genus due to its scant interest to rock gardeners. Its initial treatment in "Flora of Turkey" comprised 23 species, only 2 of which are endemic to Turkey, thus resulting in an endemism rate of 8.7 %. This changed drastically in recent decades when a wave of descriptions of local endemics increased species numbers in Turkey by nearly 100 % (up to 55, Bağci et al. 2011) and the endemism rate to >50 %.

Of special interest are monotypic endemic genera. Their isolated positions are an indication of a somewhat relictic nature. The following 17 species fall in this category: *Crenosciadium siifolium, Ekimia bornmuelleri, Microsciadium minutum, Olymposciadium caespitosum* (all Apiaceae); *Leucocyclus formosus* (Asteraceae); *Physocardamum davisii, Tchihatchewia isatidea* (Brassicaceae); *Phryna ortegioides, Thurya capitata* (Caryophyllaceae); *Cyathobasis fruticulosa, Kalidiopsis wagenitzii* (Chenopodiaceae); *Sartoria hedysaroides* (Fabaceae); *Dorystoechas hastata* (Lamiaceae); *Necranthus orobanchoides* (Orobanchaceae); *Nephelochloa orientalis, Pseudophleum gibbum, Oreopoa anatolica* (Poaceae).

The two Chenopodiaceae-species are endemics of the Central Anatolian salt steppe. *Crenosciadium* is restricted to moist open places in the upper region of Murat Dağ. Recently described *Oreopoda* is known only from a single location on Aktepe (Beydağları, southwestern Taurus Mts. 2,700 m, Parolly and Scholz 2004). The rest grows on dry, open, mostly rocky places in mid to higher altitudes (mainly 900–2,000 m). Distribution areas as a rule range between small and tiny and not a single member of this group is reaching the Euxinic region.

7.5.4 Regional Endemism in Turkey

On a global scale, local endemism shows a very strong correlation with high mountain zones. Turkey is no exception to this rule. Mill (1994) ranks the entire Taurus and the eastern Pontus among the most important centres of plant diversity in SW Asia. The most outstanding numbers in his statistics are those for SW Anatolia (mainly western Taurus, about 3,365 species, with >300 local endemics) as well as for the large area around the Anatolian Diagonal (3,200 species, with c. 390 local endemics).

The privileged position of SW Anatolia can be explained by a combination of factors which favour the development of local endemism. This begins with the discontinuous structure of the high mountains there. Between Kaz Dağ and Mendeşe Dağ there is a system of horsts and grabens, which finds its continuation in the western Taurus, where a series of isolated chains is separated by lower sedimentary basins (ovas) or abrasion zones (Atalay and Mortan 2003). A similar "patchy structure" is exhibited by the geological structure of the region. High massifs consist mainly of limestone, but there are also large areas of ultrabasic peridodites in the region, e.g. the whole Marmaris Peninsula as well as the 2,294 m high Sandras Dağ near Muğla. And finally there is a strong climatic gradient from the Mediterranean area on the coast to the Irano-Turanian sector towards central Anatolia. Annual precipitation in Antalya is 1,041 mm, but in Beyşehir north of the Taurus range only 447 mm. This endemism centre is also the most important one with respect to relict species, with 8 out of 17 monotypic genera having at least part of their distribution areas here. A "living fossil" of SW Anatolia is *Liquidambar orientalis* ("Turkish sweetgum", Hamamelidaceae), which was a constant part of the Neogene brown-coal forests of the whole of western Turkey (Gemici and Akgün 2001). Closely related forms extended into much of Europe and the Caucasus. Today *Liquidambar* is confined to some flood plains or marshy valleys in a small part of SW Anatolia (mainly around Marmaris) and the nearby island of Rhodes. Its closest relative is *L. styraciflua* from eastern N America (Photos 7.9 and 7.10).

In the adjacent section of the Taurus between Antalya and Adana concentration of local endemics remains on a similar high level. Local hotspots of particular importance are Bolkar Dağları (3,524 m) and Ermenek Valley. The latter is up to 1,000 m deep and was formed by excavation of the Ermenek River into a vast plateau of remarkably soft limestone, which was deposited during a middle Miocene transgression of the Tethys. About 50 species are more or less confined to this marvelous region (Özhatay et al. 2003), e.g. the monotypic genus *Sartoria (S. hedysaroides)*, which is closest to *Hedysarum* (Fabaceae).

The Anatolian Diagonal connects Aladağları (3,756 m) in the south with Munzur Dağları (3,462 m) in the north, thus forming a mountainous borderline between W and E Anatolia. Factors favouring general diversity and local endemism are roughly the same as in SW Anatolia. Calcareous Munzur Dağları are in a relatively isolated position in otherwise mainly siliceous (often volcanic) central Anatolia, which explains their unusually high number of local endemics. Again there are numerous outcrops of ultrabasic serpentine rocks in the region, the largest of these being Avcı Dağları (3,345 m) east of the Munzur Dağları. Of special importance for local endemism are large areas of lime marls and gypsum in the lower sections of the Taurus mountains, mainly east and south of Sivas. Monotypic *Tchihatchewia isatidea* has its centre of distribution in the northern and central parts of the Anatolian Diagonal.

The Amanus Mountains (Nur Dağları, 2,240 m) form a bridge between the Anatolian Diagonal and the "Levantine Uplands", running parallel to the eastern shore of the Mediterranean Sea into northern Israel. A total of 139 species is endemic to the Amanus, and 128 further species are subendemic (Mill 1994). One reason for these exceptionally high values might be the special position of the Amanus as an "island of humidity" between sea and desert. Precipitations exceed 1,700 mm in altitudes over 1,750 m and even summers are not completely dry here, thus enabling the survival of local *Fagus orientalis*-forests with Euxinic elements such as *Ilex colchica, Rhododentron ponticum, Tilia argentea, Prunus laurocerasus, Gentiana asclepiadea, Staphylea pinnata* etc. On the eastern slope, towards the Syrian desert, annual precipitation soon drops to <600 mm, resulting in a much more xerophilous forest vegetation. Edaphic diversity is very high, with hard limestones composing northern and central sectors, and ultra basic peridodites building up the south. The latter are covered mainly by open *Centaurea ptosimopappa-Pinus brutia* forests with a large number of local endemics (Özhatay et al. 2003).

The SE Taurus reaches its maximum heights in the Cilo Dağları (4,135 m) near the Iraqi border. Forest destruction is extremely severe in this area, but smaller remnants of dry deciduous oak forests still exist in many valleys in the southern part of this region. Floristic affinities are higher with the adjacent mountainous regions of the neighbouring countries than with the rest of the Taurus. Such a

Photo 7.9 *Liquidambar orientalis*, a living fossil of Tertiary relict swamp forests. This species is endemic to SW Anatolia, Turkey, and Rhodes, Greece (Photographed by Gerhard Pils)

Photo 7.10 *Liquidambar orientalis*, flowers and leaves (Photographed by Gerhard Pils)

"transnational" endemism region with SW Anatolia, NW Iran and N Iraq contains about 500 endemic species, which exceeds by far the other Centres of Diversity in Anatolia as delimited by Mill (1994). Examples of species confined to this extended area are *Pelargonium quercetorum* (in moist oakwoods), *Eryngium thyrsoideum, Papaver curviscapum, Scrophularia kurdica, Acantholimon petuniiflorum, Salvia atropatana* and *Colchicum kurdicum.* Recent maps showing totals of "Turkish endemics" per grid (Kutluk and Aytuğ 2001) seriously disadvantage these central regions of Turkish "Wild Kurdistan", with further negative implications if such biased diversity maps were to be used as a basis for conservation planning, as proposed e.g. by Türe and Böcük (2010). But also by using such an arbitrary delimitation c. 100 species remain as "Turkish" local endemics of the mountains of SE Anatolia. Examples are *Gypsophila hakkiarica, Eryngium bornmuelleri, Crepis hakkarica, Campanula hakkarica, Rhynchocorys kurdica, Galium zabense, Senecio davisii, Allium rhetoreanum* etc.

The NE Anatolian Center of Endemism is centred in the Colchic part of the Euxinic region, where the Pontus rises to its greatest heights in the Kaçkar Dağları (3,932 m). About 160 local species – including a lot of apomictic *Alchemilla* and *Hieracium* microspecies – are endemic to the Turkish sector (Mill 1994). Further >130 species are subendemic as they extend into Georgian territory. The edaphic diversity of this region is rather low, with predominantly igneous rocks. But there is a very steep climatic gradient between northern and southern slopes of the Kaçkar Range. Conditions on the Turkish Black Sea coast are generally very oceanic, becoming particularly moist and cloudy in the Colchic sector east of Ordu. These conditions also remained relatively stable during Pleistocene glaciations, thus enabling the survival of a considerable number of Arcto-Tertiary species (DellaSala 2010). These are typically confined to very humid forest habitats of the Colchis. A very illustrative example is *Quercus pontica*, which is morphologically distinct from the rest of the Turkish oaks. Its nearest relative is *Q. sadleriana*, a local endemic of mountain slopes on the Californian – Oregon border (Denk and Grimm 2010). Two dwarf, evergreen shrublets of the Ericaceae family have similar long-range disjunctions: *Epigaea gaultherioides* (also Georgia), with two related species in Japan und eastern N America, and *Rhodothamnus sessilifolius* (subalpine, endemic) with the only remaining species of this genus in the eastern Alps. Further fairly isolated subendemics of the region are *Picea orientalis, Osmanthus decorus* (Oleaceae), *Pachyphragma macrophylla* (a monotypic Brassicaceae), *Hypericum bupleuroides* (which reaches into Armenia), *Rhododendron smirnovii* (endemic, subalpine) and *Rh. ungernii.* Due to the mesophytic character of the Colchic vegetation, Irano-Turanian genera such as *Astragalus, Cousinia* or *Phlomis* are without local endemics there.

In the Euxinic region west of Ordu the mountains of the coastal ranges do not as a rule surpass timberline and are covered by dense, deciduous forests. Edaphic diversity remains relatively low, with schists or sediments of mainly igneous origin. Centres of local endemism are the Ilgaz Dağları south of Kastamonu (2,587 m) and Uludağ (2,543 m) south of Bursa. Both are in fairly isolated positions within their much lower surroundings. Local endemics are concentrated in treeless habitats, with some exceptions such as *Heracleum paphlagonicum*, growing by mountain streams in mixed forests of the Ilgaz Dağları. From thoroughly explored Uludağ c. 30 local endemic taxa have been described, from Ilgaz Dağları at least 8.

Uludağ is very close to Istanbul province and European Turkey. Landscape here consists of undulating hills, mostly forest covered on the Black Sea side, but transformed into agricultural steppe on the Marmara side. One would not expect any local endemic in such unspectacular scenery. But reality is not that simple in Turkey. *Colchicum micranthum*, whose delicate flowers may be found in deciduous oak forests of the Bosporus area, is endemic here, as is *Cirsium byzantinum*. The latter has a preference for moderately disturbed places, especially dry road margins, thus excluding any major ecological change as a reason for its narrow distribution area.

This holds still more true for *Onosma propontica*, the rarest endemic of the region, whose few known localities are confined to the open hillsides E and NE of Istanbul (with one locality in neighbouring Kirklareli, Baytop 2009). The plant grows here largely disjunct from its supposed closest allies in Ucraina and Dobrutscha, making it quite probable that this is a relict species (Teppner and Tuzlaci 1994).

7.5.5 Edaphic Diversity and Regional Endemism

Ultrabasic bedrocks underlying serpentine soils are distributed over large areas of Anatolia (Fig. 7.6). Their occurrence is always strongly correlated with high levels of local endemism. For instance, in the Balkans about 6 % of the total endemic flora consists of obligate serpentinophytes (Stevanović et al. 2003). Nevertheless, these special habitats have only recently received particular attention in Turkey. One of the best explored serpentine areas is Sandras Dağ (2,294 m) near Köyceğiz on the SW Anatolian coast. Vegetation there consists of open *Pinus brutia* forest on the lower slopes, which is replaced by fine stands of *P. nigra* up to the timberline at c. 2,100 m. Systematic collecting over the last 25 years has revealed a surprisingly high degree of local endemism on this mountain. In this period 15 new taxa have been described from Sandras Dağ, raising the number of local endemics from 9 to 23 (Özhatay and Kültür 2006, Pils ined.). The same applies to the serpentine outcrop of Kızıldağ near the town of Çamlık (Konya province), from where in the last decade 9 new serpentinophytes were described, 7 of which were local endemics (Aytaç and Türkmen 2011; Duran et al. 2011). Unfortunately, no precise data exist on the relative share of serpentinophytes in the endemic flora of the Amanus region, despite some recently published plant lists from the area.

Some genera of Brassicaceae, such as *Alyssum, Aethionema* (both Table 7.5), *Thlaspi, Bornmuellera* and *Cochlearia,* are well known for their resistance to high concentrations of heavy metals. The first three of these developed considerably more species in Turkey than in the whole area of Flora Europaea and are disproportionally rich in local endemics. In *Bornmuellera* (3 sp.) and *Cochlearia* (4 sp.) all Turkish species are endemics. Nickel-hyperaccumulation abilities have recently also been detected in some Turkish *Centaurea* species. This is not really a big surprise, as 15 out of 122 endemic *Centaurea* species are supposed to be restricted to serpentine (Reeves and Adigüzel 2004). Some obligate serpentinophytes are of quite isolated systematic position (e.g. *Eryngium thorifolium, Scorzonera coriacea*), thus indicating rather a long evolution time for the Anatolian serpentine flora.

Continental salt vegetation adds considerably to the floristic diversity of the endorheic basins west of the Anatolian Diagonal. The bottom of the larger of these flat basins is covered by the Big Salt Lake (Tuzgölü), which dries up to a large extent during summer. There are a number of such salty areas in the plains of Konya, Niğde and Kayseri; some of these are occupied by temporary lakes, others are covered with salt steppe. Quite a number of genera from the surrounding steppe flora have evolved local halophytic species. Examples are *Verbascum helianthemoides, V. pyroliforme, Salvia halophila, Ferula halophila, Centaurea halophila, C. tuzgoluensis, Onosma halophila, Saponaria halophila, Scorzonera hieraciifolia, S. tuzgoluensis, Cousinia birandiana, Acantholimon halophilum, Dianthus aydogduii, Taraxacum tuzgoluensis, Senecio salsugineus* and *Astragalus demirizii*. However, typical halophytic genera are also represented here by local endemics, e.g. *Limonium* (*L. anatolicum, L. iconicum, L. lilacinum, L. adilguneri*), *Petrosimonia* (*P. nigdeensis*), *Suaeda* (*S. cucullata*), and *Salsola* (*S. grandis*). Some of these local halophyts are systematically isolated to such an extent that one has to assume an old (Tertiary) origin of these habitats. The best examples are the monotypic genera *Cyathobasis* and *Kalidiopsis* (Chenopodiaceae). Other endemics show interesting long-range disjunctions, e.g. the monotypic genus *Microcnemum* (Chenopodiaceae), with *M. coralloides* ssp. *anatolicum* on very salty soils in central Anatolia and ssp. *coralloides* in Spain, or the ditypic genus *Sphaerophysa* (Fabaceae), with *S. kotschyana* in damp salty places in Central Anatolia and *S. salsula* in Central Asia (Photos 7.11, 7.12, and 7.13).

With inland salt vegetation so rich in local species, one would expect similar conditions on the Turkish coasts, the length of which is comparable to those of Italy. Salt-swamps exist in the deltas of the large rivers discharging into the Mediterranean (Çukurova near Adana, Göksu delta near Silifke and Great Menderes delta south of Izmir). But with the exception of *Tamarix duezenlii*, a very rare shrub of the Çukurova delta with close affinities to *T. arborea* in the southeast Mediterranean, endemism is virtually non-existent there. This is possibly due to the instability of these habitats, with the Mediterranean Sea drying out in the Messinian (c. 5,5 Mill BP.) and repetitive eustatic sea level depressions for at least 100 m during periods of maximum glaciations.

A still more puzzling problem is the uniformity of the halophytic coastal rock flora. To date only one endemic *Limonium* has been described from the Turkish coasts, namely *L. gueneri*, known from a single location near Kaş (SW Anatolia, Doğan et al. 2008). However, there are 102 endemic *Limonium* (micro)species in Italy, nearly exclusively growing in coastal habitats, mainly on rocks (Conti et al. 2005).

The flora of coastal sands is clearly more strongly differentiated. Local endemics of the Çukurova delta are *Medicago* (= *Trigonella*) *halophila, Bromus psam-*

Photo 7.11 *Sphaerophysa kotschyana*, endemic to damp saline habitats in Central Anatolia, Turkey (Photographed by Gerhard Pils)

mophilus and *Silene pompeipolitana. Medicago* (= *Trigonella*) *arenicola* is confined to the Antalya region and *Anthemis ammophila* is more widely distributed on the south coast. On the Black Sea coast huge dune areas existed between Kilyos and Terkos Lake (N. of Istanbul), with an estimated total surface of c. 17.5 km^2 and dunes penetrating up to 5 km into the interior, forming forest-covered hills of up to 30 m altitude (Önal 1981; Özhatay et al. 2003). Most of this area was devastated during the 1970s due to lignite extraction, but on the remaining >4 km^2 the sand flora is still exceptionally diverse, with *Centaurea kilaea*, *Isatis arenaria, Linum tauricum* ssp. *bosphori*, *Erysimum sorgerae* and *Silene sangaria* as local (sub)endemics. An interesting large range disjunction is shown by *Asperula littoralis*, growing on this part of the Black Sea Coast and with slightly different populations also on the Mediterranean coast near Antalya.

Gypsum outcrops are frequent and widespread in the Irano-Turanian part of Anatolia, covering very large areas between Çankırı and Çorum and around Sivas. They are testimonials for a hot and dry climate at the beginning of the Neogene. Remarkable gypsophytes of the Sivas gypsum hills are e.g. *Achillea sintenisii, Allium sivasicum**, *Campanula sivasica**, *Centaurea yildizii**, *Chrysocamela noeana**, *Isatis sivasica, Gypsophila heteropoda* ssp. *minutiflora**, *Onosma sintenisii, Paronychia galatica*, *Reaumurea sivasica**, *Scrophularia gypsicola**, *S. lepidota, Scorzonera aucherana, Thesium stellerioides* and *Thymus spathulifolius** (* = only on gypsum,

Photo 7.12 Saline inland steppe with *Verbascum helianthemoides* (Photographed by Gerhard Pils)

Photo 7.13 *Verbascum helianthemoides* (Photographed by Gerhard Pils)

Akpulat and Celik 2005). Some of these species are without closer relatives, indicating the high (pre Pleistocene) age of these gypsum habitats.

Neogene marls containing gypsum form large badland areas around Beypazarı (c. 120 km west of Ankara). From these barren hillsides two very local endemics have recently been described: *Verbascum gypsicola*, with a total population of less than 2,000 individuals (Vural 2009), and *Salsola grandis*. The latter grows very locally in some bedland ravines together with *Anabasis aphylla*, which is next found at the Armenian border, and *Petrosimonia nigdeensis*, a central Anatolian endemic (Freitag et al. 1999).

7.6 High Mountain Regions in Iran

Jalil Noroozi (✉)
Department of Conservation Biology, Vegetation and Landscape Ecology, Faculty Centre of Biodiversity, University of Vienna, Vienna, Austria

Plant Science Department, University of Tabriz, 51666 Tabriz, Iran
e-mail: noroozi.jalil@gmail.com

7.6.1 Physical Geography of Mountain Ranges in Iran

Iran with a total surface area of c. 1.6 million km^2 is a typical high mountain country. Almost half the country consists of high elevations, and Alborz and Zagros are the major high mountain chains (Fig. 7.7). The highest mountains are Damavand (5,671 m asl.), Alamkuh (4,850 m), Sabalan (4,810 m), and Hezar (4,465 m). More than 100 mountain peaks exceed 4,000 m.

There are glaciers in the higher elevations, i.e. Damavand, Alamkuh, Sabalan and Zardkuh (Ferrigno 1991). According to Schweizer (1972) the present snowline in Alborz, north and central Zagros, and the NW Iranian mountains lies between 4,000 and 4,200 m. In the mountains of central and southern Iran, e.g. Shirkuh, Dena, and Hezar-Lalehzar Mts. it is between 4,500 and 5,000 m. The higher elevations of Iran have a continental climate with Mediterranean precipitation regime. The annual precipitation in the higher altitudes of Alborz reaches almost 1,000 mm (Khalili 1973).

The alpine habitats are almost above 3,000 m, and are found scattered across different parts of the country (see Fig. 7.7). The Iranian alpine flora is of Irano-Turanian origin (Zohary 1973; Klein 1982, 1991; Frey et al. 1999). A conspicuous feature of this flora is the high rates of endemism. A total of 682 vascular plant species are known from alpine habitats of Iranian mountains, and of this total 394 species are endemic or subendemic to Iran (Noroozi et al. 2008). These habitats are less known, and plant species are still being discovered and described as new to

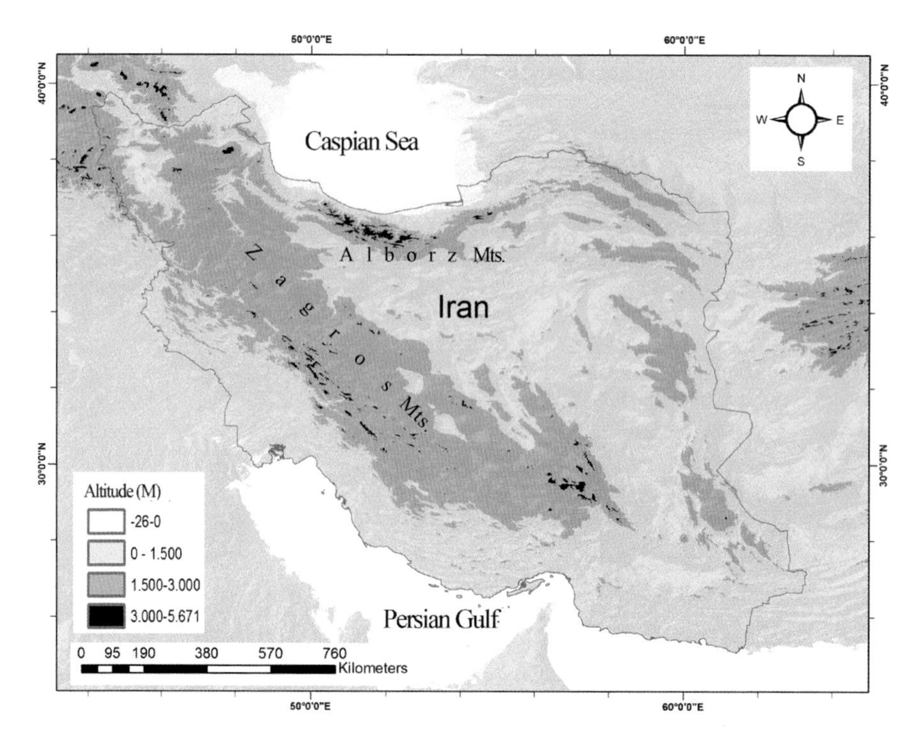

Fig. 7.7 Alborz and Zagros are the major mountain chains of Iran. The alpine habitats (*black spots*) are scattered around different parts of the country

science (e.g. Jamzad 2006; Khassanov et al. 2006; Noroozi et al. 2010a; Noroozi and Ajani 2013; Razyfard et al. 2011).

Several different vegetation types are found in the high regions of Iran. The dry slopes of the subalpine zone are usually covered with tall herbs and umbelliferous vegetation types (Klein 1987, 1988). Small patches of subalpine wetlands are found scattered on dry slopes (Naqinezhad et al. 2010). Thorn-cushion grasslands occur in alpine meadow and windswept areas, and snowbed vegetation types in snow patches or snow melting runnels (Klein 1982; Noroozi et al. 2010b). Scree vegetation types with sparse vegetation cover occupy high alpine and subnival steep slopes with a high percentage of screes and stones on the ground (Klein and Lacoste 2001; Noroozi et al. 2013).

7.6.2 Analysis of Floras and Phytosociological Investigation

To select the local endemic species, the Flora Iranica (Rechinger 1963–2012) and Flora of Iran (Assadi et al. 1988–2012) were used as the main sources.

The main data on endemism in alpine habitats of Iran and the percentage of endemism in alpine and subnival-nival zones were extracted from Klein (1991), Hedge and Wendelbo (1978) and Noroozi et al. (2008, 2011).

The rate of endemism in different habitats is based on phytosociological vegetation studies of the author in Central Alborz. This rate is measured from all character species which were recorded for each habitat. Some species which were characteristic for two different habitats were separately calculated for both habitats. For example *Astragalus macrosemius* is a character species for alpine thorn-cushion grasslands and alpine-subnival scree grounds and was thus counted separately for both habitats as a character species.

7.6.3 Endemism in Different High Mountain Areas and Habitats

Three monotypic genera are found exclusively within Iranian alpine habitats, *Elburzia* in Central Alborz (Hedge 1969), *Sclerochorton* in northern Zagros (Rechinger 1987), and *Zerdana* in central, south and southeastern Zagros (Rechinger 1968). There are also some interesting ditypic genera in these habitats, such as *Clastopus* with two species endemic to Iran (Hedge and Wendelbo 1978), *Didymophysa* with one species occurring in Iran and Transcaucasus, and another one in central Asia and Hindukush (Hedge 1968), and *Dielsiocharis* with one endemic species in Iran (Hedge 1968) and one local endemic in central Asia (Al-Shehbaz and Junussov 2003). All the above-mentioned genera belong to the Brassicaceae except *Sclerochorton*, which belongs to Apiaceae. The author has only seen *Zerdana, Clastopus* and *Didymophysa* in the wild, and all of these are restricted to scree habitats.

A total of 110 vascular plant species were considered to be rare and narrow endemic to Iranian alpine habitats and have only been recorded from one or very few locations. These species are found in Zagros, Alborz, and in the northwestern part of the country (Sahand and Sabalan Mts.). They are classified into four mountainous regions as below:

Rare species in Alborz:
Alchemilla amardica, A. rechingeri, Allium tuchalense, Astragalus aestivorum, A. herbertii, A. montis-varvashti, A. nezva-montis, Cousinia decumbens, C. harazensis, Diplotaenia damavandica, Elburzia fenestrata, Erodium dimorphum, Festuca rechingeri, Galium delicatulum, Iranecio oligolepis, Myopordon damavandica, M. hyrcanum, Nepeta allotria, N. pogonosperma, Oxytropis aellenii, O. cinerea, O. takhti-soleimanii, Paraquilegia caespitosa, Phlomis ghilanensis, Saxifraga koelzii, S. ramsarica, S. iranica, Scorzonera xylobasia, Scutellaria glechomoides, Senecio iranicus, Silene demawendica, Thlaspi maassoumi, Trachydium eriocarpum, Veronica euphrasiifolia, V. paederotae, Vicia aucheri.

Rare species in the mountains of northwestern Iran:
Astragalus azizii, A. pauperiflorus, A. savellanicus, Dianthus seidlitzii, Euphorbia sahendi, Festuca sabalanica, Nepeta sahandica, Ranunculus renzii, R. sabalanicus, Thlaspi tenue.

Rare species in north, central and southern Zagros (from Kordestan to Fars provinces):
Acantholimon eschkerense, Allium mahneshanense, Arenaria minutissima, Astragalus mahneshanensis, A. montis-parrowii, Bufonia micrantha, Chaerophyllum nivale, Cicer stapfianum, Cousinia concinna, C. eburnea, Crepis connexa, Cyclotrichium straussii, Dianthus elymaiticus, Dionysia aubrietioides, D. iranshahrii, Dracocephalum surmandinum, Erysimum frigidum, Euphorbia plebeia, Festuca iranica, Galium schoenbeck-temesyae, Jurinea viciosoi, Myopordon aucheri, Nepeta archibaldii, N. chionophila, N. iranshahrii, N. monocephala, N. natanzensis, Potentilla flaccida, Psychrogeton chionophilus, Ranunculus dalechanensis, Salvia lachnocalyx, Satureja kallarica, Sclerochorton haussknechtii, Scorzonera nivalis, Scrophularia flava, Senecio kotschyanus, Seratula melanocheila, Silene hirticalyx, Tragopogon erostris, Veronica daranica.

Rare species in southeastern Zagros (mountains of Yazd, Kerman and Baluchestan provinces):
Acantholimon haesarense, A. kermanense, A. nigricans, A. sirchense, Allium lalesaricum, Astragalus hezarensis, A. melanocalyx, A. pseudojohannis, Chaenorhinum grossecostatum, Cousinia fragilis, Dionysia curviflora, Helichrysum davisianum, Hymenocrater yazdianus, Hyoscyamus malekianus, Nepeta asterotricha, N. bornmulleri, N. rivularis, Polygonum spinosum, Rubia caramanica, Senecio eligulatus, S. subnivalis, Silene dschuparensis, Verbascum carmanicumm.

Since alpine habitats have been less well investigated than lower elevations, our knowledge about the distribution range of species living in high mountain regions is low. This means that it is more likely that new explorations at these elevations will produce new records and localities for local endemics. Nonetheless, the above mentioned species are rare, with narrow geographical and altitudinal distribution. Most of the above-mentioned species can be categorised as Endangered (EN) and Critically Endangered (CR) according to IUCN Red List criteria. However, more field studies are needed to clarify this. Some of these species are only known from type specimens. For instance, *Sclerochorton haussknechtii* has not been found for 140 years, since it was first collected. This could be because the species has become extinct in the wild or, more likely, because of a lack of information due to a scarcity of field investigations.

Based on the vegetation data of the author from Central Alborz, the highest rate of endemism occurs in alpine-subnival scree grounds (60 % taxa), followed by alpine thorn-cushion grasslands (49 % taxa), subalpine dry slopes covered with umbelliferous vegetation types (41 % taxa), alpine snowbeds (36 % taxa) and subalpine wetlands (6 % taxa).

Photo 7.14 *Astragalus macrosemius* (3,700 m) in Central Alborz with Mt. Damavand (5,671 m) in the background (Photographed by Noroozi)

Examples for endemics adapted to alpine-subnival scree grounds are: *Asperula glomerata* subsp. *bracteata, Astragalus capito, Cicer tragacanthoides, Clastopus vestitus, Crepis heterotricha, Dracocephalum aucheri, Galium aucheri, Jurinella frigida, Leonurus cardiaca* ssp. *persicus*, *Nepeta racemosa, Scutellaria glechomoides, Senecio vulcanicus, Veronica paederotae, Veronica aucheri* and *Ziziphora clinopodioides* subsp. *elbursensis.*

Examples for endemics which occur in alpine thorn-cushion grasslands are: *Acantholimon demawendicum, Allium tuchalense, Astragalus chrysanthus, A. iodotropis, A. macrosemius* (Photo 7.14), *Bufonia kotschyana, Cousinia crispa, Draba pulchella, Minuartia lineata, Oxytropis persicus, Scorzonera meyeri, Silene marschallii, Tragopogon kotschyi* and *Veronica kurdica.*

Examples of endemics which inhabit subalpine dry slopes covered with umbelliferous vegetation are: *Aethionema stenopterum, Allium derderianum, A. elburzense, Astragalus aegobromus, Cousinia adenosticta, C. hypoleuca, Echinops elbursensis, Galium megalanthum, Iris barnumae, Parlatoria rostrata, Ranunculus elbursensis* and *Rumex elbursensis.*

Examples of endemic taxa of alpine snowbeds are: *Cerastium persicum, Erigeron uniflorus* ssp. *elbursensis, Potentilla aucheriana* and *Ranunculus crymophylus.*

An example of endemics of subalpine wetlands is *Ligularia persica.*

According to Naqinezhad et al. (2010), of 323 vascular plant species recorded for the wetland flora on the southern slopes of Alborz, only 7 % are endemic and subendemic to Iran, which is consistent with our local findings. This rate of endemism is very low in comparison to other high-elevation habitats in Iranian mountains.

The percentage of endemic species for the true subnival-nival flora is 68 % (Noroozi et al. 2011), and for alpine habitats (including the subnival-nival zone) 58 % (Noroozi et al. 2008). This rate is much higher than that for the entire Iranian flora (24 %, Akhani 2006). It means that c. 23 % of Iranian endemics are concentrated in alpine habitats, suggesting that the alpine zone can be considered one of the centres of endemism. Most studies on plant biodiversity of Iranian mountains demonstrate that the proportion of narrowly distributed plant species increases consistently from low to high elevations (Klein 1991; Noroozi et al. 2008, 2011; Naqinezhad et al. 2009, 2010; Kamrani et al. 2011).

The major factors increasing the extinction risk for geographically restricted alpine species of Iran could be climate change and overgrazing. The evidence of global warming impacts on the upward shift of plant species has been demonstrated for European mountains (e.g., Grabherr et al. 1994; Walther et al. 2005; Gottfried et al. 2012; Pauli et al. 2012). The persistence of the unique cryophilic flora of Iran would be seriously threatened under the impact of ongoing global warming where the potential to migrate to appropriate habitats is very limited (Noroozi et al. 2011).

High elevations in Iran have been used as summer pastures, and overgrazing is very severe in these habitats. Thus, the high mountain flora and vegetation are seriously disturbed (personal observation).

Since alpine habitats are floristically less well researched, we have insufficient knowledge about the habitats of rare species in the various mountain ranges. We thus strongly recommend more field exploration of high mountain areas to determine the ecology, biology and conservation status of local endemics according to IUCN criteria, and to improve the protection status of the country's high mountain flora and vegetation.

7.7 The Role of Edaphic Substrate Versus Moisture Availability in Montane Endemic Plant Distribution Patterns – Evidence from the Cape Midlands Escarpment, South Africa

V. Ralph Clark (✉) • Nigel P. Barker
Department of Botany, Rhodes University, Grahamstown, South Africa
e-mail: vincentralph.clark@gmail.com

7.7.1 *Introduction: Physical Geography and Biodiversity of the Great Escarpment*

7.7.1.1 The Great Escarpment of Southern Africa

Southern Africa's macro-topography – in the simplest terms – consists of a raised, centrally dipped interior (part of the extensive African Erosion Surface, Burke and Gunnell 2008) separated from a skirting coastal plain by a steep drop-off (Partridge and Maud 1987; McCarthy and Rubidge 2005). The interior forms a saucer in cross-section, with the rim – running parallel to the coastline – forming the highest regions. This rim is generally known as the 'Great Escarpment' (Partridge and Maud 1987; Gilchrist et al. 1994; Kooi and Beaumont 1994; McCarthy and Rubidge 2005; Moore et al. 2009) and can be taken as occurring from Angola southwards through Namibia to South Africa and east and north up through Lesotho and Swaziland into eastern Zimbabwe and adjacent Mozambique (Clark et al. 2011d, Fig. 7.8). While debates on definitions, delimitations, origins and age are numerous (Birkenhauer 1991; Gilchrist et al. 1994; Matmon et al. 2002; McCarthy and Rubidge 2005; Moore and Blenkinsop 2006; Burke and Gunnell 2008), the Escarpment's effect on biodiversity has been consistent: a series of uplands and mountains with significant endemism in both plants and animals (Stuckenberg 1962; Huntley and Matos 1994; Van Wyk and Smith 2001; Steenkamp et al. 2005; Clark et al. 2011d).

7.7.1.2 The Cape Midlands Escarpment

A portion of the Great Escarpment is the Cape Midlands Escarpment, located in the Eastern Cape Province of South Africa, between 31°–32°30′S and 24°–27°E (Figs. 7.8 and 7.9). Because of deep inland incursion by the Great Fish and Great Kei River systems, through headward erosion, the Cape Midlands Escarpment is comprised of three mountain blocks: the Sneeuberg ('Snow Mountains' – 'berg' refers to 'mountain' in South Africa; Clark et al. 2009 provide a detailed overview of the Sneeuberg), the Great Winterberg–Amatolas (GWA, detailed overview in prep.), and the Stormberg (Photo 7.15a–e). The Cape Midlands Escarpment is unique in the context of the total Escarpment in that it forms a roughly circular band of mountains (rather than simply linear) bounding a large inselberg-studded plain formed from the inland advance of the Great Fish River. This Cape Midlands Basin (Photo 7.15f) is some 100 km wide and as long, while the mountains cover a total area of approximately 31,500 km^2. Although not high by world standards, the Cape Midlands Escarpment reaches reasonable heights by southern African standards: the highest peak is the Compassberg (2,504 m), followed by the Nardousberg (2,429 m – see Fig. 7.11 and Photo 7.20f), both in the Sneeuberg, and then by the Great Winterberg peak (2,369 m), in the GWA. Several other peaks in the Sneeuberg and GWA rise above 2,200 m, while the highest peak in the Stormberg is the Aasvoëlberg ('Vulture Mountain', 2,207 m). Large upland plateaux occur between

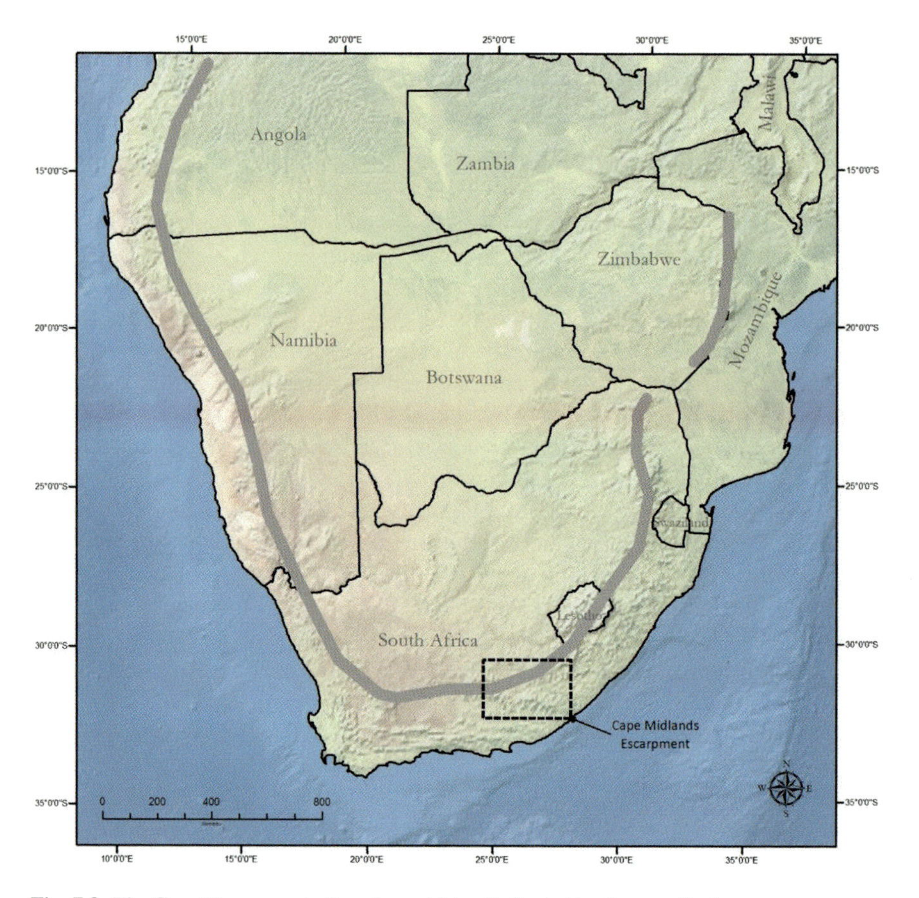

Fig. 7.8 The Great Escarpment of southern Africa (indicated by the *grey line*)

1,700 and 2,100 m. The base of the mountains varies between ca. 900 m on the coastward side to ca. 1,200 m on the inland side.

7.7.1.3 Geology and Geomorphology

The geology of the Cape Midlands Escarpment is fairly simple, mostly consisting of the arenaceous and argillaceous sediments of the Beaufort Group (Karoo Supergroup, Photo 7.16a, b, f) massively intruded by Jurassic-era Dolerites (Hill 1993; Johnson et al. 2006; Van Zijl 2006, Photo 7.16c–e). The central and eastern parts of the Stormberg are dominated by Molteno sandstones and capped sporadically by Drakensberg basalts (Johnson et al. 2006); these are excluded from the study area in favour of the western Stormberg, which has the same geology as the Sneeuberg and GWA.

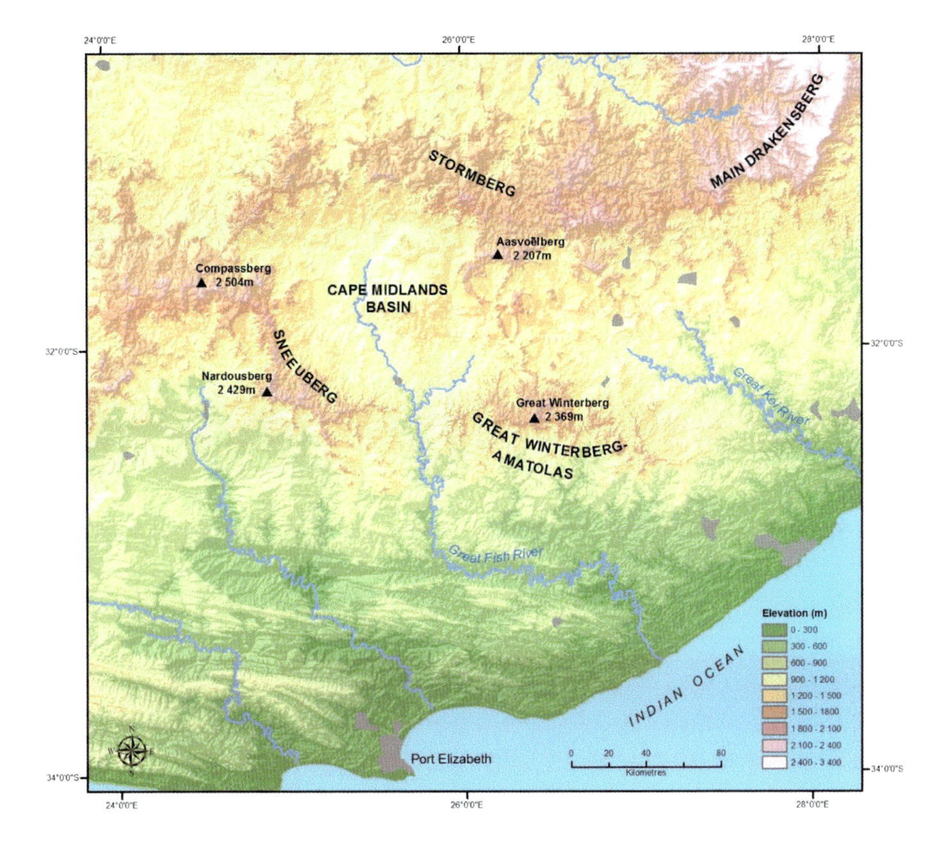

Fig. 7.9 The Cape Midlands Escarpment, South Africa

While finer details in the Karoo stratigraphy vary between the mountain blocks (Johnson et al. 2006), the general landscape features are similar: layered horizontal sediments (Photo 7.16j) intruded by dolerite either as near-vertical sill cappings (e.g. the Great Winterberg peak and the Hogsbacks) or steeply inclined sheets (e.g. the Compassberg and Nardousberg, Photo 7.16i). Throughout these mountains dolerite is the most important rock geomorphologically: while dolerite denudes rapidly in the moister eastern parts of South Africa (Brink 1983), in the drier Cape Midlands Escarpment it is much less susceptible to denudation than the Karoo sediments and has almost invariably formed the highest peaks and plateaux (Du Toit 1920; Agnew 1958; Partridge and Maud 1987, Photo 7.16k, l). Consequently almost all of the highest peaks are doleritic (Agnew 1958; Phillipson 1987; Clark et al. 2009).

7.7.1.4 Soils

Soils arising in mountainous areas differ according to *inter alia* parent material, altitude, moisture availability, mechanical processes, aspect, vegetation cover and

Photo 7.15 Representative photographs of the Cape Midlands Escarpment, Eastern Cape, South Africa. (**a**) The Nardousberg area, Sneeuberg, taken from an altitude of ca. 2,100 m, looking west, (**b**) The Great Winterberg plateau, ca. 1,700 m, west of the main Great Winterberg peak, Great Winterberg–Amatolas (GWA), (**c**) The Didima Range, with the Katberg on the far right, taken from Katberg Pass, GWA, (**d**) A view towards Aasvoëlberg, Stormberg, taken from the first plateau at 1,900 m, (**e**) The Bamboesberge on the farm 'Bamboeshoek', Stormberg, taken from the plateau at 1,900 m, (**f**) The Cape Midlands Basin as viewed looking east from the Wapadsberg, Sneeuberg (Photographed by V.R. Clark)

slope steepness (Macvicar et al. 1977; Laffan et al. 1998; Turner 2000; Burke 2002; Osok and Doyle 2004, etc.). It should be noted that while sandstone is less resistant to denudation on a landscape level – and therefore forms softer landforms – dolerite, although harder, produces better soil. Thus in the Cape Midlands Escarpment,

Photo 7.16 Typical sedimentary and igneous landforms in the Cape Midlands Escarpment. (**a**) Sandstone cliffs in the Toorberg, Sneeuberg, (**b**) Metamorphic sedimentary cliffs capped with dolerite, western slopes of the Koudeveldberge, Sneeuberg, (**c**) A dolerite batholith forming part of the Blinkberg, Sneeuberg, (**d**) A massive dolerite sill outcropping on the eastern edge of the Toorberg plateau, Sneeuberg, (**e**) Jointed, columnar dolerite forming the eastern ramparts of the main Wapadsberg, Sneeuberg, (**f**) A sedimentary spur off the Kamdebooberge, Sneeuberg, consisting largely of shale (note the poor vegetation cover), (**g**) The northern summit area of the Blinkberg, Sneeuberg, consisting of flat, shattered sandstone, (**h**) The summit plateau of the Schurfteberg, Sneeuberg, consisting of small dolerite boulders, (**i**) Dolerite sills forming typical cuesta-shaped hills in the Kikvorsberge, Sneeuberg, (**j**) Deep fluvial incision of harder sandstones forming Fenella Gorge, GWA, (**k**) A rounded sandstone hill resulting from the absence of a protective dolerite capping, taken from the nearby dolerite-capped Oppermanskop, Kikvorsberge, Sneeuberg, (**l**) A rounded sandstone and shale hill resulting from the absence of a protective dolerite capping, compared to the angular, flat-topped dolerite sill-capped plateau behind, Kamdebooberge, Sneeuberg (Photographed by V.R. Clark)

Photo 7.16 (continued)

dolerite tends to form better soils than any other substrate, forming shallow reddish loamy-clays to deeper black turf clays (Turner 2000; Clark 2010, Photo 7.17a–c). Sandstone soils on summit plateaux are generally either almost non-existent (Photo 7.16g), to poorly developed in the drier (<700 mm per annum, e.g. Sneeuberg) mountains (Photo 7.17d, e), to fairly well developed and often indistinguishable from dolerite soils in moister mountains (>700 mm per annum, e.g. on the GWA and Stormberg scarp and summit). Soils on steep slopes are typically a mixture of regolith and occasional large boulders from various parent substrates (Turner 2000). A humus-rich colluvium develops on wetter slopes where the vegetation has stabilised the loose material, and deep colluvial soils are formed on gentler slopes

(Osok and Doyle 2004, Photo 7.17g). Shale generally provides no soil, and simply weathers to a purple-grey gravel locally called 'gruis' (Photo 7.16f). Metamorphic sandstone ('baked' from contact with intruded dolerites) behaves much like shale and very little or no soil is formed. Extensive wetlands in upland areas generally host deep, humus-rich clay soils regardless of parent substrate (Photo 7.17h, i), and deep alluvial deposits occur along more permanent streams (Photo 7.17f).

7.7.1.5 Climate

A steep moisture gradient exists from south to north across each of these mountain blocks: the southern slopes and adjacent plateaux are the wettest, receiving between ca. 700 mm to in excess of 1,000 mm rainfall per annum (Phillipson 1987; Hoare and Bredenkamp 1999; Clark et al. 2009, 2011e). The northern slopes in contrast are much drier (down to ca. 400 mm), and rainfall quickly diminishes away from the south-facing scarps (Clark et al. 2011e, unpublished data). The combination of high altitude and exposure to frequent mid-latitude cyclones renders the mountains vulnerable to regular snowfalls, especially in winter (Phillipson 1987; Clark et al. 2009, 2011e).

7.7.1.6 Vegetation and Endemism

The vegetation of the Cape Midlands Escarpment is primarily montane vegetation consisting of upland grassland, azonal fynbos (the African version of heathland, consisting of temperate 'Cape elements', Levyns 1964; Oliver et al. 1983; Linder 1990; Carbutt and Edwards 2001; Galley et al. 2007; Devos et al. 2010, etc.), wetlands, scarp forest, mesic woodlands and sub-tropical thicket in a matrix of lower-altitude, arid Nama-Karoo (a shrubby semi-desert vegetation type common in the arid central-west of South Africa) and *Acacia karroo* Hayne (=*Vachellia karroo* (Hayne) Banfi & Galasso) savannoid grassland (Cook sine anno; Phillipson 1987; Palmer 1988, 1990, 1991; Mucina and Rutherford 2006; Clark et al. 2009, 2011a). Four biomes are represented on these mountains: Grassland (here including azonal fynbos as opposed to the Fynbos Biome of the Cape Floristic Region), Forest, Sub-tropical Thicket, and Nama-Karoo (Mucina and Rutherford 2006; Clark et al. 2011e).

Local plant endemism in the Cape Midlands Escarpment is high, with a current total of 88 endemic taxa (Fig. 7.10, Photos 7.18 and 7.19). Currently, the GWA has the highest number of local endemics, followed by the Sneeuberg, while a similar number is shared by all three mountain blocks. The Sneeuberg was recently described as a centre of floristic endemism (Clark et al. 2009), and based on the same standards the GWA deserves the same recognition (in prep.). The comparable number of shared endemics however suggests a case for resurrecting the old concept of a combined centre comprising all three mountain blocks, as originally described by Nordenstam (1969). Future work on the Stormberg – the

Photo 7.17 Representative soil profiles in the Cape Midlands Escarpment. (**a–c**) Dolerite-derived soils, Bankberg, Sneeuberg – note the well-developed A horizons on saprolite, and the presence of near-surface rocks from either gravitational slope processes (in **b**), or as core stones (in **c**), (**d–e**) Shallow, rocky soils on sandstone/sandstone-dolerite colluvium, Nardousberg area, Sneeuberg, (**f**) Stratified alluvial deposits along a stream-line at the base of the Blinkberg, Sneeuberg, (**g**) Deep colluvial deposits along an eroded stream-line at the base of the Blinkberg, Sneeuberg, (**h–i**) Black turf clays in two upland valley wetland systems: Blinkberg, Sneeuberg (**h**), and near the base of the Great Winterberg peak, GWA (**i**) (Photographed by V.R. Clark)

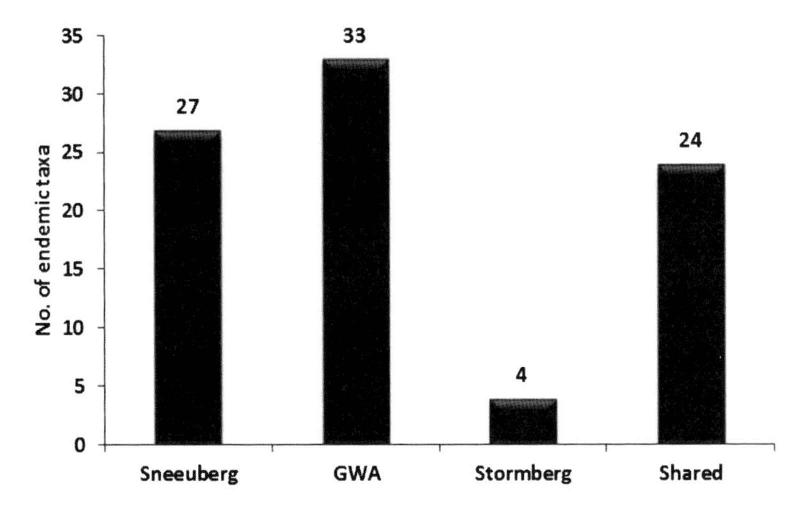

Fig. 7.10 Plant endemism in the Cape Midlands Escarpment

least well-researched mountain block, with only four known local endemics – will allow a more quantitative assessment in this regard. With approximately nine species new to science having been discovered since 2005 (Goldblatt and Manning 2007; Clark et al. 2009; Nordenstam et al. 2009; Stirton et al. 2011), the total number of Cape Midlands Escarpment endemics is likely to increase as research in these poorly explored mountains continues.

Extra-Cape Midlands phytogeographical links are mostly with the eastern Escarpment (particularly the high Drakensberg-Maluti system), west along the southern Escarpment (particularly with the Nuweveldberge), with the Albany Centre of Endemism, and loosely with the Cape Floristic Region (Cook *sine anno*, Phillipson 1987; Van Wyk and Smith 2001; Carbutt and Edwards 2006; Clark et al. 2009, 2011a, c, e).

7.7.2 Aims and Hypothesis

While the role of edaphic substrate is known to be a key player in plant speciation and ecology the world over, and no less so in South Africa (Matthews et al. 1991; Balkwill and Balkwill 1999; Carbutt and Edwards 2001; Siebert et al. 2001; Van Wyk and Smith 2001; Mucina and Rutherford 2006, etc.), this has not been explored in detail as a factor in local endemism in the Cape Midlands Escarpment, with its mosaic of sedimentary and igneous substrates. Although both dolerite and sandstone substrates are known to be rich in local endemics in other parts of southern Africa (e.g. the Hantam region, Western Cape, for dolerite endemics, and the Pondoland region, Eastern Cape, for sandstone endemics, Van Wyk and Smith 2001), this doesn't necessarily take into account the role of moisture availability in these areas (Bergh et al. 2007). The aim of this study is thus to determine

Photo 7.18 (continued)

if edaphic substrate plays a primary role in the distribution of endemics in the Cape Midlands Escarpment, or if moisture availability (represented by altitude and primary vegetation habitat as proxies, given an absence of fine-scale climate data) is a more important determining factor in their distributions (Clark et al. 2011b, c, e). We also include primary geomorphological niche as a characteristic, to determine if there are patterns in niche specialisation by endemic taxa.

We hypothesise that edaphic factors do not play a major role in the distribution of endemics, and that moisture availability plays the dominant role. The Cape Midlands Escarpment is a good region in which to examine this issue, as there are large areas dominated by both sedimentary and dolerite substrates at all altitudes and in all local moisture regimes, and the basic geology is consistent throughout the study area. Furthermore, the soils present are predominantly lithosols arising from *in situ* weathering: they are thus less likely to be contaminated by movement of soils from slope processes, except on steeper slopes where soil movement from gravitational processes is prevalent.

7.7.3 *Methods*

7.7.3.1 Selection of Endemic Taxa

The Cape Midlands Escarpment has been part of the focus of a detailed and intensive floristic study on the southern Great Escarpment since 2005 (Clark 2010; Clark et al. 2009, 2011a, b, c, e), resulting in ca. 13,000 herbarium specimens and field notes with detailed ecological data. Edaphic, lower altitude limit, primary vegetation habitat and primary geomorphological niche data were collated for 50 of the endemic taxa from this data and supplemented where necessary from available published sources. The necessary data for the other 38 endemic taxa is currently incomplete or unavailable. Four typical Cape Midlands Escarpment 'endemics' (*Ficinia compasbergensis* Drège, *Helichrysum tysonii* Hilliard, *Lessertia sneeuwbergensis* Germish. and *Ruschia complanata* L. Bolus) are now also known from the Nuweveldberge (Clark et al. 2011c) and have been excluded from this study. We have however kept *Indigofera elandsbergensis* Phillipson in the study even though there is one outlier recorded in the Main Drakensberg (Phillipson 1992).

Photo 7.18 (continued) A photographic selection of plant taxa endemic to the Cape Midlands Escarpment – Part 1: A world of *Euryops* (Asteraceae). (**a**) *E. ciliatus* B. Nord. – endemic to the Great Winterberg–Amatolas (GWA), (**b**) *E. dentatus* B. Nord. – endemic to the Sneeuberg, (**c**) *E. exsudans* B. Nord. & V.R. Clark – endemic to the Sneeuberg (discovered in 2005), (**d**) *E. galpinii* Bolus – endemic to the Sneeuberg, GWA and Stormberg, (**e**) *E. trilobus* Harv. – endemic to the Sneeuberg and Stormberg, (**f**) *E. proteoides* B. Nord. & V.R. Clark – endemic to the Sneeuberg (discovered in 2005) (Photographed by V.R. Clark, except for (**a**), Nick Helme)

Photo 7.19 A photographic selection of plant taxa endemic to the Cape Midlands Escarpment – Part 2: Other colours of endemism. (**a**) *Crassula exilis* ssp. *cooperi* (Regal) Tölken – endemic to the Sneeuberg, Great Winterberg–Amatolas (GWA) and Stormberg, (**b**) *Delosperma dyeri* L. Bolus – endemic to the Sneeuberg and GWA, (**c**) *Erica* sp. aff. *reenensis* Zahlbr. – endemic to the Sneeuberg and GWA (yet to be photographed at maturity), (**d**) *Gazania caespitosa* Bolus – endemic to the Sneeuberg, (**e**) *Garuleum tanacetifolium* (MacOwan) Norl. – endemic to the Sneeuberg and GWA, (**f**) *Geranium grandistipulatum* Hilliard & B.L. Burtt – endemic to the GWA, (**g**) *Psoralea margaretiflora* C.H. Stirt. & V.R. Clark – endemic to the Sneeuberg (discovered in 2005) (Photographed by V.R. Clark)

Fig. 7.11 Principal geomorphological habitat niches in the Cape Midlands Escarpment. (*A*) Moist scarp, (*B*) Cliffs, (*C*) Plateau, (*D*) Upland slope, (*E*) High scree, (*F*) High peak (Photo: V.R. Clark (Nardousberg, Sneeuberg))

7.7.3.2 Edaphic Substrate

Edaphic substrate options were reduced to two main categories, namely dolerite and sandstone. Sandstone is used here to include metamorphic rocks found along contact zones with dolerites. Shale was excluded from this study as it generally supports little vegetation in the study area. An attempt to classify soil types binomially (as per Macvicar et al. 1977) was not made given the complicated and often inaccurate results without specialist assistance.

7.7.3.3 Altitude

Lower limit altitude refers to the lowest altitude from which a taxon has been recorded.

7.7.3.4 Primary Vegetation Habitat

Because of the complexity of vegetation types in these mountains, primary vegetation habitats were simplified into Open and Closed Habitats:

Open Habitats: namely Montane Grassland (including azonal montane fynbos, wetlands and shrublands) and Nama-Karoo.

Closed Habitats: a lumping of the intergrading and inter-digitating varieties of woodland, thicket and forest that grade into each other with aspect and altitude.

Photo 7.20 Geomorphological habitat niches. (**a**) Moist scarp (Koudeveldberge, Sneeuberg), (**b**) Cliffs (Great Winterberg peak, Great Winterberg–Amatolas (GWA)), (**c**) Upland slope (Nardousberg area, Sneeuberg), (**d**) Plateau (Didima Range, GWA), (**e**) High scree (Koudeveldberge, Sneeuberg), (**f**) High Peak (Nardousberg) (Photographed by V.R. Clark)

7.7.3.5 Primary Geomorphological Niche

Primary geomorphological niche refers to the typical physical location of a taxon on the characteristic topography of the Cape Midlands Escarpment (Fig. 7.11, Photo 7.20). Although the three mountain blocks have distinct features unique to themselves (and various local differences within them), there is a common basic geomorphological theme. A conservative approach was adopted, with taxa placed in their least specialised niche if they occur in more than one (e.g. a taxon occurring in both Plateaux and High Peaks was placed under Plateaux).

Moist Scarps are those south- and south-east-facing slopes from the base of the Escarpment to generally the first line of cliffs. 'Moist' here differentiates these slopes from the much drier north- and west-facing scarps.

Cliffs refers to major cliff-lines at all altitudes – usually at ca. 1,400/1,500 m, then again at 1,800/2,000 m, and sometimes again at 2,300 + m on the highest peaks.

Upland Slopes refer to the higher altitude slopes (ca. >1,600 m) which are too steep to constitute plateau.

Plateaux refer to extensive flat or gently sloping regions. These are typically tiered on the Escarpment, occurring primarily at 1,200/1,300 m, again at 1,400/1,500 m, and extensively between 1,700–2,100 m.

High Screes refer to extensive steep or near-horizontal dolerite screes and boulder fields occurring above 1,800 m (often >2,000 m) – these often consist of massive stable boulder beds.

High Peaks refer to the prominent highest peaks that sit above the main summit plateaux, reaching altitudes of 2,200–2,500 m. These peaks are typically isolated from each other, are flanked with extensive High Screes, and cover a much smaller surface area than the rest of the mountain blocks.

7.7.3.6 Data Manipulation and Presentation

The data was tabulated (Table 7.6), and analysed through simple statistical means and graphical presentation using Microsoft Excel and PowerPoint. Although more complex means of statistical analysis could have been employed, a simpler approach has been deemed suitable in this study.

7.7.4 Results and Discussion

The results are shown in Figs. 7.12 and 7.13. Each point is considered separately.

7.7.4.1 Edaphic Substrate

Almost half of the taxa are edaphic ubiquitists, with approximately a third restricted to dolerite, and a smaller proportion restricted to sandstone (Fig. 7.12). There is thus no strong bias towards substrate. Edaphic indifference is easily explainable in the moister mountains as there is often a less discernible difference in the soils arising on dolerite and sandstone because of more equitable weathering and humus production. The humus content of the soil is possibly the critical factor, being as important to soil fertility in these mountains as is mineral content. In moister

Table 7.6 Selected taxa endemic to the Cape Midlands Escarpment, indicating distribution and key characteristics used in this study

Endemic taxon	Distribution	Substrate	Lower altitude limit (m)	Primary vegetation habitat	Primary geomorphological niche
Acmadenia sp. nov. aff. *sheilae* I. Williams	S	Dolerite	1,600	Montane grassland	Plateau
Alepidea macowani Dummer	S, GWA	Dolerite & sandstone	1,400	Montane grassland	Plateau
Arrowsmithia styphelioides DC.	GWA	Dolerite & sandstone	1,100	Montane grassland	Upland slope
Bergeranthus nanus A.P. Dold & S.A. Hammer	S, GWA	Dolerite & sandstone	1,300	Montane grassland	Upland slope
Brachystelma cathcartense R.A. Dyer	GWA	Sandstone	1,200	Montane grassland	Upland slope
Ceropegia macmasteri A.P. Dold	GWA	Sandstone	1,200	Montane grassland	Upland slope
Conium sp. no. 4 (Hilliard and Burtt 1985)	S, GWA	Dolerite & sandstone	1,800	Montane grassland	Cliffs
Crassula exilis subsp. *cooperi* (Regal) Tölken	S, GWA, ST	Dolerite & sandstone	1,500	Montane grassland	Cliffs
Delosperma alpinum (N.E.Br.) S.A. Hammer & A.P. Dold	GWA	Dolerite	1,800	Montane grassland	High scree
Delosperma katbergense L. Bolus	GWA	Sandstone	1,200	Montane grassland	Cliffs
Delosperma dyeri L. Bolus	S, GWA	Dolerite & sandstone	1,600	Montane grassland	Plateau
Diascia ramosa Scott-Elliot	S	Sandstone	1,200	Montane grassland	Moist scarp
Dierama grandiflorum G.J. Lewis	S	Dolerite	1,500	Montane grassland	Plateau
Drimia montana A.P. Dold & E. Brink	GWA, ST	Sandstone	2,000	Montane grassland	Upland slope
Encephalartos cycadifolius (Jacq.) Lehm.	S, GWA	Dolerite	1,400	Montane grassland	Upland slope
Erica sp. aff. *reenensis* Zahlbr.	S, GWA	Dolerite & sandstone	1,800	Montane grassland	High peak
Erica passerinoides (Bolus) E.G.H. Oliv.	S	Dolerite	1,800	Montane grassland	Plateau
Euryops ciliatus B. Nord.	GWA	Dolerite	1,600	Montane grassland	Upland slope
Euryops dentatus B. Nord.	S	Dolerite	1,400	Montane grassland	Upland slope
Euryops dyeri Hutch.	GWA	Dolerite & sandstone	1,700	Montane grassland	Upland slope
Euryops exsudans B. Nord. & V.R. Clark	S	Dolerite & sandstone	1,600	Montane grassland	Upland slope
Euryops galpinii Bolus	S, GWA, ST	Sandstone	1,800	Montane grassland	Upland slope
Euryops trilobus Harv.	S, ST	Dolerite & sandstone	1,800	Montane grassland	Upland slope
Euryops proteoides B. Nord. & V.R. Clark	S	Dolerite & sandstone	1,300	Montane grassland	Upland slope
Faurea recondita MS J. Rourke	S	Dolerite & sandstone	1,300	Montane grassland	Moist scarp
Ficinia sp. nov. aff. *gracilis* Schrad.	S, GWA	Dolerite & sandstone	1,500	Montane grassland	Upland slope

Garuleum tanacetifolium (MacOwan) Norl.	S, GWA	Dolerite & sandstone	1,600	Montane grassland	Upland slope
Gazania caespitosa Bolus	S	Dolerite	1,600	Montane grassland	Plateau
Geranium amatolicum Hilliard & B.L. Burtt	GWA	Dolerite & sandstone	1,400	Montane grassland, Closed	Upland slope
Geranium contortum Eckl. & Zeyh.	GWA	Dolerite	1,900	Montane grassland	Upland slope
Geranium grandistipulatum Hilliard & B.L. Burtt	GWA	Sandstone	1,800	Montane grassland	Upland slope
Haworthia marumiana var. *batesiana* (Uitewaal) M.B. Bayer	S	Dolerite & sandstone	1,200	Closed	Cliffs
Hermannia sneeuwbergensis MS D. Gwynne-Evans	S, GWA	Dolerite & sandstone	1,300	Montane grassland	Upland slope
Hermannia crassifolia MS D. Gwynne-Evans	S	Dolerite	1,200	Closed	Upland slope
Hermannia violacea (Burch. ex DC.) K. Schum.	S, GWA	Sandstone	1,000	Closed	Moist scarp
Hesperantha helmei Goldblatt & J.C. Manning	S	Dolerite & sandstone	2,000	Montane grassland	High peak
Hesperantha stenosiphon Goldblatt	GWA	Dolerite & sandstone	1,500	Montane grassland, Closed	Cliffs
Indigofera elandsbergensis Phillipson	GWA	Sandstone	1,600	Montane grassland	Upland slope
Indigofera magnifica MS B. Schrire	S	Dolerite	1,800	Montane grassland	Plateau
Indigofera asantasanae MS B. Schrire	S	Dolerite & sandstone	1,800	Montane grassland	Upland slope
Jamesbrittenia crassicaulis (Benth.) Hilliard	S, GWA, ST	Dolerite & sandstone	1,800	Montane grassland	Plateau
Kniphofia acraea Codd	S	Dolerite	1,600	Montane grassland	Upland slope
Lotononis alpina subsp. *multiflora* (Eckl. & Zeyh.) B.E.van Wyk	GWA	Dolerite	1,700	Montane grassland	Upland slope
Moraea reticulata Goldblatt	GWA	Dolerite	1,200	Montane grassland	Plateau
Polemannia grossulariifolia Eckl. & Zeyh.	S, GWA	Dolerite	1,500	Montane grassland	High scree
Psoralea margaretiflora C.H. Stirt. & V.R. Clark	S	Dolerite & sandstone	1,200	Montane grassland	Moist scarp
Selago bolusii Rolfe	S	Dolerite & sandstone	1,600	Montane grassland	Cliffs
Selago retropilosa Hilliard	S	Dolerite & sandstone	1,800	Montane grassland	Plateau
Sutera glandulifera Hilliard	GWA, ST	Dolerite	1,400	Montane grassland	Cliffs
Wahlenbergia laxiflora (Sond.) Lammers	S, GWA	Dolerite	1,400	Montane grassland	Upland slope

For distribution: *S* Sneeuberg, *GWA* Great Winterberg–Amatolas, *ST* Stormberg

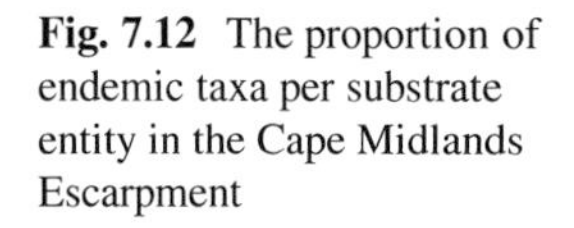

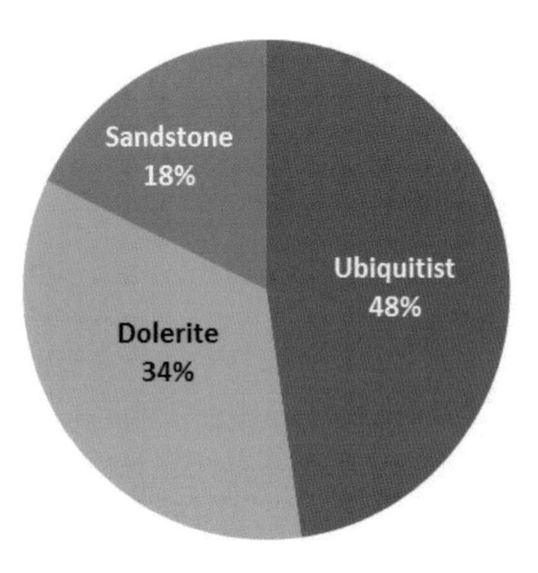

Fig. 7.12 The proportion of endemic taxa per substrate entity in the Cape Midlands Escarpment

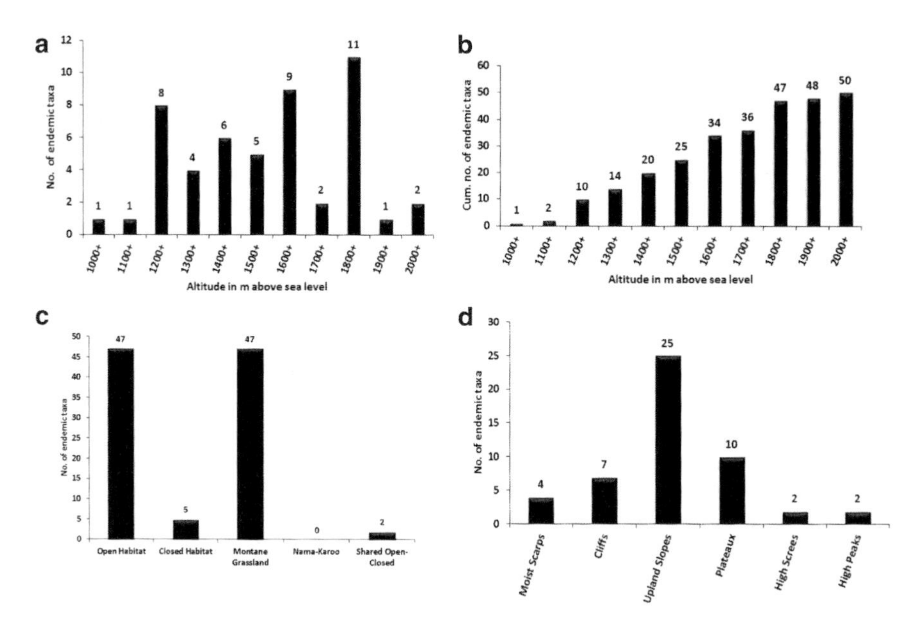

Fig. 7.13 Results for three moisture availability proxies, and primary geomorphological niche, for endemic plant taxa in the Cape Midlands Escarpment. (**a**) Minimum altitude thresholds, (**b**) accumulation of endemics with altitude, (**c**) primary vegetation habitats, (**d**) primary geomorphological niches

areas the production of plant material – and therefore humus content – is likely to be similar on both substrates, thus enriching the soil on both substrates equally well (forming a more ubiquitous A horizon). In drier areas, the deeper weathering and minerally richer dolerite (Burke 2002) encourages better vegetative cover and therefore higher humus content in the soil, thus favouring more fertile dolerite soils while sandstone soils remain comparatively stony and infertile. Thus endemics are more widespread on the richer dolerite soils. Otherwise there appears to be no significant distinction between the substrates ecologically, compared to very obvious plant specialisation on substrates with high levels of toxic heavy metals, such as the Sekhukhuneland ultramafics, the Barberton serpentines, and the Great Dyke of Zimbabwe (Balkwill and Balkwill 1999; Siebert et al. 2001; Van Wyk and Smith 2001).

7.7.4.2 Altitude

Most endemic taxa have a lower altitude limit of 1,800 m, followed closely by 1,600 m and 1,200 m respectively (Fig. 7.13a). Only two taxa have limits below 1,200 m and three above 1,800 m. Figure 7.13b indicates that there is a steady cumulative addition of endemic taxa with altitude, mostly between 1,200 and 1,800 m. Most endemics therefore occur above 1,200 m, and only three taxa are added above 1,900 m.

Lower altitude limits are easily explained as the result of increased aridity towards the base of the Escarpment in drier areas, or a change in habitat from Open to Closed (in both drier and wetter areas, discussed in *Primary vegetation habitat* below). What is surprising is the low number of endemics restricted above 2,000 + m, suggesting that the highest altitudes are not critical for endemism, or that conditions on the isolated High Peaks are not conducive to many endemic taxa (Cook *sine anno*). This is discussed further in *Primary geomorphological niche*.

7.7.4.3 Primary Vegetation Habitat

The vast majority of endemic taxa occur in Open Habitats, with only five occurring exclusively in Closed Habitats (Fig. 7.13c). Of the Open Habitat taxa, all occur in Montane Grassland and none in Nama-Karoo. Two taxa are both Open and Closed Habitat taxa, although they lean towards more open areas (forest margin, exposed cliffs in forest, etc.).

That no endemics occur in Nama-Karoo is very significant, as much of the drier leeward slopes of the mountains are dominated by Nama-Karoo and similar karroid mountain vegetation. This supports the idea that the endemics are concentrated where moisture is available, as reflected in Montane Grassland. Note however that forest, which reflects the highest rainfall in the study area, does not support many endemics. Grassland as a biome is therefore more important for endemics in these mountains than any other biome, a result that strongly supports trends in the grassland regions of South Africa (Van Wyk and Smith 2001; Mucina and Rutherford 2006). These results also overwhelmingly support Meadows and Linder's (1993) findings that montane grassland in the region is a primary, endemic-rich habitat and not a species-poor, secondary vegetation type arising from the palaeo-anthropogenic destruction of forests. The results also suggest that, being grassland taxa, most of the endemics are capable of withstanding, or even require, the periodic fires that characterise the Grassland Biome in southern Africa (Mucina and Rutherford 2006).

7.7.4.4 Primary Geomorphological Niche

Upland Slopes host by far the majority of endemic taxa, while High Screes and High Peaks host the lowest (Fig. 7.13d). Plateaux, Cliffs and Moist Scarp each host an intermediate to low number.

The dominance of Upland Slopes, followed by Plateaux, could be attributed to their majority surface area, while the other niches cover much smaller areas. It also suggests that there has been no need for these taxa to specialise into more rugged niches away from topographically gentler ones. The results also correlate well with *Altitude*: most Upland Slopes and Plateaux occur between 1,200 and 2,100 m. Similarly, Moist Scarp typically occurs below 1,200 m, and often comprises Closed Habitat with its corresponding low endemism (three of the four Moist Scarp taxa occur exclusively in Closed Habitat). Low endemism on the High Screes and High Peaks also correlates with the low addition of endemic taxa above 1,900 m. There is a definite avoidance of these highest altitudes by most taxa, a propensity also noted by Cook (*sine anno*) in the GWA, where most taxa sampled occurred between 1,500 and 2,000 m. The extreme climatic conditions prevalent on these isolated High Peaks and associated High Screes are probably the main reason for their low endemism, as could be the poor soil conditions of these habitats. The low specialisation in Cliffs, High Screes and High Cliffs in the Cape Midlands Escarpment is interesting given the impression of higher specialisation in these niches in other regions both elsewhere in South Africa and abroad (Larson et al. 2000; Pooley 2003; Van Jaarsveld 2011).

While High Peaks do not feature prominently in the results, and Cliffs only moderately, it can be noted that their role becomes more important westwards, where they compensate for increased aridity by providing moisture refugia for taxa that are more widespread in wetter areas (Larson et al. 2000; Bergh et al. 2007; Clark et al. 2009, 2011d; Clark 2010). For example, numerous endemic and more typical Afromontane taxa widespread on Upland Slopes and Plateaux in the GWA are almost completely restricted to the base of moist, south-facing Cliffs or the Highest Peaks in the drier Sneeuberg (Clark 2010). This is further augmented in the even drier Nuweveldberge (Clark et al. 2011d). It would appear then that most montane taxa will accumulate into specialised refugia only when necessary, and only require these specialist habitats in response to moisture deficits.

7.7.5 *Conclusion and Conservation Implications*

The majority of Cape Midlands Escarpment endemic taxa are edaphic ubiquitists, with dolerite preference as second, and are found in Montane Grassland on Upland Slopes above 1,200 m. Much fewer are sandstone 'specialists', and few are restricted to Closed Habitats (forest, woodland, thicket), Cliffs, High Screes or High Peaks (i.e. above 1,900 m), or occur below 1,200 m (i.e. to the base of the Escarpment). We conclude therefore that the distribution of Cape Midlands Escarpment endemics is driven primarily by moisture availability rather than edaphic substrate, and that good soil conditions in grassland on moderate topography at medium altitudes are favoured over specialised habitats such as Cliffs, High Scree and High Peaks unless moisture deficits are experienced.

Conservation implications are that the upland grasslands (and associated fynbos, wetland and montane shrublands) in the Cape Midlands Escarpment are of high conservation value and need to be well managed in order to conserve the bulk of endemic taxa, particularly against overgrazing, soil erosion and afforestation (both formal and feral). Should the effects of climate change in this region include a decrease in moisture availability, a possible concentrating of endemic taxa along the base of south-facing Cliffs and on the High Peaks can be expected. These two habitats can thus be considered the most promising refugial niches for the future and should be afforded primary conservation priority.

Acknowledgements Detailed botanical research on the Cape Midlands Escarpment has been in progress since 2005 as a PhD followed by two successive post-doctoral research fellowships and a research association at Rhodes University, Grahamstown (2010–2012). Funding for the PhD was generously provided by the South African National Research Foundation (NRF) in the form of Grant GUN 2069059 and a freestanding South African Biosystematics Initiative (SABI) grant (2006–2009), Buk'Indalo Consultancy cc (Durban, 2005–2007), the National Geographic Society Committee for Research and Exploration (Grant 8521–08), and a Dudley D'Ewes Scholarship from the Cape Tercentenary Foundation. The authors wish to thank the many land-owners for permissions to collect on private land, the Eastern Cape Department of Economic Affairs, Environment, and Tourism and CapeNature for collecting permits, and assistance from many taxonomists (see Clark 2010 for further details). Mr Tony Dold, the Curator of the Schönland Herbarium, Albany Museum, is thanked for office space and facilities. Figures 7.8 and 7.9 were kindly produced by Gillian McGregor, Department of Geography, Rhodes University. Nick Helme is thanked for permission to use Photo 7.18a, and Richard Cowling for comments on the manuscript.

7.8 Taxonomic Composition and Spatial Pattern of Endemic Seed Plant Species in Southwest China

Wenjing Yang • Jihong Huang • Keping Ma (✉)
Institute of Botany, Chinese Academy of Sciences, Beijing 100093, China
e-mail: kpma@ibcas.ac.cn

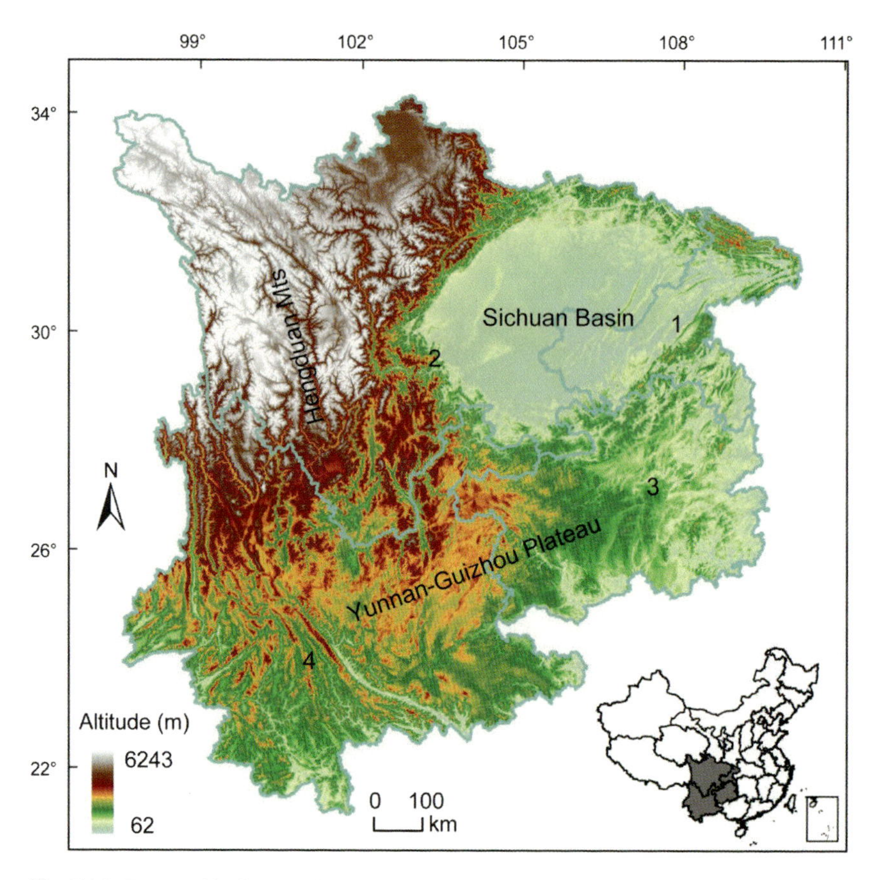

Fig. 7.14 Topographical map of southwest China. *Light blue lines* are provincial boundaries. *1* Chongqing; *2* Sichuan; *3* Guizhou; *4* Yunnan. The map in the *right bottom* shows the location of the study area. The *inset* in the *right bottom* shows islands in the South China Sea (Albers projection)

7.8.1 Physical Geography and Endemism in Southwest China

Southwest China is defined to include three provinces (Sichuan, Guizhou and Yunnan) and one municipality (Chongqing) in this study (Fig. 7.14). It lies between the easternmost edge of the Tibetan Plateau and the Central Chinese Plain and stretches over an area of 1.14 million km^2. The geology and topography of this region is complex, characterized by large mountain ranges in the west (e.g. Hengduan Moutains) with a north–south direction, and lower mountains and flat areas (e.g. Sichuan Basin) in the east. Its location in a climate transition zone and highly complex topography result in a wide range of climatic conditions, generally with subtropical climate in the Sichuan Basin and Yunnan-Guizhou Plateau, tropical climate in southernmost part of Yunnan (e.g. Xishuangbanna), and alpine climate in higher altitude areas (Pang 1996).

Southwest China is one of the botanically most diverse terrestrial regions in China (Wang 1992) as well as in the world (Myers et al. 2000). Some parts of the region fall within the Himalaya and South-Central China biodiversity hotspots (Mittermeier et al. 2005). It possesses a wide variety of vegetation types, including broad-leaved and coniferous forests, meadow, freshwater wetlands, and alpine scrub and scree communities (Wu 1980). This region has a disproportionate amount of China's overall seed plant species (c. 60 %), with 18,309 native species (Editorial Committee of Flora Reipublicae Popularis Sinicae 1959–2004), which is featured by a high level of endemism. The Hengduan Mountain region has been identified as one of the endemism centers in China (Ying et al. 1993; López-Pujol et al. 2011; Huang et al. 2012).

Because the level of endemism of an area reflects the history of diversification (speciation and extinction), studies on the geographical distribution and diversity patterns of endemic species would benefit the interpretation on phytogeographic and biodiversity patterns of the area (Qian 2001; Huang et al. 2011). Moreover, endemic species are regarded with outstanding conservation importance owing to the rarity and uniqueness of the species (Myers et al. 2000). Here, we focus on southwest China, one of the most floristic diverse regions in the world, to investigate the taxonomic composition and spatial pattern of endemic seed plant species in the region, for the purposes of improving our understanding of the phylogenetic history of the flora and providing guidance for biodiversity conservation in southwest China.

7.8.2 *Analysis of Checklists and Distribution Data*

We first compiled a checklist of native seed plant species in southwest China from a wide range of literature including Flora Reipublicae Popularis Sinicae (1959–2004) and Flora of China (Wu et al. 1994–2011). We defined that endemic species refer to the species that occur naturally in and are restricted to southwest China and this concept of endemism is based on administrative boundaries. A checklist of endemic seed plant species was then formulated based on literature review and experts' scrutiny. Information on their distribution was collated by consulting a wide range of literature and checking herbarium specimens through the Chinese Virtual Herbarium (http://www.cvh.org.cn/cms/en) and Chinese Educational Specimen Resource Center (http://mnh.scu.edu.cn/). Totally, 33,959 distribution records were obtained for this study. This region is politically divided into 391 counties, with an average size of 2,636 km^2 per county. We geo-referenced the distribution information to county level and used the county as the unit of our analysis.

We calculated the number of endemic seed plant species of six main phylogenetic groups (gymnosperms, basal angiosperms, monocots, basal eudicots, rosids, and asterids). Species were assigned to genera and families following the Catalogue of Life: Higher Plants in China (http://www.cnpc.ac.cn, Wang et al. 2011). The designation of angiosperm families to the main phylogenetic groups followed the

Table 7.7 Taxonomic richness of endemic seed plant species in southwest China

Group	No. of endemic species	No. of species	Proportion of endemics (%)
Gymnosperms	10	140	7.0
Angiosperms	4,563	18,169	25.1
Basal angiosperms	163	584	27.9
Monocots	677	3,292	20.5
Basal eudicots	577	1,912	30.2
Rosids	1,012	5,272	19.2
Asterids	2,134	7,107	30.0
Total	4,573	18,309	25.0

APG III system (APG 2009). ANOVA and Tukey's multiple comparisons (Bretz et al. 2010) were conducted to assess endemism differences in the six phylogenetic groups.

We counted the number of endemic seed plant species per county. The range size of each endemic species was calculated as the sum of areas of the counties in which they occurred. We compared the range sizes with thresholds of IUCN red list criteria to assess the conservation status of the species. The IUCN criteria suggest that species with range sizes smaller than 20,000 km^2 are vulnerable species; smaller than 5,000 km^2 are endangered species; and smaller than 100 km^2 are critically endangered species (IUCN Standards and Petitions Working Group 2008).

7.8.3 *Taxonomic Composition*

According to our data, southwest China has 4,573 endemic seed plant species, accounting for 25 % of the total seed plant species in the region (Table 7.7). Angiosperms have a significantly higher number of endemic species than gymnosperms. Among angiosperms, asteroids have the largest number of endemic species (n = 2,134; 47 % of all endemic species), followed by rosids, monocots, basal eudicots and basal angiosperms. Angiosperms also have a markedly higher proportion of endemic species than gymnosperms. The proportion of endemic species is highest in basal eudicots (30.2 %), while it is lowest in rosids (19.2 %).

On average, the proportion of endemic species is 14.5 % in each family and 14.0 % in each genus (Fig. 7.15). Families of asteroids have the highest proportion of endemic species (mean = 19.8 %), whereas those of gymnosperms have the lowest proportion (mean = 4.6 %) (Fig. 7.15). At genus level, the proportion of endemics in the main phylogenetic groups shows a similar pattern as that at family level (Fig. 7.15).

The endemic species identified in this study belong to 139 families. Among them, 17 families contain more than 100 endemic species. The top ten families in terms

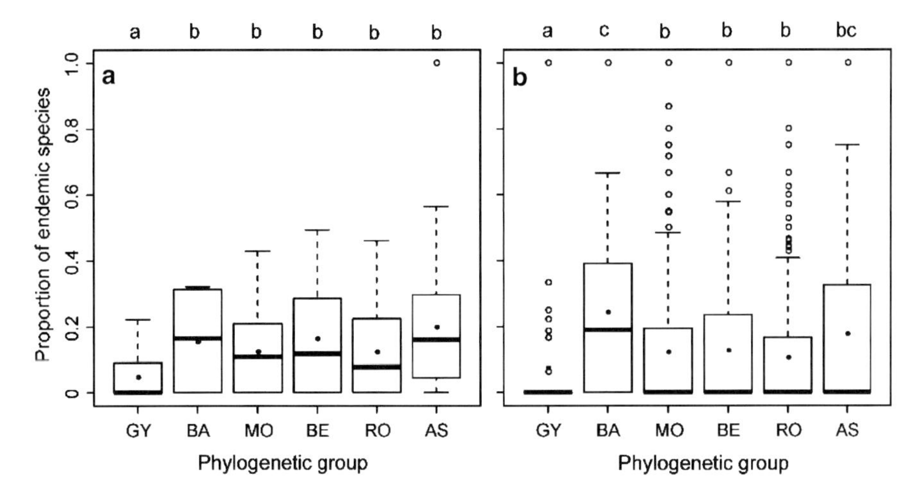

Fig. 7.15 Proportion of endemic species in each family (**a**) and genus (**b**) of the six analysed phylogenetic groups in southwest China. *GY* gymnosperms, *BA* basal angiosperms, *MO* monocots, *BE* basal eudicots, *RO* rosids, *AS* asterids. *Solid points* indicate mean values. *Letters above boxes* show the result of Tukey's multiple comparison tests

Table 7.8 Top ten families, in terms of the number of endemic seed plant species in southwest China

No.	Family	No. of endemic species	No. of species	Proportion of endemics (%)
1	Asteraceae	403	1,172	34
2	Poaceae	223	985	23
3	Lamiaceae	198	541	37
4	Scrophulariaceae	180	490	37
5	Ericaceae	179	668	27
6	Primulaceae	173	372	47
7	Ranunculaceae	156	626	25
8	Orchidaceae	148	856	17
9	Gesneriaceae	142	305	47
10	Rosaceae	142	840	17

of the number of endemic species are all large families with high species diversity (e.g. Asteraceae, Poaceae and Ericaceae). All these families have a proportion of endemics that is higher than the average value (Table 7.8). The endemic species belong to 810 genera. All top ten genera in terms of the number of endemic species are highly diversified genera (e.g. *Rhododendron*, *Pedicularis* and *Primula*) in the region, and have high proportions of endemics (Table 7.9).

Table 7.9 Top ten genera, in terms of the number of endemic seed plant species in southwest China

No.	Genus	No. of endemic species	No. of species	Proportion of endemics (%)
1	*Rhododendron*	149	479	31
2	*Pedicularis*	144	302	48
3	*Primula*	110	218	50
4	*Corydalis*	96	198	48
5	*Impatiens*	96	170	56
6	*Berberis*	77	143	54
7	*Gentiana*	67	186	36
8	*Carex*	65	230	28
9	*Begonia*	57	124	46
10	*Aconitum*	55	90	61

7.8.4 Spatial Distribution Patterns

The number of endemic species ranges from 0 to 947, with an average of 81 species per county (Fig. 7.16). Endemic seed plant species are unevenly distributed across the region. They are largely concentrated in the western part of the region especially along Hengduan Moutain ranges, whereas endemic species richness is relatively low in the east (including Sichuan Basin and Guizhou Province). Southern and southeastern Yunnan also has a high endemic richness.

Range sizes of endemic species show a Poisson distribution, varying from 719 to 523,232 km^2 (Fig. 7.17). The mean range is 30,921 km^2, while the median is 16,613 km^2. None of the species has a range of less than 100 km^2. 838 species (19.5 % of all endemics) occupy less than 5,000 km^2 and 2,439 species (56.6 % of all endemics) less than 20,000 km^2.

7.8.5 Discussion

The most derived group asterids has the largest number of endemic species, accounting for nearly half (47 %) of the total number of endemic species (Table 7.7). This is consistent with the floristic composition of endemism in East Asia and North America at genus level (Qian 2001). Endemics have generally been classified into two relative groups: palaeoendemics and neoendemics (Qian 2001; López-Pujol et al. 2011). The former refers to formerly more widespread taxa that now have restricted distributions, whereas neoendemics have arisen generally by differential evolution and have not yet spread out from their original territory (Stott 1981). The dominance of neoendemics in southwest China may be explained by the geographical and climatic histories of the region. The India-Eurasia plate collision

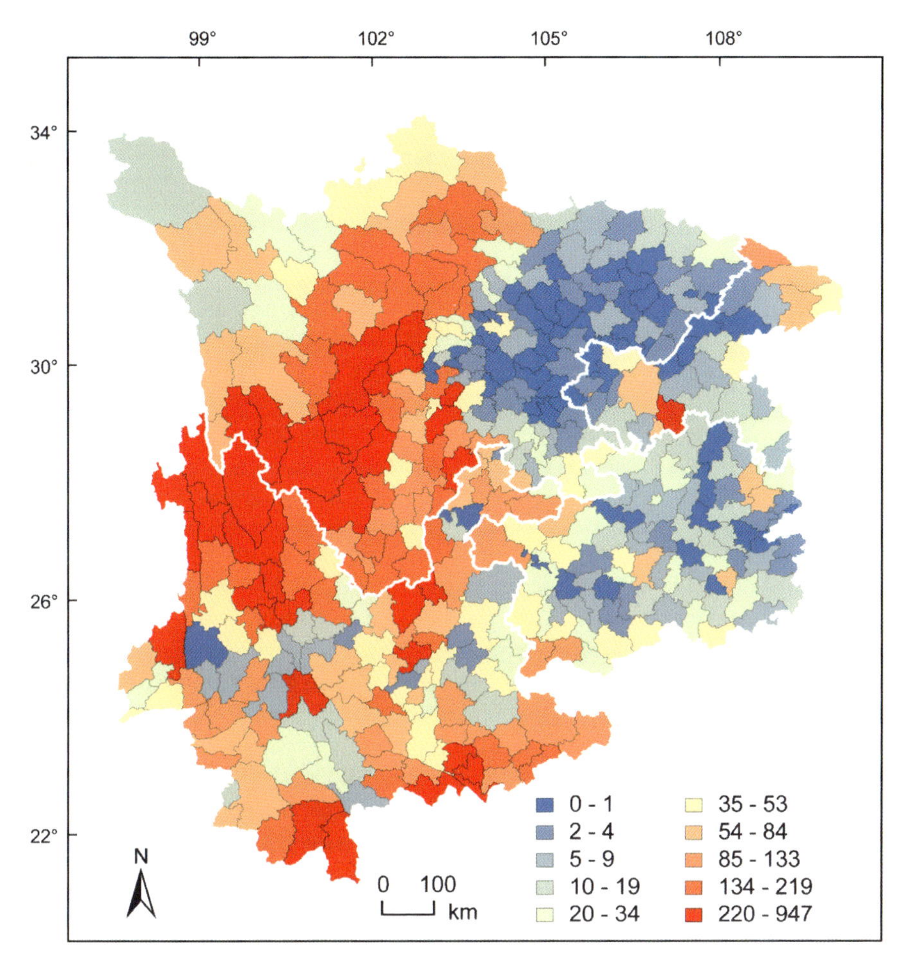

Fig. 7.16 Spatial pattern of endemic seed plant species richness in southwest China. *White lines* show provincial boundaries (Quantile classification and Albers projection)

50 million years ago has caused a major topographical transformation in southwest China (Li and Fang 1999; Zhang et al. 2000; Feng et al. 2011). The uplift of the Hengduan Mountains created a vast array of new habitats and stimulated allopatric and habitat differentiation, and ultimately gave rise to adaptive radiation (Liu and Tian 2007). Additionally, the climate became significantly colder and drier during glacial maxima. To adapt to the harsher environment, fast species differentiation occurred in many plant groups (Davis and Shaw 2001). For example, genera such as *Rhododendron*, *Pedicularis*, *Primula*, and *Corydalis*, have lineages showing adaptation to an alpine climate in southwest China (Li et al. 2007). The Hengduan Moutain region has been considered as the center of the diversification and evolutionary radiation of angiosperms (Wu 1980).

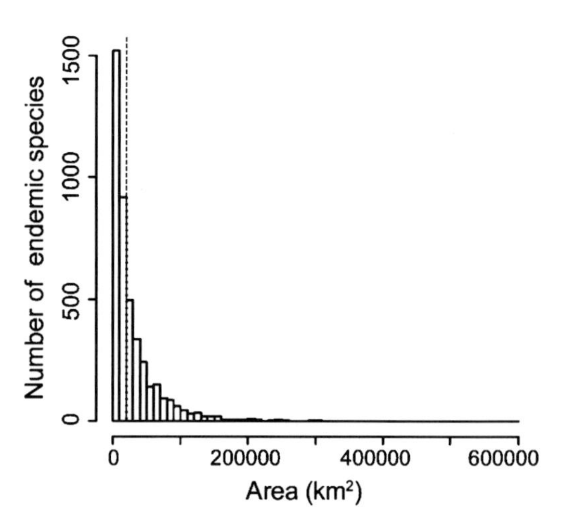

Fig. 7.17 Frequency distribution of range sizes of endemic seed plant species in southwest China. The *dashed line* is x = 20,000 km^2

The spatial pattern of endemic seed plant species richness noticeably coincides with large mountain ranges in southwest China (Figs. 7.14 and 7.16). Endemic plant species largely concentrate in the major mountain areas of the west, whereas lower mountains and flat areas in the east have low densities of endemic species. High levels of endemism in mountainous regions is probably due to the high speciation rates and low extinction rates, and especially the presence of refugia during the Pleistocene glacial episodes (Taberlet 1998; Hewitt 2000). Mountainous regions had relatively stable environmental conditions during the Quaternary, allowing plants to sustain large assemblages of a variety of living forms during glacial and interglacial periods (Hewitt 1999). Mountains also provide abundant and separated habitats where plant populations would be isolated following dispersion and colonization. Strong geographical isolation prevents genetic exchange between populations, which is a critical step for allopatric speciation and accelerates the rate of lineage diversification (Rice 1987). Moreover, geographical isolation increases the level of endemism by limiting newly evolved species to spread out from their original areas. In contrast, eco-climate conditions in lower mountains and flat areas are less stable, and are more vulnerable to various disturbances (Tribsch and Schönswetter 2003; Médail and Diadema 2009). These conditions are unfavorable for preserving endemic species.

We found that 19.5 % of all endemic species is restricted to ranges of less than 5,000 km^2 and 56.6 % to ranges of less than 20,000 km^2 (Fig. 7.17). According to the IUCN criteria (IUCN Standards and Petitions Working Group 2008; IUCN Standards and Petitions Subcomittee 2010), these species should be included in the red list either as endangered or vulnerable species. In fact, only a small proportion of the species is already in the China species red list (Wang and Xie 2004), because the selection of protected species is limited by available resources (money and effort)

(Margules and Pressey 2000). We do not find any species with range sizes smaller than 100 km^2. This is determined by the way in which we calculated range size by using 'county' as the unit area and the minimum size of the counties (115 km^2) is larger than 100 km^2. The quantification of range size would be more precise if we only include the parts that met the species' requirements of elevation and types of preferred habitats (Harris and Pimm 2007), instead of simply summing up the areas of the counties of occurrence. Nevertheless, our approach can still give a rough estimation of the range sizes of the endemic species, and would be helpful for assessing the conservation status of these species.

There are two potential sources of bias in the study. First, we use administrative boundaries to identify a species as endemic in southwest China. Ideally, endemic species are those whose distributions are delimited along natural, geographic boundaries. However, on that basis it would be difficult to identify a species as endemic species in southwest China because of the absence of data at the correct level of detail (Huang et al. 2011). Our delimitation may underestimate the levels of endemism in counties near the boundary and the levels of endemism of taxonomic groups that are narrowly distributed but not restricted in southwest China (Chen et al. 2011). Second, species distributions are often not fully documented due to the incompleteness of sampling, which would lead to an underestimation of species range sizes. We have compiled as many data as possible from herbaria and a wide range of literature to minimize this bias.

In summary, we find that the most derived group asteroids have the highest number of endemic seed plants; spatial patterns of richness of endemic seed plant species show a significant mountain character, largely coinciding with main mountain ranges in the west of the region. The overall bias in favour of asteroids and large mountain ranges results from different rates of speciation, extinction, and dispersion which primarily have been influenced by the topography, and the geological and climatic histories of the region. We also find that 19.5 % of the endemic species are restricted to ranges of less than 5,000 km^2 and 56.6 % to ranges of less than 20,000 km^2, suggesting the high conservation significance of the endemic species.

7.9 Habitats of Tertiary Relict Trees in China

Cindy Q. Tang (✉)
Institute of Ecology and Geobotany, Yunnan University, Kunming, China
e-mail: cindytang@ynu.edu.cn

Marinus J.A. Werger
Department of Plant Ecology, University of Utrecht, Utrecht, Netherlands

Masahiko Ohsawa
Institute of Ecology and Geobotany, Kunming University in China, Kunming, China

Yongchuan Yang
Faculty of Urban Construction and Environmental Engineering,
Chongqing University, Chongqing, China

Among the many endemic plant taxa in China there are a considerable number of Tertiary relict tree species. They comprise both coniferous and broad-leaved species and evergreen as well as deciduous species (Table 7.10). Of the 40 relict tree species listed in Table 7.10 about 52.5 % are conifers as against 47.5 % broad-leaved. About 86 % of the conifer relict tree species, and about 21 % of broad-leaved relict tree species, are evergreen. Thus, among the endemic Tertiary relict tree species, evergreen coniferous species and broad-leaved deciduous species form by far the largest groups, whereas deciduous coniferous and evergreen broad-leaved species are rather few.

All of these tree species now occur exclusively in a few restricted areas in China. Tertiary relict endemic (paleoendemic) species are found mainly in southern China, including Chongqing municipality, Sichuan, Hubei, Hunan, Guizhou, Guangxi, Yunnan, Zhejiang, Jiangxi, Fujian, Guangdong, and Hainan provinces, and Taiwan, but some relict coniferous tree species of *Picea*, *Pinus* and *Tsuga* are found in the cold temperate forests of China. The regions in which the relict tree species occur mostly have a rugged topography, and have provided long-term suitable habitats, enabling the survival of the relict-endemics (López-Pujol et al. 2011).

Most of this region was never covered by ice-sheets during the Last Glacial Maximum (LGM), and it is now one of the most important global Pleistocene refugia for lineages that evolved prior to the late Tertiary and Quaternary glaciations (Axelrod et al. 1998). The paleoendemic species provide a unique opportunity to understand past and recent biogeographical and evolutionary processes because of their taxonomical isolation, rarity and phylogenetic traits. Today they are highly vulnerable and even in danger of extinction because their populations in the wild have been greatly threatened by overexploitation of the land. Their conservation is of great concern for China and for the world.

A number of Tertiary relict trees occur in particular, unstable habitats within the warm, humid transitional zone between the evergreen and the deciduous broad-leaved forests of China, with some admixture of coniferous species. The well-known living fossil and relict-endemic species *Metasequoia glyptostroboides* currently survives in the wild only on unstable wet lower slopes and in stream valleys, in the border region of Hubei and Hunan provinces and Chongqing municipality in south-central China (Tang et al. 2011), and *Cathaya argyrophylla* grows on unstable slopes on Mt. Jinfo (Guan and Chen 1986), southwestern China. *Davidia involucrata* (Photo 7.21) thrives on unstable scree slopes on Mt. Emei and other mountains in Sichuan province, and cannot successfully compete with evergreen broad-leaved trees in a more favorable environment (Tang and Ohsawa 2002). We have found

Table 7.10 Examples of paleoendemic tree species with indication of leaf morphology and natural habitats in China (including the species distributed in the boundary between China and Vietnam or between China and Myanmar)

Species	Life form	Habitat
Gymnosperm		
Abies beshanzuensis M.H. Wu	Coniferous evergreen	Mountains or hills; 1,400–1,800 m
Abies fanjingshanensis W.L. Huang et al.	Coniferous evergreen	Mountain slopes; 2,100–2,350 m
Abies yuanbaoshanensis Y.J. Lu & L.K. Fu in L.K. Fu et al.	Coniferous evergreen	Mountain slopes; 1,700–2,100 m
Amentotaxus formosana H.L. Li	Coniferous evergreen	Damp, shady places, ravines, cliffs; 500–1,300 m
Cathaya argyrophylla Chun & Kuang	Coniferous evergreen	Steep exposed mountain slopes, ridges and vertical cliffs; 900–1,900 m
Calocedrus macrolepis var. *formosana* (Florin) W.C. Cheng & L.K. Fu	Coniferous evergreen	Mountain slopes, ravines, cliffs; 500–1,300 m
Cephalotaxus oliveri Masters	Coniferous evergreen	Mountain slopes; 300–1,800 m
Cryptomeria japonica var. *sinensis* Miquel	Coniferous evergreen	Deep, well-drained soils subject to warm, moist conditions; below 1,100 m
Cunninghamia lanceolata (Lambert) Hooker	Coniferous evergreen	Mountains, rocky hillsides, roadsides; 200–2,800 m
Keteleeria davidiana (Bertrand) Beissner	Coniferous evergreen	Hills, mountains, hot and dry valleys; 200–1,500 m
Picea asperata Masters	Coniferous evergreen	Mountains, river basins; 2,400–3,600 m
Pinus taiwanensis Hayata	Coniferous evergreen	Mountains, open sites and sunny ridges on sandy areas; 600–3,400 m
Pseudotaxus chienii (Cheng) Cheng	Coniferous evergreen	Gullies, cliffs at 900–1,400 m
Taiwania cryptomerioides Hayata	Coniferous evergreen	Steep slopes, 1,800–2,800 m
Thuja sutchuenensis Franchet	Coniferous evergreen	Cliffs and ridges of the deeply cleft mountain; 1,800–2,200 m
Torreya jackii Chun	Coniferous evergreen	Shady steep slopes or by streamsides; 400–1,000 m
Tsuga chinensis (Franchet) E. Pritzel	Coniferous evergreen	Mountains, valleys, river basins; 1,000–3,500 m
Glyptostrobus pensilis (Staunton ex D. Don) K. Koch	Coniferous semi evergreen	River deltas, etc., on flooded or waterlogged soil in full sun; near sea level
Metasequoia glyptostroboides Hu & W.C. Cheng	Coniferous deciduous	River valleys, rocky soil; 800–1,510 m
Pseudolarix amabilis (J. Nelson) Rehder	Coniferous deciduous	Mountain slopes; 100–1,500 m
Ginkgo biloba Linnaeus	Deciduous	Limestone areas, rocky soil, steep slopes; 800–1,200 m

(continued)

Table 7.10 (continued)

Species	Life form	Habitat
Angiosperm		
Castanopsis carlesii (Hemsley) Hayata	Broad-leaved evergreen	Mountain slopes, valleys, limestone areas; 1,000–2,300 m
Machilus leptophylla Handel-Mazzetti	Broad-leaved evergreen	Mountain slopes, valleys, limestone areas; 400–1,200 m
Hopea shingkeng (Dunn) Bor	Broad-leaved evergreen	Moist sites; 300–600 m
Lithocarpus konishii (Hayata) Hayata	Broad-leaved evergreen	Mountains, hills; 300–1,600 m
Annamocarya sinensis (Dode) Leroy	Broad-leaved deciduous	Forests along riverbanks; 200–700 m
Camptotheca acuminata Decaisne	Broad-leaved deciduous	Forest margins,streamsides; below 1,000 m
Cyclocarya paliurus (Batalin) Iljinskaya	Broad-leaved deciduous	Moist forests on mountains, 400–2,500 m
Davidia involucrata Baillon	Broad-leaved deciduous	Mountain slopes, valleys, scree slopes; 1,100–2,600 m
Emmenopterys henryi Oliver	Broad-leaved deciduous	Valleys, limestone areas; 400–1,600 m
Eucommia ulmoides Oliver	Broad-leaved deciduous	Valleys, dry ravines; 300–2,500 m
Eurycorymbus cavaleriei (H. Léveillé) Rehder & Handel-Mazzetti	Broad-leaved deciduous	Stream sides; 150–1,600 m
Fagus hayatae Palibin	Broad-leaved deciduous	Mountain ridges; 1,300–2,300 m
Fortunearia sinensis Rehder & E.H. Wilso	Broad-leaved deciduous	Mountain slopes; 800–1,000 m
Liriodendron chinense (Hemsl.) Sarg.	Broad-leaved deciduous	Mountain slopes, valleys, limestone areas; 900–1,800 m
Nyssa sinensis Oliver	Broad-leaved deciduous	Streamsides, valleys; 300–1,700 m
Ostryopsis davidiana Decaisne	Broad-leaved deciduous	Mountain slopes; 800–2,800 m
Pteroceltis tatarinowii Maxim.	Broad-leaved deciduous	Limestone areas, river and stream banks; 100–1,500 m
Rhoiptelea chiliantha Diels & Handel-Mazzetti	Broad-leaved deciduous	Hill slopes, valleys, streamside woods; 700–2,500 m
Tapiscia sinensis Oliver	Broad-leaved deciduous	Mountain slopes, valleys, streamsides; 500–2,200 m

Data sources: Fu and Jin (1992), Ying et al. (1993), and Wu and Raven (2012)

other paleoendemic trees, such as *Ginkgo biloba* (Tang et al. 2012), *Emmenopterys henryi*, and *Liriodendron chinense,* in unstable limestone habitats near creeks in the valleys of the Dalou Mountains located between Guizhou and Sichuan provinces, as well as *Taiwania cryptomerioides* on steep slopes in the Gaoligong Mountains of Yunnan and in the Central Ridge of Taiwan.

Photo 7.21 *Davidia involucrata* trees with sprouts thrive on an unstable scree slope on Mt. Emei, Sichuan (Photographed by Cindy Q. Tang)

Many of the Tertiary relict tree species, particularly the coniferous species, persist in these unstable habitats (Table 7.10), probably because the forest stands are not so dense on such sites, and thus light conditions are less limiting. Open, well illuminated stands are particularly important for the coniferous species since their photosynthetic capacity is considerably lower than that of broad-leaved species (Larcher 1976), and as a consequence the coniferous species are likely to be outcompeted in dense mixed forest stands. Furthermore, screes and other unstable substrates more frequently experience short episodes of drought in the top soil which can cause embolism in the xylem vessels. Coniferous species possess a particular type of xylem vessels that makes them less susceptible to develop xylem embolism during such drought conditions (Choat et al. 2012).

Several of the broad-leaved deciduous species are also confined to screes and unstable slopes, or to stream banks. Apart from more favourable light conditions, such sites generally have a rapid turn-over and replenishment of nutrient resources, which is beneficial for deciduous species, as they have a faster nutrient turn-over than broad-leaved evergreen species.

Effective conservation of the paleoendemic trees requires a thorough understanding of the ecological ranges and habitats of these species, as illustrated in Photo 7.21 taken from Mt. Emei, Sichuan, and Fig. 7.18.

The ecological performance and linkage to unstable habitats of the paleoendemic tree species are well exemplified by a Tertiary relict deciduous broad-leaved forest

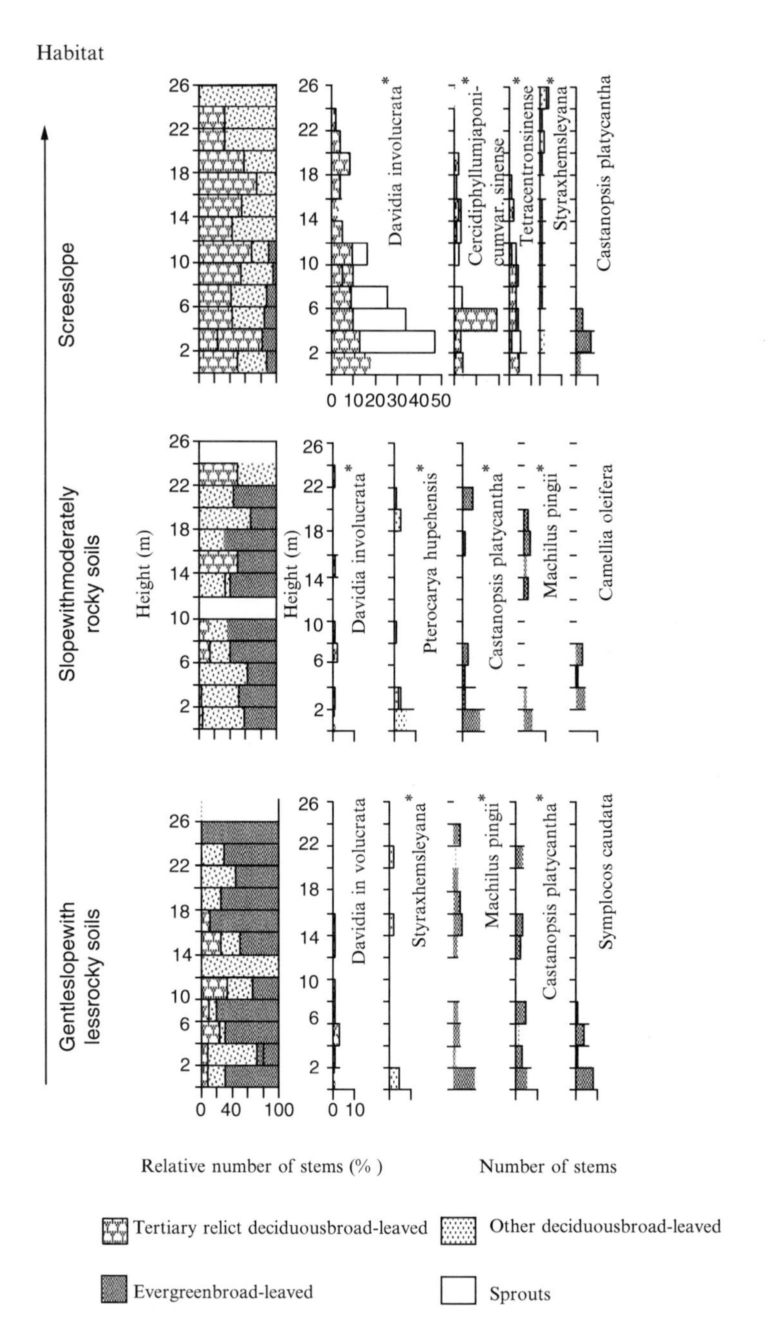

Fig. 7.18 Changes of height frequency distribution with habitat instability at altitudes around 1,600 m on the eastern slope of Mt. Emei. Dominant species are indicated by an *asterisk* (Modified from Tang and Ohsawa 2002)

growing around 1,600 m on Mt. Emei, Sichuan, on an unstable scree slope together with two other communities; one is found on moderately rocky soils and the other on a gentle slope with less rocky soils.

A comparison brings out the contrast between the unstable scree slope and the stable slope (Fig. 7.18). A site with many outcrops is characterized by an abundance of Tertiary relict broad-leaved deciduous trees, in contrast to the broad-leaved evergreen species commonly found on non-scree sites at similar altitudes. Being broad-leaved deciduous, the relict species *Davidia involucrata*, *Cercidiphyllum japonicum*, and *Tetracentron sinense* (the first one is endemic to China, the latter two are endemic to East Asia) grow well on the frequently disturbed unstable scree where new soil resources are regularly available, but where they also benefit from their high sprouting ability as an adaption to the frequent damage they are subjected to at the ground surface because of the mobility of the substrate (Photo 7.21). In contrast, evergreen broad-leaved trees, such as *Castanopsis platyacantha* and *Machilus pingii,* increase their height and canopy density on the stable habitats and outcompete the deciduous species.

The broad-leaved deciduous relict species can survive on the scree slope, probably because of frequent erosion or landslides there which keep the stands rather open. As the habitat becomes more stable, even on the scree slope, the broad-leaved deciduous relict species give way to non-relict, endemic evergreen broad-leaved species. On the gentle slopes, where relict trees are scarce, the non-paleoendemic broad-leaved evergreen *Castanopsis platyacantha* and *Machilus pingii*, with broad-leaved deciduous *Styrax hemsleyana*, dominate the forest and the canopy layer (Fig. 7.18).

Acknowledgements Sincere thanks to Professor Carsten Hobohm for his valuable suggestions, which improved the scientific quality of this chapter.

References

Ackerly DD (2009) Evolution, origin and age of lineages in the Californian and Mediterranean floras. J Biogeogr 36:1221–1233

Agnew S (1958) The landforms of the Hogsback area in the Amatola Range. Fort Hare Pap 2:1–14

Akgün F, Kayseri MS, Akkiraz MS (2007) Palaeoclimatic evolution and vegetational changes during the Late Oligocene–Miocene period inWestern and Central Anatolia (Turkey). Palaeogeogr Palaeoclimatol Palaeoecol 253:56–90

Akhani H (2006) Flora Iranica: facts and figures and a list of publications by K. H. Rechinger on Iran and adjacent areas. Rostaniha 7(2):19–61

Akpulat H, Celik N (2005) Flora of gypsum areas in Sivas in the eastern part of Cappadocia in Central Anatolia, Turkey. J Arid Environ 61:27–46

Allen MF (2009) Commentary. Bidirectional water flows through the soil-fungal-plant mycorrhizal continuum. New Phytol 182:290–293

Al-Shehbaz I, Junussov SJ (2003) *Arabidopsis bactriana* belongs to *Dielsiocharis* (Brassicaceae). Novon 13:171–172

Amigo J, Ramírez C (1998) A bioclimatic classification of Chile: woodland communities in the temperate zone. Plant Ecol 136:9–26

APG (2009) An update of the Angiosperm Phylogeny Group classification for the orders and families of flowering plants: APG III. Bot J Linn Soc 161:105–121
Arroyo MTK, Rougier D, Pérez F, Pliscoff P, Bull K (2003) La Flora de Chile central y su protección: antecedentes y prioridades para el establecimiento del Jardín Botánico Chagual. Revista Chagual 1:31–40
Aschmann H (1959) The central desert of Baja California: demography and ecology. University of California Press, Los Angeles
Assadi M, Massoumi AA, Khatamsaz M, Mozaffarian V (eds) (1988–2012) Flora of Iran, no 1–74. Research Institute of Forest and Rangelands, Tehran (in Persian)
Atalay I, Mortan K (2003) Türkiye Bölgesel Coğrafyası. Inkilap, Sirkeci
Atlas Harita Servisi (2004) Bitki Örtüsü. In: Türkiye Coğrafya Atlası. Doğan/Burda/Rizzoli, Istanbul, pp 31–32
Axelrod DI (1975) Evolution and biogeography of Madrean-Tethyan sclerophyll vegetation. Ann Mo Bot Gard 62:280–334
Axelrod DI (1978) The origin of coastal sage vegetation, Alta and Baja California. Am J Bot 65(10):1117–1131
Axelrod DI (1980) History of the maritime closed-cone pines, Alta and Baja California, Geological sciences 120. University of California Press, Berkeley, 143 p
Axelrod DI, Al-Shehbaz I, Raven PH (1998) History of the modern flora of China. In: Zhang A-L, Wu S-G (eds) Floristic characteristics and diversity of East Asia plants. Springer, New York, pp 43–55
Aytaç Z, Türkmen Z (2011) A new *Onosma* (Boraginaceae) species from southern Anatolia, Turkey. Turk J Bot 35:269–274
Bağci Y, Svran A, Dücen OD, Tutar L (2011) *Ornithogalum beyazoglui* (Hyacinthaceae), a new species from West Anatolia, Turkey. Bangladesh J Plant Taxon 18(1):51–55
Balkwill M-J, Balkwill K (1999) Characteristics, diversity, endemism, and conservation of serpentine sites in the Barberton Greenstone Belt. In: Third international conference on serpentine ecology, Post-congress tour guide. C.E. Moss Herbarium, University of the Witwatersrand, Johannesburg, 28–30 March 1999, pp 1–23
Bartlein PJ, Anderson KH, Anderson PM, Edwards ME, Mock CJ, Thompson RS, Webb RS, Webb T III, Whitlock C (1998) Paleoclimate simulations for North America over the past 21,000 years: features of the simulated climate and comparisons with paleoenvironmental data. Quat Sci Rev 17:549–585
Baytop A (2009) Notes on the flora of Istanbul. Acta Pharm Sci 51:5–8
Beard JS, Chapman AR, Gioia P (2000) Species richness and endemism in the Western Australian flora. J Biogeogr 27:1257–1268
Bergh NG, Hedderson TA, Linder HP, Bond WJ (2007) Palaeoclimate–induced range shifts may explain current patterns of spatial and genetic variation in renosterbos (*Elytropappus rhinocerotis*, Asteraceae). Taxon 56:393–408
Betancourt JL, Van Devender TR, Martin PS (1990) Packrat middens: the last 40,000 years of biotic change. University of Arizona Press, Tucson, 472 p
Billor MZ, Gibb F (2002) The mineralogy and chemistry of the chromite deposits of Southern (Kızıldağ, Hatay and Islahiye, Antep) and Tauric Ophiolite Belt (Pozantı-Karsantı, Adana), Turkey. Abstract of 9th international platinum symposium, Billings, 21–25 July
Birkenhauer J (1991) The great escarpment of southern Africa and its coastal forelands – a reappraisal. Institut für Geographie der Universität, München
Boulos L, Miller AG, Mill RR (1994) Regional overview: South West Asia and the Middle East. In: Davis SD, Heywood VH, Hamilton AC (eds) Centres of plant diversity: a guide and strategy for their conservation. Vol. 1: Europe, Africa, South West Asia and the Middle East. IUCN Publications Unit, Cambridge, pp 293–348
Bretz F, Hothorn T, Westfall P (2010) Multiple comparisons using R. CRC Press, Boca Raton
Brink ABA (1983) Engineering geology of Southern Africa, vol 3. The Karoo sequence. Building Publications, Pretoria

Bruchmann I (2011) Plant endemism in Europe: spatial distribution and habitat affinities of endemic vascular plants. Dissertation, University of Flensburg, Flensburg. www.zhb-flensburg.de/dissert/bruchmann

Burke A (2002) Properties of soil pockets on arid Nama Karoo inselbergs – the effect of geology and derived landforms. J Arid Environ 50:219–234

Burke K, Gunnell Y (2008) The African erosion surface: a continental-scale synthesis of geomorphology, tectonics, and environmental change over the past 180 million years. Mem Geol Soc Am 201:1–66

Carbutt C, Edwards T (2001) Cape elements on high-altitude corridors and edaphic islands: historical aspects and preliminary phytogeography. Syst Geogr Plant 71:1033–1061

Carbutt C, Edwards T (2006) The endemic and near-endemic angiosperms of the Drakensberg Alpine Centre. S Afr J Bot 72:105–132

Caso M, González-Abraham C, Ezcurra E (2007) Divergent ecological effects of oceanographic anomalies on terrestrial ecosystems of the Mexican Pacific coast. Proc Natl Acad Sci USA 104:10530–10535

Chen S, Ouyang Z, Fang Y, Li Z (2011) Geographic patterns of endemic seed plant genera diversity in China. Biodivers Sci 19:414–423

Choat B, Jansen S, Brodribb TJ, Cochard H, Delzon S, Bhaskar R, Bucci SJ, Feild TS, Gleason SM, Hacke UG, Jacobsen AL, Lens F, Maherali H, Martínez-Vilalta J, Mayr S, Mencuccini M, Mitchell PJ, Nardini A, Pittermann J, Pratt RB, Sperry JS, Westoby M, Wright IJ, Zanne AE (2012) Global convergence in the vulnerability of forests to drought. Nature 91(7426):752–755. doi:10.1038/nature11688

Clark VR (2010) The phytogeography of the Sneeuberg, Nuweveldberge and Roggeveldberge (Great Escarpment): assessing migration routes and endemism. Unpublished PhD thesis, Rhodes University, Grahamstown

Clark VR, Barker NP, Mucina L (2009) The Sneeuberg: a new centre of floristic endemism on the Great Escarpment, South Africa. S Afr J Bot 75:196–238

Clark VR, Barker NP, Mucina L (2011a) The Boschberg (Somerset East, Eastern Cape) – a floristic crossroads of the southern Great Escarpment. S Afr J Bot 77:94–104

Clark VR, Barker NP, Mucina L (2011b) The Roggeveldberge – notes on a botanically hot area on a cold corner of the southern Great Escarpment, South Africa. S Afr J Bot 77:112–126

Clark VR, Barker NP, Mucina L (2011c) A phytogeographic assessment of the Nuweveldberge, South Africa. S Afr J Bot 77:147–159

Clark VR, Barker NP, Mucina L (2011d) The Great Escarpment of southern Africa – a new frontier for biodiversity exploration. Biodivers Conserv 20:2543–2561

Clark VR, Barker NP, Mucina L (2011e) Taking the scenic route – the southern Great Escarpment as part of the Cape to Cairo floristic highway. Plant Ecol Divers 4:313–328

Conti F, Abbate G, Alessandrini A, Blasi C (eds) (2005) An annotated checklist of the Italian vascular flora. Palombi & Partner, Rome

Cook A (*sine anno*) The phytogeography of the Amatole mountains, Eastern Province. Department of Botany, Rhodes University, 29 pp

Cowling RM, Rundel PW, Lamont BB, Arroyo MK, Arianoutsou M (1996) Plant diversity in mediterranean-climate regions. Trends Ecol Evol 11:362–366

Crain BJ, White JW (2011) Categorizing locally rare plant taxa for conservation status. Biodivers Conserv 20:451–463

Crain BJ, White JW, Steinberg SJ (2011) Geographic discrepancies between global and local rarity richness patterns and the implications for conservation. Biodivers Conserv 20:3489–3500

Croizat L, Nelson G, Rosen DE (1974) Centers of origin and related concepts. Syst Zool 23:265–287

Dallman PR (1998) Plant life in the world's Mediterranean climates. California Native Plant Society/University of California Press, Berkeley

Davis PH (ed) (1965–1988) Flora of Turkey and the East Aegean Islands, 10 vols. University Press, Edinburgh

Davis PH (1971) Distribution patterns in Anatolia with particular reference to endemism. In: Davis PH, Harper PC, Hedge IC (eds) Plant life of South-West Asia. Botanical Society of Edinburgh, Edinburgh, pp 15–27
Davis MB, Shaw RG (2001) Range shifts and adaptive responses to quaternary climate change. Science 292:673–679
Davis SD, Heywood VH, Hamilton AC (eds) (1994) Centres of plant diversity, vol 1: Europe, Africa, South West Asia and the Middle East. IUCN Publications, Unit, Cambridge
Davis SD, Heywood VH, Herrera-MacBryde O, Villa-Lobos J, Hamilton AC (eds) (1997) Centres of Plant Diversity, vol 3: The Americas – IUCN Publications, Unit, Cambridge
Delgadillo J (1998) Florística y ecología de norte de Baja California. Universidad Autónoma de Baja California, Mexicali
DellaSala DA (ed) (2010) Temperate and boreal rainforests of the world: ecology and conservation. Island Press, Washington, DC
Denk T, Grimm GW (2010) The oaks of western Eurasia: traditional classifications and evidence from two nuclear markers. Taxon 59(2):351–366
Deutscher Wetterdienst (ed) (2010, 2002) Climate data. Unpublished CD, Hamburg
Deutscher Wetterdienst (ed) (2008) Climate data of high mountain areas. Unpublished CD, Hamburg
Devos N, Barker NP, Nordenstam B, Mucina L (2010) A multilocus phylogeny of *Euryops* (Asteraceae, Senecioneae) augments support for the "Cape to Cairo" hypothesis of floral migrations in Africa. Taxon 59:57–67
Dillon MO, Tu T, Xie L, Quipuscoa Silvestre V, Wen J (2009) Biogeographic diversification in *Nolana* (Solanaceae), a ubiquitous member of the Atacama and Peruvian Deserts along the western coast of South America. J Syst Evol 47:457–476
Djamali M, Baumela A, Brewerb S, Jackson ST, Kadereitd JW, López-Vinyallongae S, Mehreganf I, Shabaniang E, Simakovah A (2012) Ecological implications of *Cousinia* Cass. (Asteraceae) persistence through the last two glacial-interglacial cycles in the continental Middle East for the Irano-Turanian flora. Rev Palaeobot Palynol 172:10–20
Doğan U, Özel S (2005) Gypsum karst and its evolution east of Hafik (Sivas, Turkey). Geomorphology 71:373–388
Doğan M, Duman H, Akaydın G (2008) *Limonium gueneri* (Plumbaginaceae), a new species from Turkey. Ann Bot Fenn 45:389–393
Du Toit AL (1920) The Karoo dolerites of South Africa: a study in hypabyssal injection. Geol Soc S Afr Trans 23:1–42
Duran A, Dogan B, Hamzaoğlu E, Aksoy A (2011) *Scorzonera coriacea* A. Duran & Aksoy (Asteraceae, Cichorieae), a new species from South Anatolia, Turkey. Candollea 66:353–359
Editorial Committee of Flora Reipublicae Popularis Sinicae (1959–2004) Flora Reipublicae Popularis Sinicae. Science Press, Beijing
Elektrik İşleri Etüt İdaresi Genel Müdürlüğü (2012) http://www.eie.gov.tr/turkce/YEK/HES/hidroloji/havzalar.html. 12th Feb 2012
Emery-Barbier A, Thiébault S (2005) Preliminary conclusions on the Late Glacial vegetation in south-west Anatolia (Turkey): the complementary nature of palynological and anthracological approaches. J Archaeol Sci 32:1232–1251
Euro + Med (2006-) Euro + Med PlantBase – the information resource for Euro-Mediterranean plant diversity. Published on the Internet. http://ww2.bgbm.org/EuroPlusMed/. 20 Feb 2012
Ewald J (2003) The calcareous riddle: why are there so many calciphilous species in the Central European flora? Folia Geobot 38:357–366
FAO (ed) (2010) Global forest resources assessment 2010: main report. FAO forestry paper 163, Food and Agriculture Organization of the United Nations, Rome, 340 pp
Felger RS (2000) Flora of the Gran Desierto and Río Colorado of northwestern Mexico. University of Arizona Press, Tucson
Feng JM, Zhang Z, Nan RY (2011) Non-congruence among hotspots based on three common diversity measures in Yunnan, south-west China. Plant Ecol Divers 4:353–361

Ferrigno JG (1991) Glaciers of the Middle East and Africa – glaciers of Iran. In: Williams RS, Ferrigno JG (eds) Satellite image atlas of glaciers of the world, U.S. Geological Survey professional paper 1386-G. United States Government Printing Office, Washington, pp 31–47
Fontaine B, Bouchet P, Van Achterberg K, Alonso-Zarazaga MA, Araujo R, Asche M, Aspock U, Audisio P, Aukema B, Bailly N, Balsamo M, Bank RA, Barnard P, Belfiore C, Bogdanowicz W, Bongers T, Boxshall G, Burckhardt D, Camicas JL, Chylarecki P, Crucitti P, Davarveng L, Dubois A, Enghoff H, Faubel A, Fochetti R, Gargominy O, Gibson D, Gibson R, Gomez Lopez MS, Goujet D, Harvey MS, Heller K-G, Van Helsdingen P, Hoch H, De Jong H, De Jong Y, Karsholt O, Los W, Lundqvist L, Magowski W, Manconi R, Martens J, Massard JA, Massard-Geimer G, Mcinnes SJ, Mendes LF, Mey E, Michelsen V, Minelli A, Nielsen C, Nieto Nafria JM, Van Nieukerken EJ, Noyes J, Papa T, Ohl H, De Prins W, Ramos M, Ricci C, Roselaar C, Rota E, Schmidt-Rhaesa A, Segers H, Zur Strassen R, Szeptycki A, Thibaud J-M, Thomas A, Timm T, Van Tol J, Vervoort W, Willmann R (2007) The European union's 2010 target: putting rare species in focus. Biol Conserv 139:167–185
Francisco S (2001) Phylogeny and biogeography of the Arbutoideae (Ericaceae): implications for the Madrean-Tethyan hypothesis. Syst Bot 26:131–143
Franco-Vizcaino E (1994) Water regimes in soils and plants along an aridity gradient in central Baja California, Mexico. J Arid Environ 27:309–323
Freitag H, Vural M, Adıgüzel N (1999) A remarkable new *Salsola* and some new records of Chenopodiaceae from Central Anatolia, Turkey. Willdenowia 29:123–139
Frenzel B (2005) History of flora and vegetation during the quaternary North America. Prog Bot 66:409–440
Frey W, Kürschner H, Probst W (1999) Flora and vegetation, including plant species and larger vegetation complexes in Persia. In: Yarshater E (ed) Encyclopaedia Iranica 10/1. Mazda Publishers, Costa Mesa, pp 43–63
Fu L-K, Jin J-M (1992) China plant red data book rare and endangered plants. Science Press, Beijing
Galley C, Bytebier BLG, Bellstedt DU, Linder HP (2007) The Cape element in the Afrotemperate flora: from Cape to Cairo? Proc R Soc Lond B Biol Sci 274:535–543
Gams H (1938) Die nacheiszeitliche Geschichte der Alpenflora. Jahrbuch der Vereinigung zum Schutze der Alpenpflanzen und -tiere 10, pp 9–34
Garcillan PP, Gonzalez-Abraham CE, Ezcurra E (2010) The cartographers of life: two centuries of mapping the natural history of Baja California. J Southwest 52(1):1–40
Gemici Y, Akgün F (2001) Neogene and current spread of some wooden plants in Anatolia. In: Proceedings of the 2nd Balkan Botanical Congress, vol 1. Marmara Universitesi, Turkey, pp 229–238
Gilchrist AR, Kooi H, Beaumont C (1994) Post-Gondwana geomorphic evolution of southwestern Africa: implications for the controls on landscape development from observations and numerical experiments. J Geophys Res 99:12211–12228
Goldblatt P, Manning JC (2007) New species and notes on *Hesperantha* (Iridaceae) in southern Africa. Bothalia 37:167–182
Gottfried M, Pauli H, Futschik A, Akhalkatsi M, Barančok P, Benito Alonso JL, Coldea G, Dick J, Erschbamer B, Fernández Calzado MR, Kazakis G, Krajči J, Larsson P, Mallaun M, Michelsen O, Moiseev D, Molau U, Merzouki A, Nagy L, Nakhutsrishvili G, Pedersen B, Pelino G, Puscas M, Rossi G, Stanisci A, Theurillat JP, Tomaselli M, Villar L, Vittoz P, Vogiatzakis I, Grabherr G (2012) Continent-wide response of mountain vegetation to climate change. Nat Clim Change 2(2):111–115
Grabherr G, Gottfried M, Pauli H (1994) Climate effects on mountain plants. Nature 369:448
Grime JP (1977) Evidence for the existence of three primary strategies in plants and its relevance to ecological and evolutionary theory. Am Nat 111:1169–1194
Groombridge B, Jenkins MD (2002) World atlas of biodiversity: earth's living resources in the 21st century. University of California Press, Berkeley
Guan Z-T, Chen Y (1986) A preliminary study on the Cathaya mixed forest in Jinfushan, Sichuan. Acta Botanica Sinica 28:646–656 (in Chinese)

Guerrero PC, Duran AP, Walter HE (2011) Latitudinal and altitudinal patterns of the endemic cacti from the Atacama desert to Mediterranean Chile. J Arid Environ 75(11):991–997
Güldali N (1979) Geomorphologie der Türkei. – Beihefte zum Tübinger Atlas des Vorderen Orients, Reihe A, Nr. 4. Dr. Ludwig Reichert Verlag, Wiesbaden
Günay G (2002) Gypsum karst, Sivas, Turkey. Environ Geol 42:387–398
Güner A, Özhatay N, Ekim T, Başer KHC (2000) Flora of Turkey supplement 2, vol 11. University Press, Edinburgh
Harper A, Vanderplank S, Dodero M, Mata S, Ochoa J (2010) Plants of the Colonet Region, Baja California, Mexico, and a vegetation map of Colonet Mesa. Aliso 29:25–42
Harris G, Pimm SL (2007) Range size and extinction risk in forest birds. Conserv Biol 22:163–171
Hastings JR, Turner RM (1965) Seasonal precipitation regimes in Baja California, Mexico. Geogr Ann 47(4):204–223
Heads M (2009) Globally basal centres of endemism: the Tasman-Coral Sea region (south-west Pacific), Latin America and Madagascar/South Africa. Biol J Linn Soc 96:222–245
Hedge IC (1968) *Dielsiocharis* (pp 320–321), *Didymophysa* (pp 87–98) [Cruciferae]. In: Rechinger KH (ed) Flora Iranica 57. Akademische Druck u. Verlagsanstalt, Graz
Hedge IC (1969) *Elburzia*, a new genus of Cruciferae from Iran. Note R Bot Gard Edinb 29:181–184
Hedge IC (1986) Labiatae of South-West Asia: diversity, distribution and endemism. Proc R Soc Edinb 89B:23–35
Hedge IC, Wendelbo P (1978) Patterns of distribution and endemism in Iran. Note R Bot Gard Edinb 36:441–464
Hewitt GM (1999) Post-glacial re-colonization of European biota. Biol J Linn Soc 68:87–112
Hewitt G (2000) The genetic legacy of the quaternary ice ages. Nature 405:907–913
Hill RS (1993) The geology of the Graaff-Reinet area. Geological Survey, Department of Mineral and Energy Affairs, Pretoria
Hilliard OM, Burtt BL (1985) *Conium* (Umbelliferae) in southern Africa. S Afr J Bot 51:465–474
Hoare DB, Bredenkamp G (1999) Grassland communities of the Amatola/Winterberg mountain region of the Eastern Cape, South Africa. S Afr J Bot 65:75–82
Hobohm C (2008) Gibt es Ruderalpflanzen, die für Europa endemisch sind? Braunschw Geobot Arb 9:237–248
Hobohm C, Bruchmann I (2009) Endemische Gefäßpflanzen und ihre Habitate in Europa – Plädoyer für den Schutz der Grasland-Ökosysteme. RTG-Berichte 21:142–161
Huang JH, Chen JH, Ying JS, Ma KP (2011) Features and distribution patterns of Chinese endemic seed plant species. J Syst Evol 49:81–94
Huang JH, Chen B, Liu CR, Lai JS, Zhang JL, Ma KP (2012) Identifying hotspots of endemic woody seed plant diversity in China. Divers Distrib 18:673–688
Huntley BJ, Matos EM (1994) Botanical diversity and its conservation in Angola. In: Huntley BJ (ed) Botanical diversity in southern Africa. Proceedings of a conference on the conservation and utilization of southern African botanical diversity, Cape Town, September 1993, Strelitzia 1. National Botanical Institute, Pretoria, pp 53–74
Hütteroth WD, Höhfeld V (1982) Türkei: Geographie – Geschichte – Wirtschaft – Politik, 1st edn. Wissenschaftliche Buchgesellschaft, Darmstadt
IUCN Standards and Petitions Working Group (2008) Guidelines for using the IUCN Red List categories and criteria. In: Version 7.0. Prepared by the Standards and Petitions Working Group of the IUCN SSC Biodiversity Assessments Sub-Committee in August 2008
IUCN Standards and Petitions Subcomittee (2010) Guidelines for using the IUCN Red List categories and criteria. In: Version 8.1. Prepared by the Standards and Petitions Subcommittee in March 2010. Downloaded from http://intranet.iucn.org/webfiles/doc/SSC/RedList/RedListGuidelines.pdf. Nov 2012
Jamzad Z (2006) A new species and a new record from Iran. Iran J Bot 11(2):143–148
Jansson R (2003) Global patterns in endemism explained by past climatic change. Proc R Soc Lond 270:583–590

Jansson R (2009) Extinction risks from climate change: macroecological and historical insights. Biol Rep 1:44

Jetz W, Rahbek C, Colwell RK (2004) The coincidence of rarity and richness as the potential signature of history in centres of endemism. Ecol Lett 7:1180–1191

Johnson MR, Anhaeusser CR, Thomas RJ (eds) (2006) The geology of South Africa. Geological Society of South Africa/TheCouncil for Geoscience, Johannesburg/Pretori

Kamrani A, Jalili A, Naqinezhad A, Attar F, Maassoumi AA, Shaw SC (2011) Relationships between environmental variables and vegetation across mountain wetland sites, N Iran. Biologia 66:76–87

Kelly AE, Goulden ML (2008) Rapid shifts in plant distribution with recent climate change. Proc Natl Acad Sci USA 105(33):11823–11826

Kerndorf H, Pasche E (2006) *Crocus biflorus* (Liliiflorae, Iridaceae) in Anatolia (part three). Linzer Biol Beitr 38(1):165–187

Khalili A (1973) Precipitation patterns of central Elburz. Arch Met Geoph Biokl Ser B 21:215–232

Khassanov F, Noroozi J, Akhani H (2006) Two new species of *Allium* genus from Iran. Rostaniha 7(2):119–130

Klein JC (1982) Les groupements chionophiles de l'Alborz central (Iran). Comparaison avec leurs homolegues d' Asie centrale. Phytocoenologia 10:463–486

Klein JC (1987) Les pelouses xérophiles d'altitude du franc sud de l'Alborz central (Iran). Phytocoenologia 15:253–280

Klein JC (1988) Les groupements grandes ombellifères et a xerophytes orophiles: Essai de synthèse à l'échelle de la région irano-iouranienne. Phytocoenologia 16:1–36

Klein JC (2001) Endémisme à l'étage alpin de l'Alborz (Iran). Flora et Vegetatio Mundi 9:247–261

Klein JC, Lacoste A (2001) Observation sur la végétation des éboulis dans les massifs irano-touraniens: le Galietum aucheri ass. nov. de l'Alborz central (N-Iran). Documents Phytosociologiques N S 19:219–228

Kooi H, Beaumont C (1994) Escarpment evolution on high-elevation rifted margins: insights derived from a surface processes model that combines diffusion, advection, and reaction. J Geophys Res 99:12191–12209

Kraft NJB, Baldwin BG, Ackerly DD (2010) Range size, taxon age and hotspots of neoendemism in the California flora. Divers Distrib 16(3):403–413

Krebs P, Conedera M, Pradella M, Torriani D, Felber M, Tinner W (2004) Quaternary refugia of the sweet chestnut (*Castanea sativa* Mill.): an extended palynological approach. Veg Hist Archaeobot 13:145–160

Kruckeberg AR, Rabinowitz D (1985) Biological aspects of rarity in higher plants. Ann Rev Ecol Syst 16:447–479

Kutluk H, Aytuğ B (2001) Endemic plants of Turkey. In: Proceedings of the second Balkan Botanical Congress, vol 1. Istanbul, 14–18 Mayıs 2000, pp 285–288

Kutschker A, Morrone JJ (2012) Distributional patterns of the species of *Valeriana* (Valerianaceae) in southern South America. Plant Syst Evol 298(3):535–547

Laffan MD, Grant JC, Hill RB (1998) Some properties of soils on sandstone, granite and dolerite in relation to dry and wet eucalypt forest types in northern Tasmania. Tasman For 10:49–58

Larcher W (1976) Ecology of plants. Ulmer, Stuttgart

Larson DW, Matthes U, Kelly PE (2000) Cliff ecology: patterns and process in Cliff ecosystems. Cambridge University Press, Cambridge

Levyns MR (1964) Migrations and origins of the Cape flora. Trans R Soc S Afr 37:85–107

Li J, Fang X (1999) Uplift of the Tibetan Plateau and environmental changes. Chin Sci Bull 44:2117–2124

Li R, Dao Z, Ji Y, Li H (2007) A floral study on the seed plants in Northern Gaoligong Mountains in West Yunnan, China. Acta Botanica Yunnica 29:601–615

Linder HP (1990) On the relationships between the vegetation and floras of the Afromontane and the Cape regions of Africa. In: Proceedings of the 12th plenary meeting of AETFAT, symposium VII, vol 23b. Mitteilungen aus dem Institut für allgemeine Botanik, Hamburg, pp 777–790

Liu J, Tian B (2007) Origin, evolution, and systematics of Himalaya endemic genera. Newsl Himal Bot 40:20–27

Lomolino MV, Riddle BR, Whittaker RJ, Brown JH (2010) Biogeography, 4th edn. Sinauer Associates, Sunderland

López-Pujol J, Zhang FM, Sun H-Q, Ying T-S, Ge S (2011) Centres of plant endemism in China: places for survival or for speciation? J Biogeogr 38:1267–1280

Luebert F, Pliscoff P (2006) Sinopsis bioclimática y vegetacional de Chile. Editorial Universitaria, Santiago

Luebert F, Wen J (2008) Phylogenetic analysis and evolutionary diversification of *Heliotropium* sect. *Cochranea* (Heliotropiaceae) in the Atacama desert. Syst Bot 33:390–402

Macvicar CN, Loxton RF, Lambrechts JJN, Le Roux J, De Villiers JM, Verster E, Merryweather FR, Van Rooyen TH, Von M Harmse HJ (1977) Soil classification. A bionomical system for South Africa. Department of Agricultural Technical Services, Pretoria

Margules CR, Pressey RL (2000) Systematic conservation planning. Nature 405:243–253

Marticorena C, Rodríguez R (eds) (1995 ff) Flora de Chile. Universidad de Concepción, Concepción

Martorell C, Ezcurra E (2002) Rosette scrub occurrence and fog availability in arid mountains of Mexico. J Veg Sci 13(5):651–662

Matmon AM, Bierman P, Enzel Y (2002) Pattern and tempo of great escarpment erosion. Geology 30:1135–1138

Matthews WS, Bredenkamp GJ, Van Rooyen N (1991) The grassland-associated vegetation of the Black Reef Quartzite and associated large rocky outcrops in the north-eastern mountain sourveld of the Transvaal escarpment. S Afr J Bot 57:143–150

McCarthy T, Rubidge B (2005) The story of earth and life. Struik, Cape Town

Meadows ME, Linder HP (1993) A palaeoecological perspective on the origin of Afromontane grasslands. J Biogeogr 20:345–355

Médail F, Diadema K (2009) Glacial refugia influence plant diversity patterns in the Mediterranean Basin. J Biogeogr 36:1333–1345

Mihoc MAK, Morrone JJ, Negritto MA, Cavieres LA (2006) Evolución de la serie Microphyllae (Adesmia, Fabaceae) en la Cordillera de los Andes: una perspectiva biogeográphica. Revista Chilena de Historia Natural 79:389–404

Mill RR (1994) In Davis SD, Heywood VH, Hamilton AC (eds) Centres of plant diversity: a guide and strategy for their conservation, vol 1: Europe, Africa, South West Asia and the Middle East. IUCN Publications Unit, Cambridge, pp 293–348

Minnich RA (1985) Evolutionary convergence or phenotypic plasticity? Responses to summer rain by California chaparral. Phys Geogr 6:272–287

Minnich RA (2006) California climate and fire weather. In: Sugihara NG, Van Wagtendonk JW, Shaffer KE, Fites-Kaufman J, Thode AE (eds) Fire in California's ecosystems. University of California Press, Berkeley

Minnich RA (2007) California climate, paleoclimate and paleovegetation. In: Barbour MG, Keeler-Wolf T, Schoenherr AS (eds) Terrestrial vegetation of California, 3rd edn. University of California Press, Berkeley, Chapter 2

Minnich RA, Franco-Vizcaíno E (1998) Land of chamise and pines: historical descriptions of vegetation in northern Baja California, University of California publications in botany 80. University of California Press, Berkeley, pp 1–166

Mittermeier RA, Gil PR, Hoffman M, Pilgrim J, Brooks T, Mittermeier CG, Lamoreux J, da Fonseconda GAB (2005) Hotspots revisited: earth's biologically richest and most endangered terrestrial ecoregions. Cemex, Mexico City

Moore A, Blenkinsop T (2006) Scarp retreat versus pinned drainage divide in the formation of the Drakensberg escarpment, southern Africa. S Afr J Geol 109:599–610

Moore A, Blenkinsop T, Cotterill F (2009) Southern African topography and erosion history: plumes or plate tectonics. Terra Nova 21:310–315

Moreira-Muñoz A (2007) The Austral floristic realm revisited. J Biogeogr 34:1649–1660

Moreira-Muñoz A (2011) Plant geography of Chile, Plant and vegetation 5. Springer, Dordrecht

Moreira-Muñoz A, Muñoz-Schick M (2007) Classification, diversity, and distribution of Chilean Asteraceae: implications for biogeography and conservation. Divers Distrib 13:818–828
Moreira-Muñoz A, Morales V, Muñoz-Schick M (2012) Actualización sistemática y distribución geográfica de Mutisioideae (Asteraceae) de Chile. Gayana Bot 69(1):9–29
Mucina L, Rutherford MC (2006) The vegetation of South Africa, Lesotho and Swaziland, Strelitzia 19. South African National Biodiversity Institute, Pretoria, 807 pp
Mucina L, Wardell-Johnson GW (2011) Landscape age and soil fertility, climate stability, and fire regime predictability: beyond the OCBIL framework. Plant Soil 341:1–23. doi:10.1007/s11104-011-0734-x
Mulroy TW, Rundel PW (1977) Annual plants: adaptations to desert environments. Bioscience 27:109–114
Muñoz-Schick M, Moreira-Muñoz A (2000) Los Géneros Endémicos de Monocotiledóneas de Chile Continental. http://www.chlorischile.cl/Monocotiledoneas/Principalbot.htm
Muñoz-Schick M, Moreira-Muñoz A (2003a) La Flora de Chile Mediterráneo y su estado de conservación. Revista Chagual (Jardín Botànico de Santiago) 1:46–52
Muñoz-Schick M, Moreira-Muñoz A (2003b) Alstroemerias de Chile: diversidad, distribución y conservación. Museo Nacional de Historia Natural. Taller La Era, Santiago
Muñoz-Schick M, Moreira-Muñoz A (2008) Redescubrimiento de una especie de *Calceolaria* (Calceolariaceae). Gayana Botánica 65(1):111–114
Muñoz-Schick M, Moreira-Muñoz A, Villagrán C, Luebert F (2000) Caracterización florística y pisos de vegetación en los Andes de Santiago, Chile Central. Boletín Museo Nacional de Historia Natural 49:9–50
Muñoz-Schick M, Morales V, Cruzat ME, Moreira-Muñoz A (2010) Nuevo hallazgo de *Nardophyllum genistoides* (Phil.) Gray (Asteraceae) en Chile central. Gayana Botánica 67(2):234–237
Muñoz-Schick M, Morales V, Moreira-Muñoz A (2011) Validación de *Alstroemeria parvula* Phil. (Alstroemeriaceae). Gayana Botánica 68(1):110–112
Myers N (1990) The biodiversity challenge: expanded hot-spots analysis. The Environmentalist 10(4):243–256
Myers N, Mittermeier RA, Mittermeier CG, da Fonseca GAB, Kent J (2000) Biodiversity hotspots for conservation priorities. Nature 403:853–858
Naqinezhad A, Jalili A, Attar F, Ghahreman A, Wheeler BD, Hodgson JG, Shaw SC, Maassoumi AA (2009) Floristic characteristics of the wetland sites on dry southern slopes of the Alborz Mts., N. Iran: the role of altitude in floristic composition. Flora 204:254–269
Naqinezhad A, Attar F, Jalili A, Mehdigholi K (2010) Plant biodiversity of wetland habitats in dry steppes of Central Alborz. Aust J Basic Appl Sci 4(2):321–333
Nordenstam B (1969) Phytogeography of the genus *Euryops* (Compositae). A contribution to the phytogeography of southern Africa. Opera Bot 23:7–77
Nordenstam B, Clark VR, Devos N, Barker NP (2009) Two new species of *Euryops* Cass. (Asteraceae: Senecioneae) from the Sneeuberg, Eastern Cape Province, South Africa. S Afr J Bot 75:145–152
Noroozi J, Ajani Y (2013) A new alpine species of *Nepeta* sect. *Capituliflorae* (Labiatae) from Northwestern Iran. Novon. doi:10.3417/2012022
Noroozi J, Akhani H, Breckle SW (2008) Biodiversity and phytogeography of alpine flora of Iran. Biodivers Conserv 17:493–521
Noroozi J, Ajani Y, Nordenstam B (2010a) A new annual species of Senecio (Compositae-Senecioneae) from subnival zone of southern Iran with comments on phytogeographical aspects of the area. Compositae Newsl 48:4–23
Noroozi J, Akhani H, Willner W (2010b) Phytosociological and ecological study of the high alpine vegetation of Tuchal Mountains (Central Alborz, Iran). Phytocoenologia 40:293–321
Noroozi J, Pauli H, Grabherr G, Breckle WS (2011) The subnival-nival vascular plant species of Iran: a unique high-mountain flora and its threat from climate warming. Biodivers Conserv 20:1319–1338

Noroozi J, Willner W, Pauli H, Grabherr G (2013) Phytosociology and ecology of the high alpine to subnival scree vegetations of N and NW Iran (Alborz and Azerbaijan Mts.). Appl Veg Sci. doi:10.1111/avsc.12031

O'Brien B, Delgadillo Junak JS, Oberbauer T, Rebman TJ, Riemann H, Vanderplank S Rare, endangered, and endemic vascular plants of the California Floristic Province (CFP) portion of Northwestern Baja California, Mexico (in preparation for Aliso)

Oliver EGH, Linder HP, Rourke JP (1983) Geographical distribution of Cape taxa and their phytogeographical significance. Bothalia 14:427–440

Önal G (1981) Kilyos Bölgesi Kumlarının Degerlendirme Olanaklarının Araştırılması. In: Türkiye Madencilik Bilimsel ve Tektonik 7. Kongresi, 16–20, pp 319–337

Osok R, Doyle R (2004) Soil development on dolerite and its implications for landscape history in south-eastern Tasmania. Geoderma 121:169–186

Özhatay N, Kültür Ş (2006) Check-list of additional taxa to the supplement flora of Turkey III. Turk J Bot 30:281–316

Özhatay N, Byfield A, Atay S (2003) Türkiye'nin önemli bitki alanları. 88 pp + 1 CD. WWF Türkiye, Istanbul

Özhatay N, Kültür S, Aslan S (2009) Check-list of additional taxa to the supplement flora of Turkey IV. Turk J Bot 33:191–226

Özhatay N, Kültür S, Gürdal MB (2011) Check-list of additional taxa to the supplement flora of Turkey V. Turk J Bot 35:589–624

Palmer AR (1988) Vegetation ecology of the Camdebo and Sneeuberg regions of the Karoo biome, South Africa. Unpublished PhD thesis, Rhodes University, Grahamstown

Palmer AR (1990) A qualitative model of vegetation history in the eastern Cape midlands, South Africa. J Biogeogr 17:35–46

Palmer AR (1991) Asyntaxonomic and synecological account of the vegetation of the eastern Cape midlands. S Afr J Bot 57:76–94

Pang T (1996) The division of thermal belts in the tropical and sub-tropical western regions in China. Acta Geographica Sinica 51:224–229

Parolly G, Scholz H (2004) *Oreopoa* gen. novum, two other new grasses and further remarkable records from Turkey. Willdenowia 34:145–158

Partridge TC, Maud RR (1987) Geomorphic evolution of southern Africa since the Mesozoic. S Afr J Geol 90:179–208

Pauli H, Gottfried M, Dullinger S, Abdaladze O, Akhalkatsi M, Benito Alonso JL, Coldea G, Dick J, Erschbamer B, Fernández Calzado R, Ghosn D, Holten JI, Kanka R, Kazakis G, Kollár J, Larsson P, Moiseev P, Moiseev D, Molau U, Molero Mesa J, Nagy L, Pelino G, Puşcaş M, Rossi G, Stanisci A, Syverhuset AO, Theurillat J-P, Tomaselli M, Unterluggauer P, Villar L, Vittoz P, Grabherr G (2012) Recent plant diversity changes on Europe's mountain summits. Science 336:353–355

Peinado M, Alcaraz F, Delgadillo J, Aguado I (1994) Fitogeografía de la peninsula de Baja California, México. Anales Jard Bot Madrid 51:255–277

Phillipson PB (1987) A checklist of the vascular plants of the Amatola mountains, eastern Cape province/Ciskei. Bothalia 17:237–256

Phillipson PB (1992) A new species of *Indigofera* L. (Fabaceae) from the eastern Cape. S Afr J Bot 58:129–132

Pignatti S (1978) Evolutionary trends in mediterranean flora and vegetation. Vegetatio 37:175–185

Pignatti S (1979) Plant geographical and morphological evidences in the evolution of the mediterranean flora (with particular reference to the Italian representatives). Webbia 34:243–255

Pils G (1995) Die Bedeutung des Konkurrenzfaktors bei der Stabilisierung historischer Arealgrenzen. Linzer biol Beitr 27(1):119–149

Pils G (2006) Flowers of Turkey – a photo guide. Eigenverlag G. Pils, Feldkirchen

Pliscoff P, Luebert F (2008) Ecosistemas terrestres. In: CONAMA (ed) Biodiversidad de Chile, patrimonio y desafíos. Comisión Nacional del Medio Ambiente (CONAMA), Santiago, pp 74–87

Pooley E (2003) Mountain flowers: a field guide to the flora of the Drakensberg and Lesotho. Natal Flora Publications Trust, Durban
Qian H (1998) Large-scale biogeographic patterns of vascular plant richness in North America: an analysis at the generic level. J Biogeogr 25:829–836
Qian H (2001) A comparison of generic endemism of vascular plants between East Asia and North America. Int J Plant Sci 162:191–199
Rabinowitz D (1981) Seven forms of rarity. In: Synge H (ed) The biological aspects of rare plant conservation. Wiley, New York, pp 205–217
Raven PH, Axelrod DI (1978) Origin and relationships of the California flora. University of California Press, Berkeley
Razyfard H, Zarre S, Fritsch RM (2011) Four new species of *Allium* (Alliacea) from Iran. Ann Bot Fenn 48(4):352–360
Rechinger KH (ed) (1963–2012) Flora Iranica 1–179. Akademische Druck- u. Verlagsanstalt und Naturhistorisches Museum Wien, Graz/Wien
Rechinger KH (1968) *Zerdana* (pp. 307–308) [Cruciferae]. In: Rechinger KH (ed) Flora Iranica 57. Akademische Druck u. Verlagsanstalt, Graz
Rechinger KH (1987) *Zerdana* (pp 307–308) [Cruciferae]. In: Rechinger KH (ed) *Flora Iranica*, no 162. Akademische Druck u. Verlagsanstalt, Graz
Reeves RD, Adigüzel N (2004) Rare plants and nickel accumulators from Turkish serpentine soils, with special reference to *Centaurea* species. Turk J Bot 28:147–153
Rice WR (1987) Speciation via habitat specialization: the evolution of reproductive isolation as a correlated character. Evolut Ecol 1:301–314
Richerson PJ, Lum K (2008) Patterns of plant species diversity in California: relation to weather and topography. Am Nat 116(4):504–536
Riemann H, Ezcurra E (2007) Endemic regions of the vascular flora of the peninsula of Baja California, Mexico. J Veg Sci 18:327–336
Roberts N, Reed JM, Leng MJ, Kuzucuoglu C, Fontugne M, Bertaux J, Woldring H, Bottema S, Black S, Hunt E, Karabiyikoglu M (2001) The tempo of Holocene climatic change in the eastern Mediterranean region: new high-resolution crater-lake sediment data from central Turkey. Holocene 11:721–736
Rull V (2004) Biogeography of the 'Lost World': a palaeoecological perspective. Earth-Sci Rev 67:125–137
Rundel PW, Bowler PA, Mulroy TW (1972) A fog-induced lichen community in Northwestern Baja California, with two new species of *Desmazieria*. Byrol 75:501–508
Schmithüsen J (1956) Die räumliche Ordnung der chilenischen Vegetation. Bonner Geographische Abhandlungen 17:1–86
Schweizer G (1972) Klimatisch bedingte geomorphologische und glaziologische Züge der Hochregion vorderasiatischer Gebirge (Iran und Ostanatolien). In: Troll C (ed) Landschaftökologie der Hochgebirge Eurasiens. Franz Steiner Verlag GMBH, Wiesbaden, pp 221–236
Shreve F (1936) The transition from desert to chaparral in Baja California. Madroño 3:257–264
Shreve F (1951) Vegetation of the Sonoran desert, no 591. Carnegie Institution of Washington, Washington, DC. Reprinted as vol 1, Shreve L, Wiggins IL. Vegetation and flora of the Sonoran desert, vol 1. Stanford University Press, Stanford
Siebert SJ, Van Wyk AE, Bredenkamp GJ (2001) Endemism in the flora of ultramafic areas of Sekhukhuneland, South Africa. S Afr J Sci 97:529–532
Smith TB, Wayne RK, Girman DJ (1997) A role for ecotones in generating rainforest biodiversity. Science 276:1855–1857
Smith TB, Saatchi S, Graham C, Slabbekoorn H, Spicer G (2005) Putting process on the map: why ecotones are important for preserving biodiversity. In: Purvis A, Gittleman J, Brooks T (eds) Phylogeny and conservation, Cambridge University Press, Cambridge, UK, pp 166–197
Sorrie BA, Weakley AS (2010) Coastal plain vascular plant endemics: phytogeographic patterns. Castanea 66(1–2):50–82

Squeo FA, Arancio G, Gutiérrez JR (eds) (2001) Libro Rojo de la Flora Nativa y de los Sitios Prioritarios para su Conservación: Región de Coquimbo. Ediciones Universidad de La Serena, La Serena

Squeo FA, Arancio G, Gutiérrez JR (eds) (2008) Libro Rojo de la Flora Nativa y de los Sitios Prioritarios para su conservación: Región de Atacama. Ediciones Universidad de La Serena, La Serena, pp 137–163

Stebbins GL (1980) Rarity of plant species: a synthetic viewpoint. Rhodora 82:77–86

Stebbins GL, Major J (1965) Endemism and speciation in the California flora. Ecol Monogr 35:2–35

Steenkamp Y, Van Wyk AE, Smith GF, Steyn H (2005) Floristic endemism in southern Africa: a numerical classification at generic level. In: Friis I, Balslev H (eds) Plant diversity and complexity patterns: local, regional and global dimensions, vol 55. Biologiske Skrifter, Copenhagen, pp 253–271

Stevanović V, Tan K, Iatrou G (2003) Distribution of the endemic Balkan flora on serpentine I. – obligate serpentine endemics. Plant Syst Evol 242(1):149–170

Stirton CH, Clark VR, Barker NP, Muasya AM (2011) *Psoralea margaretiflora* (Psoraleeae, Fabaceae): a new species from the Sneeuberg centre of floristic endemism, Eastern Cape, South Africa. Phytokeys 5:31–38

Stott P (1981) Historical plant geography. Allen & Unwin, London

Stuckenberg BR (1962) The distribution of the montane palaeogenic element in the South African invertebrate fauna. Ann Cape Prov Mus 2:190–205

Taberlet P (1998) Biodiversity at the intraspecific level: the comparative phylogeographic approach. J Biotechnol 64:91–100

Tang CQ, Ohsawa M (2002) Tertiary relic deciduous forests on a humid subtropical mountain, Mt. Emei, Sichuan, China. Folia Geobotanica 37:93–106

Tang CQ, Yang Y, Ohsawa M, Momohara A, Hara M, Cheng S, Fan S (2011) Population structure of relict *Metasequoia glyptostroboides* and its habitat fragmentation and degradation in south-central China. Biol Conserv 144:279–289

Tang CQ, Yang Y, Ohsawa M, Yi S-R, Momohara A, Su W-H, Wang H-C, Zhang Z-Y, Peng M-C, Wu Z-L (2012) Evidence for the persistence of wild *Ginkgo biloba* (Ginkgoaceae) populations in the Dalou mountains, southwestern China. Am J Bot 99(8):1408–1414

Teppner H, Tuzlaci E (1994) *Onosma propontica* Aznavour (Boraginaceae, Lithospermeae). Stapfia (Linz) 37:77–83

Thorne RF (1993) Phytogeography. In: Flora of North America Editorial Committee (ed) Flora of North America North of Mexico 1. Oxford University Press, New York, pp 132–153

Thorne JH, Viers JH (2009) Spatial patterns of endemic plants in California. Nat Area J 29(4):344–366

Tribsch A, Schönswetter P (2003) Patterns of endemism and comparative phylogeography confirm palaeoenvironmental evidence for Pleistocene refugia in the Eastern Alps. Taxon 52:477–497

Türe C, Böcük H (2010) Distribution patterns of threatened endemic plants in Turkey: a quantitative approach for conservation. J Nat Conserv 18:296–303

Turner DP (2000) Soils of KwaZulu-Natal and Mpumalanga: recognition of natural soil bodies. Unpublished PhD thesis, University of Pretoria, Pretoria

Tutin TG, Burges NA, Chater AO, Edmondson JR, Heywood VH, Moore DM, Valentine DH, Walters SM, Webb DA (1996a) Flora Europaea vol 1: Psilotaceae-Platanaceae, 2nd edn (Reprint, first published 1993). Cambridge University Press, Cambridge

Tutin TG, Heywood VH, Burges NA, Moore DM, Valentine DH, Walters SM, Webb DA (1996b) Flora Europaea vol 2: Rosaceae-Umbelliferae (Reprint, first published 1968). Cambridge University Press, Cambridge

Tutin TG, Heywood VH, Burges NA, Valentine DH, Walters SM, Webb DA (1996c) Flora Europaea vol 3: Diapensiaceae-Myoporaceae (Reprint, first published 1968). Cambridge University Press, Cambridge

Tutin TG, Heywood VH, Burges NA, Valentine DH, Walters SM, Webb DA (1996d) Flora Europaea vol 4: Plantaginaceae-Compositae (and Rubiaceae) (Reprint, first published 1976). Cambridge University Press, Cambridge

Tutin TG, Heywood VH, Burges NA, Valentine DH, Walters SM, Webb DA (1996e) Flora Europaea, vol 5: Alismataceae-Orchidaceae (Reprint, first published 1980). Cambridge University Press, Cambridge

Underwood EC, Viers JH, Klausmeyer KR, Cox RL, Shaw MR (2009) Threats and biodiversity in the mediterranean biome. Divers Distrib 15:188–197

Valencia R, Pitman N, León-Yánez S, Jørgensen PM (eds) (2000) Libro Rojo de las Plantas Endémicas del Ecuador 2000. Publicaciones del Herbario QCA, Ponticicia Universidad Católica del Ecuador, Quito

Valiente-Banuet A, Verdú M (2007) Facilitation can increase the phylogenetic diversity of plant communities. Ecol Lett 10:1029–1036

Valiente-Banuet A, Rumebe AV, Verdú M (2006) Modern quaternary plant lineages promote diversity through facilitation of ancient tertiary lineages. Proc Natl Acad Sci USA 103(45):16812–16817

van der Werff H, Consiglio T (2004) Distribution and conservation significance of endemic species of flowering plants in Peru. Biodivers Conserv 13:1699–1713

Van Jaarsveld EJ (2011) Cremnophilous succulents of southern Africa: diversity, structure and adaptions. Unpublished PhD thesis, University of Pretoria, Pretoria

Van Wyk AE, Smith GF (2001) Regions of floristic endemism in Southern Africa. Umdaus Press, Hatfield

Van Zijl JSV (2006) Physical characteristics of the Karoo sediments and mode of emplacement of the dolerites. S Afr J Geol 109:329–334

Vandergast AG, Bohonak AJ, Hathaway SA, Boys J, Fisher RN (2008) Are hotspots of evolutionary potential adequately protected in southern California? Biol Conserv 141(6):1648–1664

Vanderplank SE (2011a) The flora of Greater San Quintín, Baja California, Mexico. Aliso 29:65–106

Vanderplank SE (2011b) Rare plants of California in Greater San Quintín, Baja California, Mexico. In: Willoughby JW, Orr BK, Schierenbeck K, Jensen N (eds) Proceedings of the CNPS conservation conference: strategies and solutions, California Native Plant Society, Sacramento, 17–19 Jan 2009, pp 381–387

Verdú M, Valiente-Banuet A (2008) The nested assembly of plant facilitation networks prevents species extinctions. Am Nat 172:751–760

Verdú M, Dávila P, García-Fayos P, Flores-Hernández N, Valiente-Banuet A (2003) 'Convergent' traits of Mediterranean woody plants belong to pre-mediterranean lineages. Biol J Linn Soc 78:415–427

Viers JH, Thorne JH, Quinn JF (2006) CalJep: a spatial distribution database of CalFlora and Jepson plant species. San Franc Estuary Watershed Sci 4(1):1–18

Viruel J, Catalan P, Segarra-Moragues JG (2012) Disrupted phylogeographical microsatellite and chloroplast DNA patterns indicate a vicariance rather than long-distance dispersal origin for the disjunct distribution of the Chilean endemic *Dioscorea biloba* (Dioscoreaceae) around the Atacama desert. J Biogeogr 39(6):1073–1085

Vural M (2009) Endemik bitkiler. Published on the internet, http://www.xn--nallhan-ufb.org.tr/nallihan_hakkinda_detay.php?no=14. 23 Feb 2012

Walther GR, Beißner S, Burga CA (2005) Trends in upward shift of alpine plants. J Veg Sci 16:541–548

Wang H (1992) Floristic geography. Science Press, Beijing

Wang S, Xie Y (eds) (2004) China species red list. Higher Education Press, Beijing

Wang L, Zhang Y, Xue N, Qin H (2011) Floristics of higher plants in China. Report from catalogue of life: higher plants in China database. Plant Divers Resour 33:69–74

Whittaker RJ, Araujo MB, Paul J, Ladle RJ, Watson JEM, Willis KJ (2005) Conservation biogeography: assessment and prospect. Divers Distrib 11:3–23

Wu Z (1980) The vegetation of China. Science Press, Beijing

Wu CY, Raven PH (eds) (2012) Flora of China. Flora of China@efloras.org. http://www.efloras.org. Accessed online on 9 Aug 2012

Wu Z, Raven PH, Hong D (1994–2011) Flora of China. Missouri Botanical Garden Press, St. Louis

Ying T, Zhang Y, Boufford DE (1993) The endemic genera of seed plants of China. Science Press, Beijing

Zhang D, Li F, Bian J (2000) Eco-environmental effects of the Qinghai-Tibet Plateau uplift during the quaternary in China. Environ Geol 39:1352–1358

Zohary M (1973) Geobotanical foundations of the Middle East 2. Fischer, Stuttgart

Zuloaga F, Morrone O, Belgrano M (eds) (2008) Catálogo de las plantas vasculares del cono sur (Argentina, southern Brazil, Chile, Paraguay y Uruguay). Monographs in systematic botany from the Missouri Botanical Garden 107, 3 vols. Missouri Botanical Garden, St. Louis

Part IV
Endemism in Vascular Plants

Chapter 8
Synthesis

Carsten Hobohm, Sula E. Vanderplank, Monika Janišová, Cindy Q. Tang, Gerhard Pils, Marinus J.A. Werger, Caroline M. Tucker, V. Ralph Clark, Nigel P. Barker, Keping Ma, Andrés Moreira-Muñoz, Uwe Deppe, Sergio Elórtegui Francioli, Jihong Huang, Jan Jansen, Masahiko Ohsawa, Jalil Noroozi, Miguel Pinto da Silva Menezes de Sequeira, Ines Bruchmann, Wenjing Yang, and Yongchuan Yang

8.1 Definition

Animals and plants that have small ranges are called endemics. *Small* is a relative term. Ultimately, every taxon can be called *endemic* in an area that includes its entire range. We use the term *endemic* for any taxonomic category entirely restricted to a given geographical or biogeographical unit such as a locality, island, region, mountain range, country, grid cell etc.

The vascular plant species *Stapelianthus madagascariensis* is endemic e.g. to the Indian Ocean Islands, to Madagascar, to the South of Madagascar, and also to the dry spiny bush in the South of Madagascar.

C. Hobohm (✉) • U. Deppe • I. Bruchmann
Ecology and Environmental Education Working Group, Interdisciplinary Institute of Environmental, Social and Human Studies, University of Flensburg, Flensburg, Germany
e-mail: hobohm@uni-flensburg.de

S.E. Vanderplank
Department of Botany & Plant Sciences, University of California, Riverside, CA, USA
e-mail: sula.vanderplank@gmail.com

M. Janišová
Institute of Botany, Slovak Academy of Sciences, Banská Bystrica, Slovakia

C.Q. Tang
Institute of Ecology and Geobotany, Yunnan University, Kunming, China
e-mail: cindytang@ynu.edu.cn

G. Pils
HAK Spittal/Drau, Kärnten, Austria
e-mail: gerhardpils@yahoo.de

M.J.A. Werger
Department of Plant Ecology, University of Utrecht, Utrecht, The Netherlands

C.M. Tucker
Department of Ecology and Evolutionary Biology, University of Toronto, Toronto, ON, Canada

C. Hobohm (ed.), *Endemism in Vascular Plants*, Plant and Vegetation 9,
DOI 10.1007/978-94-007-6913-7_8,

8.2 Causes

A taxon can be endemic for both environmental and biological reasons. Environmental factors include, for example, dispersal barriers such as salt water or mountain ranges, unique geological and soil conditions such as those supporting serpentine habitat isolates, and unique combinations of ecological conditions such as those in high mountain regions. Examples of biological causes of endemism include reduced seed or fruit production, reduced pollination opportunities outside of the taxon's range, or limited dispersal opportunities. Other positive and negative species interactions, such as competition, maintain the border of the endemic's range.

8.3 Meaning/Perception

Endemism is a subject of growing interest and importance in biogeography, ecology and nature conservation management. The terms *endemism* and *endemic*, formerly limited to scientific circles, are now becoming wider currency amongst

V.R. Clark • N.P. Barker
Department of Botany, Rhodes University, Grahamstown, South Africa
e-mail: vincentralph.clark@gmail.com

K. Ma • J. Huang • W. Yang
Institute of Botany, Chinese Academy of Sciences, Beijing, China
e-mail: kpma@ibcas.ac.cn

A. Moreira-Muñoz
Instituto de Geografía, Pontificia Universidad Católica de Chile, Santiago, Chile
e-mail: asmoreir@uc.cl

S. Elórtegui Francioli
Facultad de Ciencias de la Educación, Pontificia Universidad Católica de Chile, Santiago, Chile

J. Jansen
Institute for Water and Wetland Research, Radboud University, Nijmegen, The Netherlands
e-mail: jan.jansen@science.ru.nl

M. Ohsawa
Institute of Ecology and Geobotany, Kunming University in China, Kunming, China

J. Noroozi
Department of Conservation Biology, Vegetation and Landscape Ecology,
Faculty Centre of Biodiversity, University of Vienna, Vienna, Austria

Plant Science Department, University of Tabriz, 51666 Tabriz, Iran
e-mail: noroozi.jalil@gmail.com

M.P. da S.M. de Sequeira
Centro de Ciências da Vida, Universidade da Madeira, Funchal, Portugal
e-mail: miguelmenezessequeira@gmail.com

Y. Yang
Faculty of Urban Construction and Environmental Engineering, Chongqing University,
Chongqing, China

non-scientists in many countries and in many languages and are quoted in environmental protection laws and nature conservation measures all over the world.

The perception of *endemism* is geographically biased. Knowledge of endemism on marine islands that are popular tourist regions, for example, is often much better than in inaccessible terrestrial regions that lack any infrastructure. We hypothesise that this disparity in perception is related to the uneven distribution of scientific, economic and political effort, including tourism and scientific excursions.

Addressing this imbalance in perception as well as the general lack of knowledge on the endemism phenomena is vital for nature conservation management theory and practice. The education of young people is the foundation on which the future survival of species depends.

8.4 Quantification

Endemism can be quantified in different ways. The most popular measures are the absolute number (E) and the proportion (E/S) of endemic species. The number of endemics (E) in a region reflects the environmental and biological history including speciation processes and extinction events. The proportion (level, percentage) of endemics (E/S) is an indicator for geographic separation, genetic isolation and/or uniqueness of ecological conditions. Islands and archipelagos often have high rates of endemism. However, this does not necessarily mean that they have many endemics per unit area (density). For example, the density of endemics in the Eastern Arc Mountains and Southern Rift, Africa, is similar or a little higher than the density of endemics on the Hawaiian Islands. The rate of endemism, however, is c. 90 % in Hawaii but only 30 % in the Eastern Arc Mountains and Southern Rift. In this regard, regions with special habitat types, such as the area supporting the dry spiny bush in SW Madagascar, or unique ecological conditions, such as regions with serpentine soils, are very like oceanic islands; they are often ecologically isolated and covered by a vegetation with high levels of endemism.

Other, recently developed, parameters such as Range-Size Rarity (RSR) or Phylogenetic Analyses of Endemism (PE), when combined with the recent progress in molecular genetics, are likely to result in important innovations in landscape ecology and nature conservation practices.

8.5 Systematic Groups

Endemism develops over evolutionary timescales. Different phylogenetic groups exhibit different amounts of endemism within a given region. In general, systematic groups of vascular plants with high total species numbers also have high numbers of endemics. In many non-tropical regions, asteroids or Asteraceae is the taxonomic group with the highest number of both species and endemic taxa. Like the Orchidaceae, for example, this family is characterised by a high proportion of taxa that have wind-dispersed seeds. Thus, even though long-distance dispersal is a normal phenomenon, the two richest families in the world harbour very many species that are restricted to small regions. Factors other than mode of dispersal, such as pollination mode or competition with other taxa, can also limit the expansion of a taxon and explain its spatial restrictedness. Most taxa of Orchidaceae and Asteraceae are characterised by insect pollination. In Fabaceae and Myrtaceae, large plant families with many endemic species (e.g. in Australia), insect pollination is also common but wind dispersal of seeds is not.

Gymnosperms were the first seed-bearing plants on earth; they are much older than angiosperms. China has the richest flora of gymnosperms in the world – more than a quarter, and more than a tenth of the world's gymnosperms are endemic here. The proportion of Tertiary relict trees in central and southern China that are gymnosperms is higher than that of putative old angiosperm tree taxa, even though the overall diversity of gymnosperm trees is much lower than that of angiosperm trees. However, the family Asteraceae harbours more endemic taxa in China than any other plant family.

8.6 Genetics

Taxa that are restricted to small regions generally have a lower genetic diversity than more widespread congeners. Lower heterozygosity and genetic variation implies higher inbreeding levels and less ability to adapt to changing environmental conditions.

Reduced pollination, low fruit sets, and other problems of reproduction are often ecologically and not genetically controlled. However, most endemic vascular plant taxa do not show any problems with reproduction in their original habitat.

8.7 Increase of Endemism Over Time

Most, but not all, taxa start (and finish) their existence – branch of the phylogenetic tree – with quite a small range.

The high diversity of vascular plant taxa including endemics at regional scales seems to be related to long, uninterrupted evolutionary processes under relatively stable climate conditions and/or conditions that favour high evolutionary speed. Evolutionary speed depends on biological and environmental parameters: productivity, warmth, light and high precipitation rate might be relevant factors.

Most of the regions with high plant diversity are characterised by high environmental heterogeneity and habitat diversity. We assume that environmental heterogeneity (different aspect, gradient, substrate, water and nutrient conditions etc.) stimulates and promotes speciation. Furthermore, especially under changing conditions, environmental heterogeneity can ensure the local or regional survival of (endemic) species: under changing conditions, having a sufficient number of patches ensures that changes in one patch will not necessarily affect other patches and that there is therefore little or no net change over the landscape or ecoregion. For example, almost all the mid- or high altitudes of the Andes, Himalaya, Alps and other high mountain ranges were covered by Pleistocene glaciers, but the higher zones currently harbour many endemic vascular plants which had survived in the valleys or at lower altitudes during the height of the glaciations. Stability at this scale is nevertheless a stochastic phenomenon which involves local and regional processes, such as catastrophes, gap dynamics, successions and vertical displacement of the biota.

Geographical separation increases genetic isolation and thereby favours speciation processes. It also decreases dispersal opportunities. Many oceanic islands far from the mainland have a small species pool and gene pool; a small gene pool resulting from reduced dispersal and a small population size is a less favourable precondition for the evolution of new species. Thus, separation both promotes and limits speciation.

8.8 Decrease in Endemism Over Time – The Need for Conservation

Habitat loss, habitat degradation, habitat fragmentation, disturbance of habitat dynamics and biological resource use and pollution have all been identified as major drivers of species decline. Other factors such as climate change are of minor importance at the moment.

Nature conservation management needs to focus more on those threatened species which are restricted to small regions. Securing appropriate habitat conditions, including maintenance of traditional and sustainable resource use, is often the best protection from extinction.

Fortunately, while many groups of animals (e.g. small insects or large vertebrates such as whales) cannot be effectively safeguarded in zoos, almost all vascular plants could be protected in botanical gardens and indeed in horticulture in general. Only if plants have problems with reproduction or if they have very special ecological demands might conservation of these plant taxa become difficult. Having said that, gardening should be the exception and the last step adopted for the protection of threatened and endemic plant species.

8.9 Biogeography – Climate Zones

As expected, we can confirm that endemism (E, E/S, and other values) is high in the tropics, especially in the wet tropics, e.g. in tropical rainforests. However, other tropical landscapes and habitat types, such as those in extremely arid regions, are poor in endemic vascular plants. Endemism in most water bodies and wetlands in the tropics is also relatively low.

Endemism is also high in (not extremely) arid regions such as Namaqualand, parts of the Atacama or Socotra, for instance, and in regions with a Mediterranean climate, particularly in the more or less open shrub and dwarf-shrub formations. Other subtropical regions are also rich in endemics, for example in Southern Africa, Madagascar and the Brazilian Cerrado with their thickets, heaths, scrub and savanna.

In most temperate regions, endemism is low compared to subtropical and tropical climates but high in comparison to boreal-arctic zones. The highest endemism in temperate regions of the Holarctic is found in rocky habitats, screes, grasslands, heaths and other shrubby habitats. Endemism is low in water bodies and wetlands, as in water bodies and wetlands of most regions and climate zones of the world.

Endemism in boreal-arctic regions is very low, something which seems clearly related to glaciation cycles. However, to date this apparent fact has been explained by different and competing hypotheses. These take into account various historical and contemporary parameters and processes, environmental and biological constraints, and also include evolutionary, ecological and genetic aspects.

8.10 Biogeography – Altitudinal Zones

Endemism is normally higher in mountainous regions and lower in the lowlands. Central America, northern South America and South-West China, with their complex geology and high mountain ranges, are some of the most botanically diverse regions of the world. The *rate* of endemism (proportion of endemics) usually increases with altitude, whereas the belt with the highest *number* of endemics is found at mid- or low altitudes in mountain ranges.

However, exceptions to the rule also occur. Because of the Pleistocene glaciations, high latitude mountain ranges are also poor in endemics. Some flatland areas, notably the maritime chaparral of Baja California, the lowlands of Ecuador, the South-West of the Iberian Peninsula, various lowland regions of southern Africa, Madagascar and Australia are characterised by a high diversity of vascular plants and show rich endemism in combination with low or intermediate environmental heterogeneity.

8.11 Biogeography – Substrates

Vascular plant diversity and endemism in northern and Central Europe is higher on substrates with intermediate or high pH-values, although acidic substrates cover much larger areas in these regions. In other regions, the opposite is true e.g. in the Mediterranean climate regions of South-Western Australia and in the Cape Floristic Province. In these areas, most endemics occur on the very poor and acidic soils which represent the majority of substrates.

Serpentine soils, with high levels of heavy metals, low levels of nutrients, intermediate or high pH-values, and a low Ca/Mg ratio, often show high levels of endemism – serpentine endemics. Regions or countries with appreciable areas of serpentine outcrops, soils and serpentine endemism are e.g. eastern North America, Cuba, Brazil, Aegean Islands, Turkey, Zimbabwe, southern Africa, New Caledonia.

Thus, in some regions endemism is higher on dominant substrates, in other regions on non-dominant substrates.

8.12 Biogeography – Habitats

The richness of endemics associated with a certain habitat type can be discussed in the context of zonal, extrazonal and azonal vegetation types. *Zonal* vegetation types are controlled by climate. Such vegetation types sometimes have outlier occurrences in adjacent climatic zones but on special soils or landscape conditions; these are extrazonal. Azonal vegetation types are determined by local edaphic or hydrographic factors, such as aquatic vegetation, swamps, fens, bogs, salines, but also sparse vegetation of steep rock faces or mobile screes, etc. *Orobiomes* are the vegetation belts of mountain ranges, and *pedobiomes* are controlled by extensive edaphic factors.

The relationship between zonality and endemism is not clear. Some azonal vegetation types are very rich in endemics, e.g. on serpentine soils, which often give rise to sparse associations with many endemic plants. Rocky habitats of subalpine and alpine orobiomes which are strongly affected by both climate and substrate normally also harbour many endemic plants. In contrast, azonal vegetation types such as aquatic vegetation, or bogs, are normally poor in endemics. Different zonal vegetation types also differ remarkably with respect to endemism.

At the moment it is impossible to quantify habitat-related endemism in vascular plants at global scales, i.e. we do not know whether or not more endemic vascular plant taxa on earth occur in forest/woodland or in (semi-) open habitat types and landscapes, nor do we know whether or not most endemics are woody or herbaceous plants. And we do not know the proportion of annuals, perennials, epiphytes, shrub or tree species which have small distributional ranges.

However, for a few regions we can relate endemism to the array of ecological conditions and to habitat composition. The highest proportion of endemics associated with forest can be found e.g. in wet tropical regions of Middle America and northern South America (Ecuador), Madagascar, New Caledonia, and also on islands such as the Juan Fernandez and Madeira Islands.

In most Mediterranean climate regions the largest number of endemics occurs in heath, garigue, fynbos, kwongan, Chilean matorral, chaparral or thornbush including sparse units with a low vegetation cover on rocky ground.

We assume that rocks, screes, scrub, dwarf shrub habitats and grasslands of tropical, subtropical and temperate mountain ranges are in general relatively rich in endemics, whereas water bodies and wetlands, for example are normally poor.

8.13 Biogeography – Case Study Baja California/CFP

The California Floristic Province exhibits lower endemism than most other Mediterranean regions except for the Central Chile Hotspot. The formation of the arid deserts of Baja California and California over the last 10,000 years resulted in the emergence of several new ecotones where speciation is promoted via either 'synclimatic' or 'anticlimatic' migration. In Baja California the consistent presence of a stable fog zone has created a climate refugium through deep time that has both conserved ancient lineages and promoted the radiation of new lineages as water availability from precipitation has fluctuated. Whereas endemism normally increases with elevation, we see a narrow coastal band of very narrowly endemic plants at elevations below 500 m. Although fossil evidence for this region is very limited, it appears that chaparral was historically dominant, with the modern coastal scrub species previously existing as understorey elements. The coastal scrub we see today may be a very recent habitat type; this supports the notion that many of the micro-endemic plants found here are neo-endemics. The paleo-endemic taxa in the region are more often long-lived woody chaparral or tree species, although some genera appear to have radiated more recently as well. For example, *Arctostaphylos australis* is a putative palaeo-endemic taxon; however several seeder-*Arctostaphylos* species appear to have radiated recently as a result of changes in disturbance patterns, primarily shifting fire regimes with long-term weather patterns.

8.14 Biogeography – Case Study Ecuador, South America

Ecuador is one of the most endemic- and species-rich countries in the world, with two thirds of the country's area falling into two Biodiversity Hotspots: the Tropical Andes Hotspot and the Tumbes-Chocó-Magdalena Hotspot.

The main vegetation belts, in sequence from the coast to the highest mountains, are: mangroves, rain forest, wet to dry and deciduous forest, scrub paramo, other paramo types, rock and scree habitats.

A third of the endemics are in the Orchidaceae, less than 10 % belong to the Asteraceae. More than one third of the endemics are epiphytes, making these the largest group. The second largest group are shrubs or dwarf shrubs, followed by herbs, trees, lianas and vines, and finally others such as hydrophytes. In the regions of continental Ecuador most endemics (more than three quarters) are found in forest or woodland, with a minority occurring in paramo vegetation, including scrub paramo in the subalpine belt and other habitat types, such as scrub or thicket, in the dry parts of the Andes or in near-coastal South-West Ecuador.

The high number of endemics in such a small country occurs in a context of high spatial heterogeneity combined with more or less stable climate conditions over long time-periods. The relatively low endemism of the Amazonian part of Ecuador is most likely related to the lower topographical, geological and climatic heterogeneity there.

Many endemics are threatened by the same human activities in Ecuador as in many other parts of the wet tropics: deforestation and destruction of habitats by the expansion of arable land, settlements and roads, the extraction of oil, and so on. The potential threat from global warming cannot be quantified.

8.15 Biogeography – Case Study Europe

In Europe, rock and scree areas harbour the highest number of endemic vascular plant taxa. The second largest group of endemics is found in grassland, followed by scrub and heath, forest, arable land, ruderal and urban habitats, coastal/saline habitats, freshwater habitats, and finally mires and swamps.

In general, the diversity of endemic taxa increases from North to South and from flatlands with low environmental heterogeneity to mountain regions with high heterogeneity. Notwithstanding the general trend, the distribution patterns of wetland endemics in Europe do not show a clear North-South gradient. The highest wetland endemism occurs in temperate western Europe, including the Alps, with very low endemism in some southern, eastern and northern regions of Europe.

The East of Europe shares many taxa with the West of Asia. Many landscapes, habitat types and ecological conditions are quite similar to the West and East of the border between Europe and Asia. The southern part of the continental border, in particular, is artificially defined. Thus, the marginal regions bordering Asia necessarily have fewer endemics than comparable regions close to the Atlantic Ocean or Mediterranean Sea.

However, some of the most serious nature conservation problems concern grassland ecosystems. This is because most European grassland habitats are threatened by both abandonment – the cessation of livestock grazing or mowing – and intensification/ploughing. In Europe, the use of farmland is heavily influenced by payments made under European Union schemes. Changing these policies is essential, with payments being shifted from continuing to support productivity to supporting biodiversity and other environmental objectives. In the absence of such a change, the European Union will fail to achieve its environmental goals, specifically the halting of biodiversity loss.

8.16 Biogeography – Case Study Great Escarpment, Southern Africa

Southern Africa includes both tropical and sub-tropical regions of endemism, with the highest concentration in the exceptionally rich Cape, fynbos of the south-western Cape, which has a mediterranean climate and fringes the adjacent dry lands with sharp ecotones. High endemism is also associated with the Great Escarpment and its coastal forelands. While many factors affect endemism in this region, the strongest cause of endemism appears to be long-term climatic stability from reliable winter rainfall in the southern and western areas and a stable moisture regime from orographic rainfall and the ameliorating effects of altitude on the Escarpment itself. While substrate does play an important role in local endemism on e.g. ultramafic and ultrabasic geologies, it appears to be localised on specific substrates as a secondary causal factor or in response to high soil toxicity.

8.17 Biogeography – Case Study Madagascar

Madagascar is the fourth largest and most likely the oldest island in the world. Its environmental heterogeneity and habitat diversity is low or intermediate in most parts of the island. The outstanding plant and animal richness of Madagascar can be explained by the age of the geological surface, by the fact that Madagascar is of continental origin and by its long and uninterrupted evolutionary history. The level of endemism among the indigenous vascular flora is calculated as four-fifths or more. The flora is predominantly woody, and most endemics occur in forest ecosystems, woodlands and thickets.

Madagascar has a peripheral ring of lowland to montane forest surrounding an upland that is covered by fire-simplified grassland and savanna. The largely degraded and depauperated open landscapes are dominated by a few cosmopolitan fire-adapted grasses. The remaining primary vegetation, species richness and endemism are therefore concentrated in near-coastal rather than inland regions.

The human population is still growing, and man is destroying the natural habitats even in protected areas. The island's problems include burning, livestock grazing, deforestation and the selective cutting of particular tree species, e.g. of *Diospyros* (black timber), *Uapaca* spp. (red timber), and other tree species in the remaining humid forest areas. The impact of global warming or invasive plants currently seems to be low or insignificant.

8.18 Biogeography – Case Study China

In the rugged mountainous subtropical regions of central and southern China, a considerable number of Tertiary relict trees have survived. These regions were never covered by a single large ice-sheet; during the Pleistocene the climate was less severe than in other parts of the Northern Hemisphere, where it was cold and dry and glaciations were extensive. Since surviving the cold period, the relicts have had to contend with other, apparently often more competitive species. In the subtropical zone, most phylogenetically primitive taxa, which are classified as palaeoendemics, that were once widespread across the Northern Hemisphere (exemplified by the gymnosperms *Ginkgo biloba*, *Glyptostrobus pensilis*, *Metasequoia glyptostroboides*, *Cathaya argyrophylla*, *Cunninghamia lanceolata*, *Taiwania cryptomerioides*, and the angiosperms *Camptotheca acuminata*, *Davidia involucrata*, *Eucommia ulmoides*, and *Tapiscia sinensis*) now grow only in unstable habitats in central and southern China; on steep slopes, scree slopes, rocky or limestone areas, stream sides, and river deltas. Those relict trees are pioneer, shade-intolerant, and long-lived species. Natural disturbances may play a chief role in their persistence in unstable habitats where competition from other plants is limited. A broad zone of modern tree species on the more stable ground creates a soft boundary that does not allow long- or intermediate-distance dispersal of the older taxa. It might be that regional and local climate conditions such as combinations of annual dry/wet and hot/cold periods also prevent the successful dispersal of the endemics.

In SW China the most derived group of asteroids has the largest number of endemic species, accounting for nearly half of the total number of endemic species. This is consistent with the floristic composition of endemism in East Asia and North America at the genus level. Endemic plant species in SW China are largely concentrated in the major mountains of the west, whereas lower mountains and flat areas in the east have a low density of endemic species. The high endemism in the mountainous regions is probably due to the high speciation and low extinction rate, and in particular to the presence of refugia during the Pleistocene glacial episodes.

Glossary and Abbreviations

Alien plant (neophyte) Plant taxon which is not indigenous; it comes from another place.

Apoendemic Geographically restricted neo-polyploid taxon that is derived from a widely distributed diploid taxon (Favarger and Contandriopoulos 1961).

The categorization of Favarger and Contandriopoulos 1961 focuses on relationships between endemics and wide-spread vicariant relatives, which gives space for interpretation. Often it is not clear which taxon ancestor and which taxon descendent is. It may be that the ancestor has become extinct in the past and another descendent seems to be the ancestral relative of an endemic taxon (Keener 1983).

Area of endemism fundamental unit in historical biogeography and conservation biology. An area of endemism can be defined by different conditions and characteristics, e.g. history, size, no overlap with another area of endemism, two or more endemic taxa with more or less congruent ranges (Giokas and Sfenthourakis 2008; Linder 2001).

asl. above sea level

Boundary, hard and soft boundaries Barrier for dispersal of diaspores. High mountain ranges, glaciated pole regions and oceans are usually defined as *hard boundaries*. The probability to cross these boundaries are very small for most diaspores of vascular plant taxa. However, a boundary for one species might be a dispersal medium for another. For example, sea water is a hard boundary for many species but some hydrochorous species of coastal habitats use currents as dispersal vectors (e.g. *Cocos nucifera*). Here we define differences of habitats in terrestrial or freshwater environments as *soft boundaries*. For many grassland plants e.g. a broad forest zone may be insurmountable, similar to marine environments. For many species bound to forest a broad zone of grassland or arable land might be insurmountable as well.

CBD Convention on Biological Diversity (Rio 1992); one of the first international nature conservation strategies including a focus on endemic species.

C. Hobohm (ed.), *Endemism in Vascular Plants*, Plant and Vegetation 9,
DOI 10.1007/978-94-007-6913-7, © Springer Science+Business Media Dordrecht 2014

Characteristic taxon If more than 50 % of the individuals or populations of a taxon inhabit a biogeographical unit, the taxon is defined as characteristic to this region (sensu Van Opstal et al. 2000).

Coastal, brackish and saline habitats These habitat types comprise maritime sands, salt marshes, inland salt steppe, coastal scrub and thicket, coastal dunes, coastal rocks and cliffs (cf. EUNIS classification of habitats, Davies et al. 2004).

Continental, marine, oceanic island *Marine islands* originated as *oceanic islands* within the sea and have never been part of the mainland – often: *atolls* or *volcanic islands* – or as *continental islands* which formerly were part of a continent. All islands in the central parts of the oceans are oceanic islands, all very large islands are continental islands (cf. Gillespie and Clague 2009).

However, a marine island is surrounded by sea water which in most cases is saltwater (exception: Baltic Sea with many parts of brackish water). In contrary, islands in lakes and rivers of the terrestrial land are normally surrounded by freshwater.

Cross-border endemics Endemics distributed on both sides of artificial boundaries are not of the same state of awareness than e.g. national endemics. However, a taxon occurring in different countries must not be widely distributed. According to our list of endemic vascular plants in Europe (EvaplantE) more than thousand taxa occur in only 2 of 42 regions in Europe which in most cases are defined by artificial borders. Most of them cannot be found in the scientific context of endemism even if many of them have very small ranges. For example, most endemics of mainland Spain most likely have larger areas than Iberian endemics which occur in Portugal plus Spain.

The term is first used in this book.

Cultivated, agricultural, horticultural and domestic habitats Different habitat types which are strongly influenced by human activities belong to this group of habitats including cultivated ground, cereal fields, cornfields, rice fields, fallow land, habitats dominated by weeds, waste places, disturbed ground, roadsides, margins of food paths, artificial walls, places beside walls, and other urban habitats such as parks and gardens (cf. EUNIS classification of habitats, Davies et al. 2004).

Diversification Rate Hypothesis The speciation rate depends on climate (Zobel et al. 2011; Goldie et al. 2010; Mittelbach et al. 2007). However, the underlying processes are still incompletely understood. We assume that the speciation rate in wet tropical regions is higher than in dry or cold regions because of a faster molecular evolution which most likely depends on the length of the growing season, on water, temperature and energy supply.

Endemic A taxon restricted to a (small) geographical region is endemic to this region. Following De Candolle (1820) and Dhar (2002; see also e.g. Zobel et al. 2011) we use the term *endemic* for any taxonomic category entirely restricted to a given biogeographical unit (e.g. locality, island, region, mountain range, country, grid cell etc.).

"Endemic through ignorance" A taxon which occurs in a certain geographical area but is not recognized outside; thus, the taxon is labelled as endemic (Stebbins and Major 1965; Fiedler 1986). Normally, the checklist of all taxa in a region is increasing during time and during increasing intensity of scientific work. In contrary, the number of endemics is decreasing whenever one of the taxa is found outside a region and changes its label from *endemic* to *non-endemic*. Gentry (1986: 158) stated that even, "*the revised Hawaiian Flora now in preparation will substantially reduce the number of* (endemic; C. H.) *species recognized*." Due to the realization of former overestimations the number of endemic vascular plants e.g. of the Andes Hotspot has been modified from 20,000 to 15,000 (cf. Mittermeier et al. 1999, 2005) and for Crimea from 250 to 300 to less than 150 (Yena 2007). Five species previously reported as endemics of the Drakensberg Alpine Centre are now known to occur also in the Sneeuberg, South Africa (Clark et al. 2009).

Extinction, extinct Disappearance of a species from the globe or of a species within a certain region. Many regional Red Lists give information about *extinct* species which still exist in other regions. Many fossils have become extinct by normal evolution. This means that the phenotype disappeared because one or several other descendants evolved from the ancestors. Thus, the terms *extinction* and *extinct* are ambiguous and should be clarified in every case.

Flagship species Many endemic species such as the *Panda* in China have been labelled as flagship species. Flagship species are living in a vulnerable habitat or threatened environment. The survival of the species depends on the existence of a special habitat or environmental conditions.

Forest, woodland and other wooded land Forests are habitat types dominated by trees (crown in the majority of cases interlocking). Trees in contrary to shrubs are woody plants which normally are taller than 10 m (according to FAO taller than 5 m) with a single stem. Forests comprise virgin forests, forests which are used for timber production, and also tree plantations with a dense crown and typical forest-herb layer, but exclude olive groves, agro forestry and plantations with ploughed, unnaturally grazed or mowed ground (cf. EUNIS classification of habitats, Davies et al. 2004).

Gene pool All the genes in all the individuals of a species, subspecies or population.

Genome Genetic composition of an organism or species. There is no clear difference between genome and gene pool.

Grassland As defined here, grasslands include different (non-woody) habitats dominated by graminoids and forbs which are normally lower than 1(2) m tall, including e.g. meadows, pastures, steppes, non-woody savannas, grassland-connected fringe communities, and megaforb communities, but excluding arable land dominated by Poaceae (cereals, maize).

Habitat Place or locality which can be characterized by abiotic and biotic factors. In contrary to *biome* a habitat is related to local scales. The term *vegetation unit* is more or less scale-independent and focuses on biotic factors rather than abiotic factors.

Habitat-group Being aware of the difficulties and biases which result from the different habitat terminologies in various languages or from different national regulations and standards of classification we here distinguish physiognomic habitat-groups which correspond well with those defined by the Habitat Directive of the European Commission and several other classification systems (e.g. Davies et al. 2004; Song and Xu 2003).

Hardy Weinburg equilibrium Frequency of genes in a population which might be expected if there would be perfect random mixing of genes.

Heath and scrub These are woody habitats which are of lower height than forest or woodland (<5 m tall). The habitats include woody garigue, sclerophyllous scrub and thickets dominated by woody plants, also wood-margins, hedges and openings in forests.

We here include also succulent-dominated vegetation units which are of shrubby habit (such as *Euphorbia canariensis*-communties of the Canaries).

Holoschizoendemic A mature, stabilized, and diversified schizoendemic taxon that occupies a maximum area but is restricted by habitat, hard boundaries or both (Keener 1983).

Hotspot The concept of *Biodiversity Hotspot* as developed at the end of the 1980s by Myers (1988, 1990; Mittermeier et al. 1999, 2005) is currently one of the leading approaches in global nature conservation strategies. The amount of endemics is one of the main features of Biodiversity Hotspots. Orme et al. (2005) distinguished *species richness hotspots, threat hotspots* and *endemism hotspots*. They analysed maps of avian diversity and demonstrated that hotspots of species richness, threat and endemism do not show the same geographical distribution patterns. Moreover, they found out that endemism hotspots contained a greater proportion of overall species richness than did the species richness hotspots and a greater proportion of threatened species than did the threat hotspots.

Immigration, colonisation Movement of individuals into an area.

Inland surface water, wetland Inland waters comprise standing and running waters and connected habitats, including e.g. ponds, margins of pools and lakes, minerogenic springs, wet streamsides, river banks, also wet to dry pioneer-vegetation on river banks and seasonally flooded ground (cf. EUNIS classification of habitats, Davies et al. 2004).

Island, isolated habitat Land surrounded by water, or water surrounded by land, or a group of trees in a desert (for example).

IUCN International Union for the Conservation of Nature.

Keystone species In relation to its biomass a keystone species has a disproportionate high impact on the ecosystem. The identification of keystone and umbrella species and the characterization of flagship or target species makes nature conservation planning much easier than concerning millions of species in an equal measure.

Kryptoendemic Taxon which is genetically separated but morphologically equal or similar in comparison with another taxon. The term was introduced by Stebbins and Major (1965) who analysed the flora of California (cf. Fiedler 1986).

Mires, bogs and fens These habitats include bogs, fens, swamps, swamp-springs, moorland, damp marshy ground and peaty soils covered by characteristic vegetation.

National park Large nature conservation area defined and controlled by the national government. In former times wilderness and the total reduction of human influences was the primary goal. Nowadays the idea of what a national park should be is changing due to the fact, that some impact of man can influence and stabilise the species composition in a positive manner. This has also to do with the fact that *Homo sapiens* is indeed part of nature.

Neoendemic Relatively young endemic taxon that originated in the late Pleistocene or Holocene period. In Europe many *Oenothera* species occurred after colonisation of the genus in modern times as a result of recombination processes. These sometimes called "homeless" new species are both neoendemics and neophytes.

Neoschizoendemic A relatively young schizoendemic taxon that is restricted geographically because of its age, neither because of ecological reasons nor because of a limited distribution potential. Normally the size of area is increasing with time (Keener 1983).

NGO Non-governmental organisation – any group which is not under direct governmental control.

Old stable landscape (OSL) OSLs "*are regions which have experienced prolonged tectonic quiescence older than the Late Cenozoic*" ... "*and which have not experienced large scale glacial scouring during Plio-Pleistocene.*" These "*landscapes are found in regions over evolutionary time scales characterised by either an equable (non-seasonal) temperature regime*" ... "*or are still highly dynamic (on year-to-year time scales), yet characterised by predictable climatic dynamics over very long evolutionary scales. The latter regions have evolutionary old (and well stabilised) seasonal precipitation regimes. Along the dimension of Fire regime predictability, there are areas for which fire is a highly unpredictable occurrence (e.g. many deserts, cool temperate and boreal forests) and areas with highly predictable fire regimes*" (Mucina and Wardell-Johnson 2011: 13).

Fire is one reason for the reduction of biomass, disturbance of vegetation, and promoting gap dynamics. Other factors such as grazing work in the same direction. However, in the context of endemism and species richness in total the continuity of climate, ecological conditions – including disturbance of vegetation and reduction of biomass – and the age of a landscape (geomorphology and general composition of habitat-groups) is discussed as one of the main important drivers of species composition and endemism at regional scales.

Palaeoendemic Relatively old endemic taxon. Palaeoendemic taxa are evolutionary older than Quaternary period, in many cases morphologically uniform and taxonomically isolated ancient relics with obscure origins (Keener 1983; Favarger and Contandriopoulos 1961).

Patroendemic Ancient element of a restricted paleoendemic taxon that is ancestral to a widespread polyploid taxon (Favarger and Contandriopoulos 1961).

Polymorphism Range of morphological types within a species or subspecies (except differentiation between male and female).

Radiation Burst of evolution during which many different descendants from one or a few ancestors evolved. *Adaptive* means the way in which a system such as a morphology successfully relates to a given environment. Often the term *adaptive* is more related to plausibility than to empirical data or statistics.

Range Size "*No standard methodology exists for measuring the sizes of species' geographic ranges*" (Gaston 1996: 197). The measured size of a geographic range depends on the spatial resolution of the occurrence data that are analysed. Often the geographic area is given as number of quadrates on a grid system from which a taxon is recorded. If the grid system is based on longitudes and latitudes than grid cells from different latitudes differ in size. This fact causes numerical biases and the need of corrections which are often ignored.

Rapoport's Rule Rapoport (1982) states that geographical ranges of animals and plants are generally smaller towards the equator than towards higher latitudes.

Rarity Rarity and commonness depend on spatial scale (cf. Soulé 1986). A taxon may locally be rare but distributed worldwide, another one may locally be characterised by high density but is endemic to a region.

In general, reasons for rarity are threesome: a rare habitat or rare successional stage (1), reduced gene flow and/or reduced production of diaspores or low dispersal of diaspores (2), or an inadequate relationship between biological demands and environment (3).

Red Data Book, Red List Internationally or nationally recognised list – formerly published as book, today often as electronic version – of endangered species of a certain systematic group within a region, nation or whole world. The IUCN Red List of Threatened Species (cf. Walter and Gillett 1998; Baillie et al. 2004, www.iucnredlist.org) is the most comprehensive red list for the whole world which is regularly updated.

Rocky habitats These habitats include screes, alpine moraines, rocky and stony ground with scarce vegetation, and caves. In most cases the vegetation cover is low. According to the EUNIS classification of habitats (Davies et al. 2004) rocks and screes belong to the category *inland unvegetated or sparsely vegetated habitats*.

Savanna, savannah This ecosystem type is characterized by trees or groups of trees with an open canopy surrounded by grasses and herbaceous plants. The origin of the word most likely is not African but Caribbean (cf. Spanish *sabána*).

Non-woody savanna belong to grassland s.l., wooded savannas to woodlands (cf. EUNIS classification of habitats; Davies et al. 2004).

Schizoendemic Diploid or polyploid taxon that originated from diploid ancestors. Often gradual speciation is resulting in a progressive divergence of morphologically and genetically different taxa (Favarger and Contandriopoulos 1961).

Smolenice Grassland Declaration (EDGG Bulletin 7; June 2010) In Europe the second largest group of endemic vascular plants – more than 1,200 species groups (apomictic taxa), species, and subspecies – are associated with grassland (s. l.).

The signatories of the Smolenice Declaration call for a strong and comprehensive Convention on Grassland Conservation in Europe within the framework of the Pan-European Landscape and Biodiversity Strategy, to secure the future of grasslands which provide vital ecosystem services to human society, are home to biodiversity, sources of natural beauty and cultural values.

More than 320 scientists, politicians and other persons interested in nature conservation from more than 40 countries have signed the declaration until 2012 (see on the internet).

Species Pool Hypothesis The hypothesis justifies *local diversity patterns through large-scale historical factors* (Zobel et al. 2011: 252). Conditions which have been more abundant in time or space throughout evolutionary history favour high species richness in a habitat.

Strict endemic Endemic restricted to a single habitat type (cf. Keener 1983).

Subendemic taxon If more than 75 % of the populations belong to a given biogeographical unit we call the taxon subendemic (sensu Rabitsch and Essl 2009).

Target Species According to the concept of Ozinga and Schaminée (2005) a target species can be characterised by a common set of criteria (such as legal protection status, threat status and geographical distribution). Most target species listed in Ozinga and Schaminée (2005) for Europe are restricted to Europe.

Umbrella species Protection and survival of an *umbrella species* guarantees the survival of other species in the same habitat if they are interconnected by the same food web or depend on the same environmental conditions or habitat type.

Woodland Like the terms *forest* or *grassland* woodland is an often used but inconsistently defined umbrella-term for different habitat types. Very often the term includes both forest and more open tree stands. According to White (1983) woodlands are open stands of trees at least 8 m tall with a canopy cover of 40 % or more (cf. also EUNIS classification of habitats; Davies et al. 2004).

References

Baillie JEM, Hilton-Taylor C, Stuart N (2004) 2004 IUCN red list of threatened species: a global species assessment. IUCN, Gland/Cambridge

Clark VR, Barker NP, Mucina L (2009) The Sneeuberg: a new centre of floristic endemism on the Great Escarpment, South Africa. S Afr J Bot 75:196–238

Davies CE, Moss D, Hill MO (2004) EUNIS habitat classification, revised 2004. Report to European Environment Agency, European Topic Centre on Nature Protection and Biodiversity. http://eunis.eea.europa.eu/upload/EUNIS_2004_report.pdf

De Candolle AB (1820) Essai elementaire de geographie botanique. In: Dictionnaire des sciences naturelles, vol 18. Flevrault, Strasbourg, pp 1–64

Dhar U (2002) Conservation implications of plant endemism in high-altitude Himalaya. Curr Sci 82:141–148

Favarger C, Contandriopoulos J (1961) Essai sur lèndimism. Berichte der Schweizerischen Botanischen Gesellschaft 71:384–406

Fiedler PL (1986) Concepts of rarity in vascular plant species, with spezial reference to the genus Calochortus Pursh (Liliaceae). Taxon 35(3):502–518

Gaston KJ (ed) (1996) Biodiversity: a biology of numbers and difference. Blackwell Science, Oxford et al

Gentry AH (1986) Endemism in tropical versus temperate plant communities. In: Soulé ME (ed) Conservation biology: the science of scarcity and diversity. Sinauer Associates, Inc.-Publ, Sunderland, pp 153–182

Gillespie RG, Clague DA (eds) (2009) Encyclopedia of islands. University Press of California, Berkeley

Giokas S, Sfenthourakis S (2008) An improved method for the identification of areas of endemism using species co-occurrences. J Biogeogr 35:893–902

Goldie X, Gillman L, Crisp M, Wright S (2010) Evolutionary speed limited by water in arid Australia. Proc Biol Sci 277:2645–2653

Keener CS (1983) Distribution and biohistory of the endemic flora of the Mid-Appalachian Shale Barrens. Bot Rev 49:65–115

Linder HP (2001) Plant diversity and endemisms in sub-Saharan tropical Africa. J Biogeogr 28:169–182

Mittelbach GG, Schemske DW, Cornell HV et al (2007) Evolution and the latitudinal diversity gradient: speciation, extinction and biogeography. Ecol Lett 10:315–331

Mittermeier RA, Gil PR, Hoffman M, Pilgrim J, Brooks T, Mittermeier CG, Lamoreux J, da Fonseconda GAB (2005) Hotspots revisited: Earth's biologically richest and most endangered terrestrial ecoregions. Cemex, Mexico City

Mittermeier RA, Myers N, Mittermeier CG, Gil PR (1999) Hotspots: Earth's biologically richest and most endangered terrestrial ecoregions. Cemex, Mexico City

Mucina L, Wardell-Johnson GW (2011) Landscape age and soil fertility, climate stability, and fire regime predictability: beyond the OCBIL framework. Plant Soil 341:1–23. doi:10.1007/s11104-011-0734-x

Myers N (1988) Threatened biotas: hotspots in tropical forests. Environmentalist 8:1–20

Myers N (1990) The biodiversity challenge: expended hotspots analysis. Environmentalist 10: 243–256

Orme CDL, Davis RG, Burgess M, Eigenbrod F, Pickup N, Olson VA, Webster AJ, Ding TS, Rasmussen PC, Rigely RS, Stattersfield AJ, Bennett PM, Blackburn TM, Gaston KJ, Owens IPF (2005) Global hospots of species richness are not congruent with endemism or threat. Nature 436:1016–1019

Ozinga WA, Schaminée JHJ (eds) (2005) Target species – species of European concern. Alterra-report 1119, 193 pp

Rabitsch W, Essl F (eds) (2009) Endemiten – Kostbarkeiten in Österreichs Pflanzen- und Tierwelt. Umweltbundesamt, Wien

Rapoport EH (1982) Areography: geographical strategies of species. Bergamon Press, New York

Song Y-C, Xu G-S (2003) A scheme of vegetation classification of Taiwan, China. Acta Bot Sin 45(8):883–895

Soulé ME (1986) Conservation biology: the science of scarcity and diversity. Sinauer Associates Inc., Sunderland

Stebbins GL, Major J (1965) Endemism and speciation in the California flora. Ecol Monogr 35: 2–35

Van Opstal AJFM, Brandwijk T, van Duuren L, Schaminée JHJ (2000) Endemic and characteristic plant species in Europe: Part 1 Northern Europe, Rapport IKC Natuurbeheer 53. Landbouw, Natuurbeheer en Visserij, Wageningen, p 92

Walter KS, Gillett HJ (eds) (1998) 1997 IUCN red list of threatened plants. Compiled by the World Conservation Monitoring Centre. IUCN, Gland/Cambridge

White F (1983) The vegetation of Africa: a descriptive memoir to accompany the UNSECO/AEFTAT/UNSO vegetation map of Africa. United Nations Educational, Scientific and Cultural Organization, Paris

Yena AV (2007) Floristic endemism in the Crimea. Fritschiana 55:1–8

Zobel M, Otto R, Laanisto L, Naranjo-Cigala A, Pärtel M, Fernandez-Palacios JM (2011) The formation of species pools: historical habitat abundance affects current local diversity. Glob Ecol Biogeogr 20:251–259

Index

C. Hobohm (ed.), *Endemism in Vascular Plants*, Plant and Vegetation 9,
DOI 10.1007/978-94-007-6913-7, © Springer Science+Business Media Dordrecht 2014

C

I

J

N

Q

R

T